Lecture Notes in Mathematics

Edited by A. Dold and [illegible]mann

1340

S. Hildebrandt D. Kinderlehrer
M. Miranda (Eds.)

Calculus of Variations and Partial Differential Equations

Proceedings of a Conference held in Trento, Italy
June 16–21, 1986

Springer-Verlag
Berlin Heidelberg New York London Paris Tokyo

Editors

Stefan Hildebrandt
Mathematisches Institut, Universität Bonn
Wegelerstr. 10, 5300 Bonn, Federal Republic of Germany

David Kinderlehrer
School of Mathematics, University of Minnesota
206 Church St. SE, Minneapolis, MN 55455, USA

Mario Miranda
Centro Internazionale per la Ricerca Matematica
38050 Povo-Trento, Italy

Mathematics Subject Classification (1980): 35A15, 35G20, 35J20, 35J25, 35J65, 35Q10

ISBN 3-540-50119-3 Springer-Verlag Berlin Heidelberg New York
ISBN 0-387-50119-3 Springer-Verlag New York Berlin Heidelberg

Library of Congress Cataloging-in-Publication Data.
Calculus of variations and partial differential equations: proceedings of a conference, held in Trento, Italy, June 16–21, 1986 / S. Hildebrandt, D. Kinderlehrer, M. Miranda, eds. p.cm. – (Lecture notes in mathematics; 1340) Includes bibliographies.
ISBN 0-387-50119-3 (U.S.)
1. Calculus of variations – Congresses. 2. Differential equations, Partial – Congresses.
I. Hildebrandt, Stefan. II. Kinderlehrer, David. III. Miranda, Mario, 1937–. IV. Series: Lecture notes in mathematics (Springer-Verlag); 1340.
QA3.L28 no. 1340 [QA315] 510 s–dc 19 [515'.64] 88-20180

Printing and binding: Druckhaus Beltz, Hemsbach/Bergstr.
2146/3140-543210

PREFACE

A delight and a privilege is before us: we have gathered from many nations and across many generations to dedicate this conference and its proceedings to Hans Lewy. The scientific achievements of Professor Lewy are well established. None of us cease to be influenced by the depth and imagination of his scientific vision, by what we may correctly refer to as his genius. Equally important, we have continued to receive our own inspiration from *il suo gusto ed il suo garbo* - his taste and his style - in his choice of scientific issues and their resolution. His constant dedication to the encouragement of young mathematicians stands before us as a model. We are grateful for this opportunity to demonstrate our esteem and appreciation of Hans Lewy, the man and the scientist.

Stefan Hildebrandt

David Kinderlehrer

Mario Miranda

Conference Committee

ACKNOWLEDGEMENTS

We wish to thank all the participants for their contribution to the success of the meeting.

We express our gratitude to the Comitato Nazionale per le Scienze Matematiche of the Consiglio Nazionale delle Ricerche and to the Centro Internazionale per la Ricerca Matematica of the Istituto Trentino di Cultura for their invaluable support.

Special thanks are due to Augusto Micheletti for his help in running the conference, and for his wonderful job in the typing of all the manuscripts.

Stefan Hildebrandt

David Kinderlehrer

Mario Miranda

Conference Committee

LIST OF PARTICIPANTS

- Hans Lewy (Berkeley, California, USA)

- Emilio Acerbi (Scuola Normale Superiore, Pisa, Italy)
- Giovanni Alessandrini (Istituto di Matematica, Università Firenze, Italy)
- Luigi Ambrosio (Scuola Normale Superiore, Pisa, Italy)
- Gabriele Anzellotti (Dipartimento di Matematica, Univ. Trento, Povo, Italy)
- Alberto Arosio (Dipartimento di Matematica, Univ. Parma, Italy)
- Abbas Bahri (Departement de Mathématiques, Ecole Polytechnique Palaiseau, France)
- Claudio Baiocchi (Dipartimento di Matematica, Univ. Pavia, Italy)
- Michele Balzano (S.I.S.S.A., Trieste, Italy)
- Martino Bardi (Seminario Matematico, Univ. Padova, Italy)
- Giovanni Bassanelli (Dipartimento di Matematica, Univ. Trento, Povo, Italy)
- Hugo Beirão da Veiga (Dipartimento di Matematica, Univ. Trento, Povo, Italy)
- Enrico Bernardi (Dipartimento di Matematica, Univ. Bologna, Italy)
- Antonio Bove (Dipartimento di Matematica, Univ. Bologna, Italy)
- Haim Brezis (Analyse Numérique, Univ. Pierre et Marie Curie, Paris, France)
- Giuseppe Buttazzo (Scuola Normale Superiore, Pisa, Italy)
- Elio Cabib (Istituto di Meccanica Teorica e Applicata, Univ. Udine, Italy)
- Michele Carriero (Dipartimento di Matematica, Univ. Lecce, Italy)
- Filippo Chiarenza (Dipartimento di Matematica, Univ. Catania, Italy)
- Maurizio Chicco (Istituto di Matematica, Univ. Genova, Italy)
- Franco Conti (Scuola Normale Superiore, Pisa, Italy)
- Gianni Dal Maso (S.I.S.S.A., Trieste, Italy)
- Anneliese Defranceschi (S.I.S.S.A., Trieste, Italy)
- Ennio De Giorgi (Scuola Normale Superiore, Pisa, Italy)
- Michele Emmer (Dipartimento di Matematica, Univ. Roma I, Italy)
- Robert Finn (Department of Mathematics, Stanford Univ., California, USA)

- Hisao Fujita-Yashima (Scuola Normale Superiore, Pisa, Italy)
- Paul R. Garabedian (Courant Institute, New York Univ., USA)
- Nicola Garofalo (Dipartimento di Matematica, Univ. Bologna, Italy)
- Maria Giovanna Garroni (Dipartimento di Matematica, Univ. Roma I, Italy)
- Fabio Gastaldi (Istituto di Analisi Numerica, Pavia, Italy)
- Andrea Gavioli (Dipartimento di Matematica, Univ. Modena, Italy)
- Mariano Giaquinta (Istituto di Matematica Applicata, Univ. Firenze, Italy)
- Enrico Giusti (Istituto di Matematica, Univ. Firenze, Italy)
- Robert Gulliver (School of Mathematics, Univ. of Minnesota, Minneapolis, USA)
- Erhard Heinz (Mathematisches Institut, Univ. Göttingen, Fed. Rep. of Germany)
- Mimmo Iannelli (Dipartimento di Matematica, Univ. Trento, Povo, Italy)
- Fritz John (Courant Institute, New York Univ., USA)
- Ermanno Lanconelli (Dipartimento di Matematica, Univ. Bologna, Italy)
- Peter Laurence (Mathematics Department, Pennsylvania State University, University Park, USA)
- Antonio Leaci (Dipartimento di Matematica, Univ. Lecce, Italy)
- Francesco Leonetti (Dipartimento di Matematica, Univ. L'Aquila, Italy)
- Marco Longinetti (I.A.G.A., Firenze, Italy)
- Enrico Magenes (Dipartimento di Matematica, Univ. Pavia, Italy)
- Giovanni Mancini (Dipartimento di Matematica, Univ. Trieste, Italy)
- Pierangelo Marcati (Dipartimento di Matematica, Univ. L'Aquila, Italy)
- Silvana Marchi (Dipartimento di Matematica, Univ. Parma, Italy)
- Elvira Mascolo (Istituto di Matematica, Univ. Salerno, Italy)
- Umberto Massari (Dipartimento di Matematica, Univ. Ferrara)
- Albert Milani (Dipartimento di Matematica, Univ. Torino, Italy)
- Enzo Mitidieri (Dipartimento di Matematica, Univ. Trieste, Italy)
- Giuseppe Modica (Istituto di Matematica Applicata, Univ. Firenze, Italy)
- Stefano Mortola (Dipartimento di Matematica, Univ. Pisa, Italy)
- Umberto Mosco (Dipartimento di Matematica, Univ. Roma I, Italy)
- Venkatesha Murthy (Dipartimento di Matematica, Univ. Pisa, Italy)
- Roberta Musina (S.I.S.S.A., Trieste, Italy)

- Wei Ming Ni (School of Mathematics, Univ. of Minnesota, Minneapolis, USA)
- Louis Nirenberg (Courant Institute, New York Univ., USA)
- Tullia Norando (Dipartimento di Matematica, Politecnico Milano, Italy)
- Pirro Oppezzi (Istituto Matematico, Univ. Genova, Italy)
- Gabriella Paderni (S.I.S.S.A., Trieste, Italy)
- Luigi Pepe (Dipartimento di Matematica, Univ. Ferrara, Italy)
- Carlo Pucci (Istituto di Matematica, Univ. Firenze, Italy)
- Andrea Pugliese (Dipartimento di Matematica, Univ. Trento, Povo, Italy)
- Marco Sabatini (Dipartimento di Matematica, Univ. L'Aquila, Italy)
- Sandro Salsa (Dipartimento di Matematica, Univ. Milano, Italy)
- Donato Scolozzi (Dipartimento di Matematica, Univ. Pisa, Italy)
- Paolo Secchi (Dipartimento di Matematica, Univ. Trento, Povo, Italy)
- Raul Serapioni (Dipartimento di Matematica, Univ. Trento, Povo, Italy)
- James Serrin (School of Mathematics, Univ. of Minnesota, Minneapolis, USA)
- Dan Socolescu (Institut für Angew. Mathematik, Univ. Karlsruhe, Fed. Rep. of Germany)
- Joel Spruck (Department of Mathematics, Univ. Massachusetts, Amherst, USA)
- Italo Tamanini (Dipartimento di Matematica, Univ. Trento, Povo, Italy)
- Ermanna Tomaini (Dipartimento di Matematica, Univ. Ferrara, Italy)
- Franco Tomarelli (Dipartimento di Matematica, Univ. Pavia, Italy)
- Mario Tosques (Dipartimento di Matematica, Univ. Pisa, Italy)
- Luciano Tubaro (Dipartimento di Matematica, Univ. Trento, Povo, Italy)
- Alberto Valli (Dipartimento di Matematica, Univ. Trento, Povo, Italy)
- Giorgio Vergara Caffarelli (Dipartimento di Matematica, Univ. Pisa, Italy)
- Epifanio Virga (Dipartimento di Matematica, Univ. Pisa, Italy)
- Maria Antonia Vivaldi (Dipartimento di Matematica, Univ. Roma I, Italy)
- Anna Zaretti (Dipartimento di Matematica, Politecnico Milano, Italy)

and the Conference Committee:

- Stefan Hildebrandt (Mathematisches Institut, Univ. Bonn, Fed. Rep. of Germany)
- David Kinderlehrer (School of Mathematics, Univ. of Minnesota, Minneapolis, USA)
- Mario Miranda (Dipartimento di Matematica, Univ. Trento, Povo, Italy)

CONTENTS

GLOBAL SOLVABILITY OF SECOND ORDER EVOLUTION EQUATIONS IN BANACH SCALES

Alberto Arosio
Dipartimento di Matematica - Università
Via dell'Università
43100 Parma (Italy)

Dedicated to Hans Lewy

0. Introduction

Let us consider the abstract Cauchy problem

$$(0.1)\quad \begin{cases} u'' + A(t)\, u = 0 \quad (t > 0)\ , \\ u(0) = u_0, \quad u'(0) = u_1\ , \end{cases}$$

where $A(t)$ is a family of non-negative symmetric operators in a Hilbert space H.

Problem (0.1) is the abstract version of the mixed problem for some partial differential equations of evolutionary type, not necessarily of kovalevskian type; see [AS2].

In [AS2] it is proved that problem (0.1) is globally well-posed in the Banach scale generated by a suitable operator.

Here we pose the question of the global solvability of (0.1) in a general (not necessarily generated by an operator) Banach scale, and we give a first result in this direction (Theorem 3.3).

1. Global solvability of (0.1) in a Sobolev-type framework

We recall that a Hilbert triplet (V,H,V') is composed from a reflexive Banach space V, a Hilbert space H and a dense continuous embedding of V into H: denoting by V' the antidual of V and identifying H' with H via the Riesz isomorphism, we have the following chain of dense and continuous embeddings

$$V \subset H \subset V'\ .$$

We shall denote by $|\ |$ and $(\,,\,)$ respectively the norm and the inner product in H, and by $<\,,\,>$ the antiduality between V and V'. We note that $<\,,\,>$ coincides with $(\,,\,)$ on $H \times V$.

For each $t \geq 0$, we assume that $A(t) \in B(V,V')$, the space of bounded linear operators from V into V', and moreover that A(t) is symmetric and non-negative, i.e. that for each $v,w \in V$ we have

$$<A(t)v,w> = \overline{<A(t)w,v>} \quad , \tag{1.1}$$

$$<A(t)v,v> \geq 0 \quad . \tag{1.2}$$

Thus, $V \times H$ is the natural space in which one can ask whether (0.1) is well-posed (the equation is then understood in a variational sense, cf. [LM]). The answer to such a question is in the affirmative under two assumptions. The first one has an algebraic character: A(t) must be coercive on V, uniformly on compact time intervals. The second one concerns the time regularity of the map $A(\bullet)$: it must be locally of bounded variation. More precisely, we have

Theorem 1.1 ([A2], cf. [DJ], [DT]). *Assume that* (1.1) *holds, and that for each* $T > 0$

$$<A(t)v,v> \geq c(T)\ \|v\|_V^2 \quad (c(T) > 0),\ \forall\, v \in V, 0 \leq t \leq T \quad , \tag{1.3}$$

$$A \in BV([0,T], B(V,V')) \quad . \tag{1.4}$$

Then for every u_0 in V and u_1 in H there exists a unique solution of (0.1)

$$u \in C^0([0,+\infty[,V) \cap C^1([0,+\infty[,H) \quad .$$

Regularity results may also be proved: under suitable further assumptions the solution belongs in fact to $C^1([0,+\infty[,V) \cap C^2([0,+\infty[,H)$, see [A2], [A3] and the references quoted there. For further regularity, see [K].

If the assumptions (1.3) and (1.4) are somewhat relaxed, then even the *local* solvability in $V \times H$ may, in general, fail to be true. Actually any Sobolev-type framework is no more adequate, as it is shown by two delicate counterexamples, due to F. Colombini, E. De Giorgi and S. Spagnolo [CDS] and F. Colombini and S. Spagnolo [CS2], for the special case

$$\begin{cases} u'' + a(t)\, A_0 u = 0 \quad (t > 0) \quad , \\ u(0) = u_0, \quad u'(0) = u_1 \quad , \end{cases} \tag{1.5}$$

where A_0 is a positive, symmetric isomorphism of V onto V', and $a(t)$ is a continuous non-negative function. More precisely, [CDS] ([CS2]) exhibited a function $a(t) \geq 1$ ($a(t) \geq 0$) which is Hölder continuous for each exponent strictly less than 1 (infinitely differentiable), and a pair of initial data

$$u_0,\ u_1 \in D_\infty(A_0) \stackrel{\text{def}}{=} \bigcap_{k=1}^{\infty} D(A_0^k) \ ,$$

such that no local solution to (1.5) exists.

2. A preliminary result: the global solvability of (1.5) in the analytic-type framework

The above counterexamples are straightforward abstract version of results proved by [CDS] and [CS2] for the hyperbolic equation

$$\text{(2.1)} \qquad \frac{\partial^2 u}{\partial t^2} - a(t)\frac{\partial^2 u}{\partial x^2} = 0 \quad (x \in \mathbf{R},\ t > 0) \ ,$$

under the periodic boundary condition $u(x,t) = u(x + 2\pi, t)$.

On the other hand, [CDS] provided also a positive result (1): if the initial data are *analytic* 2π-periodic functions, then the Cauchy problem for (2.1) admits a classical (C^2) solution for all time.

In order to generalize such a result to the abstract Cauchy problem (1.5), we need to introduce the following definition, cf. [LM]:

Definition 2.1. Let $B : D(B) \subset H \to H$ be a closed linear operator. A vector v in $D_\infty(B)$ is a B-*analytic vector* iff there exists some $r > 0$ such that

$$\sum_{j=0}^{\infty} \left(|B^j v| \frac{r^j}{j!} \right)^2 < \infty \ .$$

The supremum of admissible r in (2.2) is called the B-analyticity radius of v, and is denoted by $r_B(v)$.

Theorem 2.1 ([AS2], cf. [CDS], [A1], [AS1]). *Assume that* A_0 *is a positive symmetric isomorphism of* V *onto* V', *and that* $a(t)$ *is a continuous non-negative function.*

Then for each pair (u_0,u_1) of $A_0^{1/2}$-analytic vectors (2) there exists a unique global solution to (1.5). The solution and its derivative are $A_0^{1/2}$-analytic vectors for all time, with

$$\min\left\{ r_{A_0^{1/2}}(u(t)),\ r_{A_0^{1/2}}(u'(t)) \right\} = \text{constant} \quad (\forall\, t > 0) \ .$$

Remark 2.1. The above theorem is a preliminary step to solve [AS1] for all time the Cauchy problem for the *nonlinear* equation

$$(2.2) \qquad u'' + m(\langle A_0u,u\rangle)\, A_0u = 0 \quad (t > 0) ,$$

where the function $m(\cdot)$ is merely assumed to be continuous and non-negative, and A_0^{-1} is a compact operator in H. This improves the results of S. Bernstein [Be] and S.I. Pohozaev [P], and answers affirmatively a question raised by J.L. Lions [L]. Of course, uniqueness for the Cauchy problem for (2.2) may fail.

We note that some authors (G.F. Carrier [C], R. Narasimha [N]) interpreted (2.2) for $m(\rho) = c^2 + \varepsilon\rho$, $\varepsilon > 0$, as a nonlinear approximate model for the transversal vibrations of an elastic string.

3. The global solvability of (0.1) in the Banach scale generated by an operator

The above investigations show that if we want to relax the assumptions (1.3) and (1.4) in the study of (0.1), then we must introduce a richer framework: an analytic-type framework. Such a tool is provided by the notion of scale of Banach spaces, or Banach scale.

Definition 3.1. A *Banach scale* is a family of Banach spaces $(X_r)_{0<r<R}$ $(0 < R \leq \infty)$, with norms $|\ |_r$, such that

$$X_r \subset X_s \text{ with } |\ |_s \leq |\ |_r \text{ whenever } s \leq r .$$

We set

$$r(v) =^{def} \sup\{r : v \in X_r\} .$$

The number $r(v)$ is called the radius of v (in the scale (X_r)).

Definition 3.2. A Banach scale (X_r) is *dense* (in itself) iff X_r is dense in X_s whenever $s \leq r$.

Definition 3.3. A Banach scale (X_r) is *compatible* with the Hilbert triplet (V,H,V') iff $X_r \subset V$ with continuous inclusion for each r, and X_{r_0} is dense in H for some r_0.

The set of B-analytic vectors is, in an obvious way, a Banach scale:

Definition 3.4. Let $B : D(B) \subset H \to H$ be a closed linear operator. We set

$$X_r(B) =^{def} \left\{ v \in D_\infty(B) : |v|_r =^{def} \left(\sum_{j=0}^{\infty} \left(|B^j v| \frac{r^j}{j!} \right)^2 \right)^{1/2} < \infty \right\}.$$

We say that the scale $(X_r(B))$ is *generated* by the operator B. It may be proved (see [AS2]) that such a scale is dense and compatible with (V,H,V') if B is a self-adjoint operator in H with $D(B) \subset V$.

Definition 3.5. Let (X_r) be a dense Banach scale, compatible with (V,H,V'). A *radius function* for (0.1) is any continuous non-increasing function $y : [0,T(y)[\rightarrow]0,\infty[$, $0 < T(y) \leq \infty$, with the following property: whenever the initial data u_0 and u_1 have a radius in (X_r) larger or equal than $y(0)$, then (0.1) admits a unique solution u in the class $C^0([0,T(y)[,V) \cap C^1([0,T(y)[,H)$, which moreover satisfies

$$u \in \bigcap_{s<y(T)} W^{2,1}([0,T],X_s) \quad \text{for each} \quad T \in]0,T(y)[\tag{3.1}$$

(so the radius of $u(t)$ and $u'(t)$ are estimated from below by $y(t)$).

Definition 3.6. Let (X_r) be a dense Banach scale, compatible with (V,H,V'). The Cauchy problem (0.1) is *locally (globally) well-posed* in (X_r) iff for each $r_0 > 0$ there exists a radius function y for (0.1) with $y(0) = r_0$ (and y is defined for all time).

Theorem 2.1 may be rewritten in this terminology as follows:

Theorem 2.1'. *The Cauchy problem* (1.5) *is globally well-posed in the Banach scale* $(X_r(A_0^{1/2}))$, *with radius functions constant in time.*

The linear abstract version of the Cauchy-Kovalevskaya theorem, due to T. Yamanaka [Y] and L.V. Ovsjannikov [O], may be generalized to the case of second order equations, yielding the

Theorem 3.1 [C]. *Let* (X_r) *be a dense Banach scale, compatible with* (V,H,V'). *Assume that*

i) A(t) *is of analytic order* ≤ 2 *in the scale* (X_r), *i.e.* A(t) *maps* X_r *into* X_s *whenever* $s < r$, *with*

$$| A(t)v |_s \leq c(T)(r-s)^{-2} \, | v |_r \quad (s < r, 0 \leq t \leq T) \ ;$$

ii) *for each* $v \in X_r$, *the map* $A(\cdot)v$ *is a strongly measurable* X_s*-valued function, whenever* $s < r$.

Then (0.1) is *locally* well-posed in the scale (X_r), with radius functions which die down linearly.

We formulate the following

Open problem. To find compatibility assumptions between the operator map $A(\bullet)$ and a dense Banach scale (X_r) compatible with (V,H,V'), which ensure that (0.1) is *globally* well-posed in (X_r).

In [AS2] a partial answer has been provided for the case of the Banach scale generated by an operator:

Theorem 3.2 [AS2]. *Assume that* $A \in C^0([0,+\infty[,B(V,V'))$.

Let $B : V \to H$ *be a linear operator such that*

i) *the graph norm of* B *is equivalent to the initial one of* V;

ii) $A(t)$ *has Sobolev order* ≤ 2 *with respect to* B, *i.e. for each* $T > 0$

$$| A(t)v | \leq c(T) (| v | + | Bv | + | B^2 v |) \quad (0 \leq t \leq T) \ ;$$

iii) $A(t)$ *maps the space of* B*-analytic vectors into* $D_\infty(B)$;

iv) *for each* B*-analytic vector* v *and each* $j \in N$, *the map* $B^j A(\bullet)v$ *is a strongly measurable* H*-valued function.*

Let $R > 0$ *be such that*

v) *the scale* $(X_r(B))_{0<r<R}$ *is dense, and compatible with* (V,H,V');

vi) $$\left|\left(B^j A(t) - A(t) B^j\right) v\right| \leq K\, j < A(t)B^j v, B^j v>^{1/2} + K^2 (j+2)!\, R^{-j} \sum_{h=0}^{j} \frac{R^h}{h!} |B^h v| \ .$$

Then (0.1) is globally well-posed in the scale $(X_r)_{0<r<R}$ with radius functions which die down exponentially $(\sim \exp(-3Kt))$.

The above theorem generalizes some of the results on hyperbolic partial differential equations of second order due to E. Jannelli [J], which improved the earlier results of F. Colombini and S. Spagnolo [CS1] and J.-M. Bony and P. Schapira [BS], which in turn had improved a classical result of S. Mizohata [M].

4. The global solvability of (0.1) in a general Banach scale

We provide here a second result in the direction of solving the above Open Problem, concerning the special case (1.5).

Theorem 3.3. *Assume that* A_0 *is a positive symmetric isomorphism of* V *onto* V', *and that* a(t) *is a continuous non-negative function.*

Let (X_r) *be a dense Banach scale, compatible with* (V,H,V'). *For each* r, *assume that*

i) $A_0^{1/4}$ *maps* X_r *into* X_s *whenever* $s<r$;

ii) X_s *is a Hilbert space with inner product* $(\,,)_s$, *and* $A_0^{1/4}$ *is symmetric with respect to* $(\,,)_s$ *on the subspace* X_r, *whenever* $s<r$;

iii) *there exists* $\frac{d}{ds}|v|_s^2$ *as a continuous function on* $X_r\times]0,r[$, *with*

$$|A_0^{1/4}v|_s^2 \le c(r)\frac{d}{ds}|v|_s^2 \quad on \quad]0,r[\quad (\forall\, v\in X_r)\,.$$

Then (1.5) is *globally* well-posed in the scale (X_r), with radius functions constant in time.

Proof. Let $r_0 =^{def} \min\{r(u_0),r(u_1)\}$, and let $r_1>r_0$.

If $s<r<r_1$, integrating iii) we get

$$\begin{aligned}(r-s)\,|A_0^{1/4}v|_s^2 &\le \int_s^r |A_0^{1/4}v|_\rho^2\,d\rho\\ &\le c(r_1)\left(|v|_r^2-|v|_s^2\right)\\ &\le c(r_1)\,|v|_r^2\end{aligned}$$

for each $v\in X_{r_1}$ hence for each $v\in X_r$ since the scale (X_r) is assumed to be dense. Then condition i) of Theorem 3.1 holds true in the subscale $(X_r)_{0<r<r_1}$ whence (1.5) is locally well-posed in such a scale.

Moreover, there exists $T=T(r_0,r_1)>0$ with the property that if the radii of the initial data are larger than or equal to r_1, then the solutions belong to $C^2([0,T],X_{r_0})$. If we provide a suitable a priori estimate of such solutions, then, using in an essential way the fact that the scale (X_r) is assumed to be dense in itself, the local existence result of Theorem 3.1 may be improved to a *global* one. Such an estimate will be deduced from the study of a certain energy function, which we now introduce.

Let $b\in C^1([0,T])$, $b(t)>0$, and let $y(t)$ be an absolutely continuous non-increasing function on $[0,T]$, $0<y(t)<r_0$ (we will choose b and y later on). Let us define the energy function E(t) by setting

$$E(t) =^{def} b(t)\,|A_0^{1/2}u(t)|^2_{y(t)} + |u'(t)|^2_{y(t)} \quad (0 \le t \le T)\ .$$

Since $A_0{}^{1/4}$ is $(\ ,\ )_r$-symmetric, we have

$$(3.2) \quad E'(t) = b'\,|A_0^{1/2}u|^2_y + 2\mathrm{Re}(bA_0u + u'',u')_y + y'\partial E/\partial y$$

$$\le |b'|\,|b|^{-1}E + 2\,|b-a|\,|(A_0u,u')_y| + y'\partial E/\partial y$$

$$\le |b'|\,|b|^{-1}E + |b-a|\,|b|^{-1/2}\left(b\,|A_0^{3/4}u|^2_y + |A_0^{1/4}u'|^2_y\right) + y'\partial E/\partial y$$

$$\le |b'|\,|b|^{-1}E + \left(c(r_0)\,|b-a|\,|b|^{-1/2} + y'\right)\partial E/\partial y\ .$$

We can now choose the functions b and y. For a fixed $\varepsilon > 0$, one can choose [CDS] th function b in such a way that

$$c(r_0)\int_0^T |b-a|\,|b|^{-1/2}dt < \varepsilon\ .$$

If we denote by $y(t)$ the function defined by

$$y(t) =^{def} r_0 - \varepsilon - c(r_0)\int_0^t |b-a|\,|b|^{-1/2}dx \quad (0 \le t \le T)\ ,$$

it follows from (3.2) that

$$(3.3) \qquad E'(t) \le c(r_0,T,\varepsilon)\,E(t) \quad (0 \le t \le T)\ ,$$

$$(3.4) \qquad y(t) \ge r_0 - 2\varepsilon \quad (0 \le t \le T)\ .$$

By the classical argument of Gronwall, (3.3) gives an a priori estimate of the energy function E, hence of $|u'|_y$, and, in turn, of $|u|_y$. Then arguing as in Step 6 of Theorem 3.2 of [AS2], we can deduce, using the assumption that the scale (X_r) is dense in itself, that the solution to (1.5) exists on the whole interval $[0,T]$ as an $X_{r_0-2\varepsilon}$-valued $W^{2,1}$ function.

Since $\varepsilon > 0$ is arbitrary, we have proved that the constant map r_0 is a radius function for (1.5) on $[0,T]$. By considering (1.5) as a sequence of Cauchy problems on $[T_{n-1}, T_n]$, one can complete the proof.

Remark 3.1. As a simple application of Theorem 3.3, we get an easier proof of Theorem 2.1'. Indeed if we choose

$$X_r = D(\exp(rA_0^{1/2})), \quad |v|_r = \left(\sum_{j=0}^{\infty} |A_0^{j/4} v|^2 \frac{(2s)^j}{j!} \right)^{1/2},$$

then the scale (X_r) is dense and compatible with (V,H,V'), since $A_0^{1/2}$ is a positive self-adjoint operator with $D(A_0^{1/2}) = V$ [AS2], and it is easy to check the other assumptions of Theorem 3.3. Now, such a scale is equivalent (with respect to (3.1)) to the scale generated by $A_0^{1/2}$.

Notes

(1) Since we are not interested in generalizing results based on the phenomenon of dependence domains, we report here from [CDS], and later on from [CS1], [CS2], [J], [BS], [M], only results relevant to the topic.

(2) The realization of A_0 in H happens to be a positive self-adjoint operator in H, hence it admits a unique positive self-adjoint square root $A_0^{1/2}$.

Acknowledgements. The author was partially supported by the Italian Ministero della Pubblica Istruzione. The author is a member of G.N.A.F.A. (C.N.R.).

References

[A1] **A. Arosio**, Asymptotic behaviour as $t \to +\infty$ of the solutions of linear hyperbolic equations with coefficients discontinuous in time (on a bounded domain), J. Differential Equations 39 (1981), no. 2, 291-309.

[A2] **A. Arosio**, Linear second order differential equations in Hilbert space. The Cauchy problem and asymptotic behaviour for large time, Arch. Rational Mech. Anal. 86 (1984), no. 2, 147-180; Equations differentielles operationnelles lineaires du deuxieme ordre: problème de Cauchy et comportement asymptotique lorsque $t \to +\infty$, C.R. Acad. Sci. Paris, Ser. I, 295 (1982), 83-86.

[A3] **A. Arosio**, Abstract linear hyperbolic equations with variable domain, Ann. Mat. Pura Appl. 135 (1983), 173-218.

[AS1] **A. Arosio and S. Spagnolo,** Global solutions of the Cauchy problem for a non-linear hyperbolic equation, Nonlinear Partial Differential Equations and their applications, College de France, Seminar 6, 1-26; H. Brezis and J.L. Lions ed., Research Notes Math. 109, Pitman, Boston 1984.

[AS2] **A. Arosio and S. Spagnolo,** Global existence for abstract evolution equations of weakly hyperbolic type, J. Math. Pures Appl. 65 (1986), 1-43.

[Be] **S. Bernstein,** Sur une classe d'equations fonctionnelles aux derivees partielles, Izv. Akad. Nauk SSSR, Ser. Mat. 4 (1940), 17-26.

[BS] **J.-M. Bony and P. Schapira,** Existence et prolongement des solutions holomorphes des equations aux derivees partielles, Invent. Math. 17 (1972), 95-105.

[C] **L. Cardosi,** Evolution equations in scales of abstract Gevrey spaces, Boll. Un. Mat. Ital., Ser. VI, 4-C, no. 1 (1985), 379-406.

[Ca] **G.F. Carrier,** On the non-linear vibration problem of the elastic string, Quart. Appl. Math. 3 (1945), 157-165; A note on the vibrating string, ibidem 7 (1949), 97-101.

[CDS] **F. Colombini, E. De Giorgi and S. Spagnolo,** Sur les equations hyperboliques avec des coefficients qui ne dependent que du temps, Ann. Scuola Norm. Sup. Pisa Cl. Sci. (4) 6 (1979), 511-559.

[CS1] **F. Colombini and S. Spagnolo,** Second order hyperbolic equations with coefficients real analytic in space variables and discontinuous in time, J. Analyse Math. 38 (1980), 1-33.

[CS2] **F. Colombini and S. Spagnolo,** An example of a weakly hyperbolic Cauchy problem not well posed in C^∞, Acta Math. 148 (1982), 243-253.

[DJ] **V.I. Derguzov and V.A. Jakubovič,** Existence of solutions of linear Hamilton equations with unbounded operator coefficients, Dokl. Akad. Nauk SSSR 151 (1963), no. 6, 1264-1267 (transl.: Soviet Math. Dokl. 4 (1963), 1169-1172.

[DT] **L. De Simon and G. Torelli,** Linear second order differential equations with discontinuous coefficients in Hilbert spaces, Ann. Scuola Norm. Sup. Pisa Cl. Sci. (4) 1 (1974), 131-154.

[J] **E. Jannelli,** Weakly hyperbolic equations of second order with coefficients real analytic in space variables, Comm. Partial Differential Equations 7 (1982), 537-558.

[K] **T. Kato,** Linear and quasi-linear equations of evolution of hyperbolic type, Hyperbolicity, C.I.M.E. Cortona, 1976, Liguori, Napoli (1977), 127-192.

[LM] **J.L. Lions and E. Magenes,** Problèmes aux limites non homogènes et applications, Dunod, Paris 1968.

[M] **S. Mizohata,** Analyticity of solutions of hyperbolic systems with analytic coefficients, Comm. Pure Appl. Math. 14 (1961), 547-559.

[N] **R. Narasimha,** Nonlinear vibration of an elastic string, J. Sound Vibration 8 (1968), 134-146.

[O] **L.V. Ovsjannikov,** A singular operator in a scale of Banach spaces, Dokl. Akad. Nauk SSSR 163 no. 4 (1965), 819-822 (transl.: Soviet Math. Dokl. 6 (1965), 1025-1028).

[P] **S.I. Pohožaev,** On a class of quasilinear hyperbolic equations, Mat. Sbornik 96 (138), no. 1 (1975), 152-166 (transl.: Math. USSR Sbornik 25, no. 1 (1975), 145-158).

[Y] **T. Yamanaka,** Note on Kowalevskaja's system of partial differential equations, Comment.

ON THE INCOMPRESSIBLE LIMIT OF THE COMPRESSIBLE NAVIER-STOKES EQUATIONS

Hugo Beirão da Veiga
Dipartimento di Matematica - Università di Trento
38050 Povo (Italy)

Dedicated to Hans Lewy

1. Introduction. In papers [2], [3], we proved a set of general and quite complete results concerning the stationary, compressible, heat-conductive, Navier-Stokes equations (an overview of some of those results can be found as well in reference [5]). Here, for convenience, we will consider the simplified system

$$\text{(1.1)}\qquad \begin{aligned} &-\mu\,\Delta u - \upsilon\nabla\,\mathrm{div}\,u + \rho\,(u\cdot\nabla)u + \nabla p(\rho) = \rho f\,,\\ &\mathrm{div}(\rho u) = 0\,,\quad \text{in } \Omega\,,\\ &u_{|\Gamma} = 0\,,\end{aligned}$$

which describes the barotropic motion of a compressible viscous fluid in an open bounded domain Ω of the euclidean space $\mathbf{R}^n$, for arbitrarily large n. It is assumed here that Ω lies (locally) on one side of its boundary Γ, a C^2 manifold. As usual, we will use the notation $(v\cdot\nabla)u = \sum_i v_i(\partial u/\partial x_i)$.

In equation (1.1), $u(x) = (u_1(x),\ldots,u_n(x))$ is the velocity field, $\rho(x)$ is the density of the fluid, $f(x)$ is the assigned external force field, and $p = p(\rho)$ is the pressure. In order to avoid technicalities, we assume that the coefficients $\mu > 0$, and $\upsilon > -\mu$, are constants.

Since the total mass of fluid is given, we impose on $\rho(x)$ the following condition:

$$\text{(1.2)}\qquad \bar\rho \equiv (1/|\Omega|)\int_\Omega \rho(x)dx = m\,,$$

or equivalently

$$\text{(1.3)}\qquad \bar\rho \equiv (1/|\Omega|)\int_\Omega \sigma(x)dx = 0\,,$$

where $m > 0$ is given, and $\sigma(x)$ is defined by $\rho(x) = m + \sigma(x)$.

The function $p(\rho)$ is defined, and has a Lipschitz continuous first derivative $p'(\rho)$ in a neighborhood $[M-\ell, m+\ell]$ of m. Without loss of generality, we assume here that $0 < \ell \leq m/2$. We set $k = p'(m)$, we assume that $k > 0$, and we define $\omega(\sigma)$ by setting $p'(m+\sigma) = k + \omega(\sigma)$.

Moreover, we set $T=\sup|\omega(\sigma)|$, and $S=\sup|(\omega(\sigma)-\omega(\tau))/(\sigma-\tau)|$.

Before describing some of the results proved in [2], [3], we introduce some notation:

We denote by $W^{j,p}$, j an integer, $1<p<+\infty$, the Sobolev space $W^{j,p}(\Omega)$, endowed with the usual norm $\|\ \|_{j,p}$, and by $|\ |_p$, $1\leq p\leq+\infty$, the usual norm in $L^p=L^p(\Omega)$. Hence, $\|\ \|_{0,p}=|\ |_p$. For convenience, we also use the symbol $W^{j,p}$ to denote the space of vector fields v in Ω such that $v_i\in W^{j,p}(\Omega)$, $i=1,2,\ldots,n$. This convention applies to all the functional spaces and norms used here. For $j\geq 1$ we define $W_0^{j,p}=\{v\in W^{j,p}: v=0 \text{ on } \Gamma\}$. Note that $W_0^{j,p}=W^{j,p}\cap W_0^{1,p}$, is not the closure of $\mathcal{D}(\Omega)=C_0^\infty(\Omega)$ in $W^{j,p}$. Furthermore, we set $\overline{W}^{j,p}=\{\tau\in W^{j,p}:\overline{\tau}=0\}$, $\overline{W}_0^{j,p}=W_0^{j,p}\cap\overline{W}^{j,p}$, $j\geq 1$, where in general $\overline{\phi}$ denotes the mean value of $\phi(x)$ in Ω. Moreover, for vector fields, we define

$$W_{0,d}^{j,p}=\{v\in W_0^{j,p}: \operatorname{div} v=0 \text{ on } \Gamma\},\quad j\geq 2.$$

Finnally, $\mathcal{L}(X,Y)$ denotes the Banach space of bounded linear operators from X into Y. Moreover, $\mathcal{L}(X)\equiv\mathcal{L}(X,X)$.

The symbols c, c_i, $i\geq 0$, will denote positive constants depending at most on Ω,n,p. The symbol c may be utilized (even in the same equation) to indicate distinct constants.

One has the following result (cfr. [3], [5]; for related results, cfr. [2], [10], [11]):

Theorem 1.1. *Let* $p>n$. *There exist positive constants* c'_0, c'_1, *such that if* $f\in L^p$, *and*

$$|f|_p\leq c'_0\,, \tag{1.4}$$

then there exists a unique solution $(u,\rho)\in W_0^{2,p}\times W^{1,p}$ *of problem* (1.1), (1.2) *in the ball*

$$\|u\|_{2,p}+\|\rho-m\|_{1,p}\leq c'_1\,, \tag{1.5}$$

where the constants c'_0, c'_1 *depend only on* $\Omega,n,p,\mu,\upsilon,k,m,\ell,T,S$.

In reference [3] we proved a similar existence result in $W^{j,p}$ under the assumption $j+2>n/p$, where $j\geq -1$, $p\in\,]1,+\infty[$ (theorem 1.1 corresponds to $j=-1$). Moreover, in reference [3] the results were proved for the heat-conductive case.

We remark that in the presence of arbitrarily large external forces f, the above problem is (in general) not solvable; see [3], section 5.

In papers [2], [3], we have also studied the incompressible limit of a family of compressible barotropic flows. We assume that $p_\lambda(\rho)$ is a family of state functions depending on a parameter λ, such that $k_\lambda \equiv p'_\lambda(m) \to +\infty$ as $\lambda \to +\infty$, and we prove that the solution of the incompressible Navier-Stokes equations (1.7) is the limit of the solution of the compressible Navier-Stokes equations (1.6), as $\lambda \to +\infty$. Here, for the readers convenience, we will assume that $p_\lambda(\rho)$ has the particular form [1]

$$p_\lambda(\rho) = \lambda^2 p(\rho) \quad , \quad \lambda \in [1,+\infty[,$$

where $p(\rho) \in C^{1,\alpha}([m-(\ell/k), m+(\ell/k)])$ for fixed $\alpha \in]0,1]$ and $\ell > 0$. As above, we set $k=p'(m)$, and we assume that $k>0$. Without loss of generality, we assume that $\ell<(mk)/2$. We define $\omega(\sigma)$ as above, and we denote by S a non-negative real number such that $|\omega(\sigma)| \le S|\sigma|^\alpha$.

One has the following result:

Theorem 1.2. *Let* $p>n$ *be fixed, and let the above assumptions on the family of state functions* $p_\lambda(\rho)$ *hold. Then, there exist positive constants* c'_2, c'_3, *depending at most on* $\Omega, n, p, \mu, \upsilon, m, \ell, k, S$, *such that if* $f \in L^p$, $|f|_p \le c'_2$, *then the following statements hold:*

(i) *for each* λ , *the problem*

$$\text{(1.6)} \qquad \begin{aligned} &-\mu \Delta u_\lambda - \upsilon \nabla \operatorname{div} u_\lambda + \rho_\lambda (u_\lambda \cdot \nabla) u_\lambda + \nabla p_\lambda(\rho_\lambda) = \rho_\lambda f \ , \\ &\operatorname{div}(\rho_\lambda u_\lambda) = 0 \ \text{ in } \Omega, \\ &(u_\lambda)_{|\Gamma} = 0, \quad \overline{\rho_\lambda} = m, \end{aligned}$$

has a unique solution $(u_\lambda, \rho_\lambda) \in W_0^{2,p} \times W^{1,p}$ *in the ball* $\|u_\lambda\|_{2,p} \le c'_3$, $\|\rho_\lambda - m\|_{1,p} \le c'_3/k_\lambda$.

(ii) *Let* $\lambda \to +\infty$. *Then* $u_\lambda \to u_\infty$ *weakly in* $W_0^{2,p}$, $\operatorname{div} u_\lambda \to 0$ *weakly in* $W_0^{1,p}$, $\rho_\lambda \to m$ *strongly in* $W^{1,p}$, $\nabla p_\lambda(\rho_\lambda) \to \nabla\pi$ *weakly in* L^p, *where* $(u_\infty, \nabla\pi)$ *is the unique solution of the incompressible Navier-Stokes equations*

[1] In this case, $1/\lambda$ may be viewed as the Mach number.

$$
(1.7)\qquad
\begin{aligned}
&-\mu\,\Delta\, u_{\infty}+m(u_{\infty}\cdot\nabla)u_{\infty}+\nabla\pi(x)=mf\,,\\
&\operatorname{div} u_{\infty}=0 \quad in \quad \Omega\,,\\
&(u_{\infty})_{|\Gamma}=0\,.
\end{aligned}
$$

A similar result was proved for $n=3, j=0, p=2$, in reference [2]. The proof applies, without difficulty, to the case $W^{j,p}$, if $j+2 > n/p$, $\Gamma \in C^{3+j}$, $p(\rho) \in C^{3+j}$; see [6], for complete calculations.

In the sequel, we will concentrate our attention on Theorem 1.2.

2. A priori estimates. The incompressible limit.

By setting

$$
(2.1)\qquad \rho_{\lambda}(x)=m+\sigma_{\lambda}(x)\,,
$$

equation (1.6) becomes

$$
(2.2)\qquad
\begin{aligned}
&-\mu\,\Delta\, u_{\lambda}-\upsilon\nabla \operatorname{div} u_{\lambda}+k_{\lambda}\nabla\sigma_{\lambda}=F(f,u_{\lambda},\sigma_{\lambda})\,,\\
&m \operatorname{div} u_{\lambda}+u_{\lambda}\cdot\nabla\sigma_{\lambda}+\sigma_{\lambda}\operatorname{div} u_{\lambda}=0 \qquad \text{in } \Omega,\\
&(u_{\lambda})_{|\Gamma}=0,\quad \overline{\sigma_{\lambda}}=0,
\end{aligned}
$$

where $k_{\lambda}=\lambda^2 k$, and

$$
(2.3)\qquad F(f,u_{\lambda},\sigma_{\lambda})=(m+\sigma_{\lambda})[f-(u_{\lambda}\cdot\nabla)u_{\lambda}-(k_{\lambda}/k)\,\omega\,(\sigma_{\lambda})\nabla\sigma_{\lambda}]\,.
$$

In the sequel we denote by c', c'_i, $i \geq 0$, positive contants depending at most on $\Omega,n,p,\mu,\upsilon,m,k,\ell,S$. The symbol c' may be utilized (even in the same equation) to indicate different constants. We denote by c_3 a positive constant such that

$$
(2.4)\qquad |\tau|_{\infty} \leq c_3\, \|\tau\|_{1,p}\,,\quad \forall\, \tau \in \overline{W}^{1,p}\,.
$$

Part (i) on theorem 1.2 is proved as follows: we start by assuming that $f \in L^p$ verifies

(2.5) $\quad |f|_p \leq c'_2$,

and that $(v_\lambda,\tau_\lambda) \in W_{0,d}^{2,p} \times \overline{W}^{1,p}$ verifies

(2.6) $\quad \|v_\lambda\|_{2,p} \leq c'_3 , \quad k_\lambda \|\tau_\lambda\|_{1,p} \leq c'_3$,

and we prove that the linear system

$$
(2.7) \qquad \begin{aligned} &-\mu\Delta u_\lambda - \upsilon\nabla \operatorname{div} u_\lambda + k_\lambda \nabla\sigma_\lambda = F(f,v_\lambda,\tau_\lambda) , \\ &m \operatorname{div} u_\lambda + v_\lambda \cdot \nabla\sigma_\lambda + \sigma_\lambda \operatorname{div} v_\lambda = 0, \end{aligned}
$$

has a unique solution $(u_\lambda,\sigma_\lambda) \in W_{0,d}^{2,p} \times \overline{W}^{1,p}$. This first steep will be not discussed here. However, it is worth noting that the stationary linear system (2.7) is not elliptic, in the sense of Agmon, Douglis, and Nirenberg [1].

For each fixed f which satisfies (2.5), we denote by T_f the map $T_f : (v_\lambda,\tau_\lambda) \rightarrow (u_\lambda,\sigma_\lambda)$, defined on the ball

$$B = \{(v_\lambda,\tau_\lambda) \in W_{0,d}^{2,p} \times \overline{W}^{1,p}, \text{ such that (2.6) holds}\},$$

where $(u_\lambda,\sigma_\lambda)$ is the solution of problem (2.7).

The second steep on the proof of part (i) consists of showing that for each f verifying (2.5), one has

(2.8) $\quad T_f(B) \subset B$,

provided c'_2 and c'_3 are sufficiently small. This is the main tool, in order to prove that T_f has a fixed point in B. Clearly, such a fixed point is a solution of (2.2). This solution is unique in B, since T_f is a contraction on B, with respect to the $W_0^{1,2} \times L^2$ norm; see [2], [3].

In the sequel, we will concentrate our attention on point (2.8). We start by assuming that $c'_3 \leq \ell/c_3$. This guarantees that $m/2 \leq m+\sigma_\lambda(x) \leq (3/2)m$, $\forall\, x \in \Omega$.

On the calculations which follow, we abbreviate the notations, by setting $v_\lambda=v$, $u_\lambda=u$, $\tau_\lambda=\tau$, $\sigma_\lambda = \sigma$.

By taking into account the definition (2.3), one easily shows that

$$|F(f,v,\tau)|_p \leq (3m/2)\,(|f|_p + c\|v\|^2_{1,p}) + (k_\lambda/k) \quad S\,|\tau|_\infty^\alpha\,\|\tau\|_{1,p}\ .$$

In particular,

$$(2.9) \qquad |F(f,v,\tau)|_p \leq c'(|f|_p + \|v\|^2_{2,p} + \|k_\lambda\,\tau\|_{1,p}^{1+\alpha}),$$

since $k_\lambda \geq k$, and c' may depend on k. For convenience, we set $F=F(f,v,\tau)$.

By applying the divergence operator to both sides of equation $(2.7)_1$, by applying the operator $[(\mu+\upsilon)/m]\Delta$ to both sides of equation $(2.7)_2$, and by adding the corresponding results side by side, one gets

$$(2.10) \qquad \Delta(k_\lambda\sigma) + [(\mu+\upsilon)/mk_\lambda]\, v \cdot \nabla[\Delta(k_\lambda\sigma)] = G(F,v,\sigma,\tau),$$

where

$$(2.11) \qquad G(F,v,\sigma,\tau) = \operatorname{div} F(f,v,\tau) - [(\mu+\upsilon)/m]\,[2\,\Sigma_{i,j}\,(D_i v_j)(D^2_{ij}\sigma) + \Delta v\cdot\nabla\sigma + \\ + \Delta(\sigma \operatorname{div} v)].$$

The reader can easily verify that

$$(2.12) \qquad \|G\|_{-1,p} \leq |F|_p + c'\|v\|_{2,p}\,\|\sigma\|_{1,p}\ .$$

On the other hand, the solution $\Delta(k_\lambda\sigma)$ of problem (2.10) verifies the estimate

$$(2.13) \qquad \|\Delta(k_\lambda\sigma)\|_{-1,p} \leq (c_0/2)\,\|G\|_{-1,p}\ ,$$

provided

$$(2.14) \qquad c\,[(\mu+\upsilon)/mk_\lambda]\,\|v\|_{2,p} \leq 1/2\ ,$$

where c_0 and c an positive constants, depending only on Ω,n,p. This result will be discussed in section 3. Condition (2.14) holds, if the constant c'_3 (cfr. equation $(2.6)_1$) is sufficiently small.

Furthermore, by applying the divergence operator to both sides of equation $(2.7)_1$, one gets

$-(\mu+\upsilon)\,\Delta\,\mathrm{div}\,u = -\Delta(k_\lambda\sigma) + \mathrm{div}\,F$. Since $(\mathrm{div}\,u)_{|\Gamma} = 0$ (this follows from equation $(2.7)_2$), one gets

$$\text{(2.15)} \qquad \|\mathrm{div}\,u\|_{1,p} \le c' \left(\|\Delta(k_\lambda\sigma)\|_{-1,p} + |F|_p\right) .$$

Finally, by using well known estimates for the linear Stokes problem $-\mu\Delta u + k_\lambda\nabla\sigma = F + \upsilon\nabla\mathrm{div}\,u$ in Ω, with the boundary condition $u_{|\Gamma} = 0$, one proves that

$$\text{(2.16)} \qquad \|u\|_{2,p} + \|k_\lambda\sigma\|_{1,p} \le c' \left(|F|_p + \|\mathrm{div}\,u\|_{1,p}\right) .$$

From (2.9), (2.12), (2.13), (2.15), (2.16), one deduces that the solution $(u_\lambda,\sigma_\lambda)$ of problem (2.7) verifies the estimate

$$\text{(2.17)} \qquad \|u_\lambda\|_{2,p} + \|k_\lambda\sigma_\lambda\|_{1,p} \le c' \left(|f|_p + \|v\|^2_{2,p} + \|k_\lambda\tau\|_{1,p}^{1+\alpha}\right) + c'\left(k^{-1}\|v\|_{2,p}\,\|k_\lambda\sigma\|_{1,p}\right) .$$

By putting $c'_2 = (c'_3)^2$, and by choosing c'_3 small enough, one proves that u_λ and σ_λ verify the estimates (2.6). Hence, (2.8) holds.

■

As shown before, for each $\lambda \in [1,+\infty[$, the problem (1.6) has a unique solution (u_λ,ρ_λ) in the ball

$$\text{(2.18)} \qquad \|u_\lambda\|_{2,p} \le c'_3 , \quad \|\rho_\lambda - m\|_{1,p} \le c'_3/k_\lambda ,$$

provided $|f|_p \le c'_2$. The estimate (2.18) shows that there exists $u_\infty \in W_0^{2,p}$ such that $u_\lambda \to u_\infty$ weakly in $W_0^{2,p}$, and that $\rho_\lambda \to m$ strongly in $W^{1,p}$, as $\lambda \to +\infty$. The convergence of the whole sequence u_λ follows from the uniqueness of the solution u_∞ of problem (1.7), for small data.

From equation $(1.6)_2$ it follows that $\mathrm{div}\,u_\lambda \to 0$ in L^p. Moreover, $\mathrm{div}\,u_\lambda \to 0$ weakly in $W_0^{1,p}$, since $\|u_\lambda\|_{2,p}$ is uniformely bounded. Clearly, $\mathrm{div}\,u_\infty = 0$.

Finally, by passing to the limit as $\lambda \to +\infty$ in equation $(1.6)_1$, one gets

$$\text{(2.19)} \qquad \lim \nabla p_\lambda(\rho_\lambda) = \mu\Delta u_\infty + m[f - (u_\infty\cdot\nabla)u_\infty] ,$$

weakly in L^p, since: $\Delta u_\lambda \to \Delta u_\infty$, and $-\upsilon\nabla \operatorname{div} u_\lambda \to 0$, weakly in L^p; $\rho_\lambda \to m$, strongly in L^p; $\rho_\lambda(u_\lambda\cdot\nabla)u_\lambda \to m(u_\infty\cdot\nabla)u_\infty$, strongly in L^p. Obviously, the limit of the sequence $\nabla p_\lambda(\rho_\lambda)$ must be of the form $\nabla\pi(x)$, where $\pi \in \overline{W}^{1,p}$. This proves that $(u_\infty,\nabla\pi)$ is a solution of $(1.7)_1$.

3. The stationary transport equation.

Here we show that the solution $\Delta(k_\lambda\sigma)$ of problem (2.10) verifies the estimate (2.13), provided that the coefficient v satisfies (2.14). This result is a consequence of the following theorem[2]:

Theorem 3.1: *Let* $p \in]n, +\infty[$, *and let* $v=(v_1,\dots,v_n)\in W^{2,p}$ *be a vector field in* Ω *such that* $(v\cdot\upsilon)_{|\Gamma}=0$. *Then, the equation*

$$\lambda u + v\cdot\nabla u = f \tag{3.1}$$

has a solution $u \in W^{-1,p}$, *for each* $f \in W^{-1,p}$, *provided*

$$\lambda > c\,\|v\|_{2,p}\,, \tag{3.2}$$

where $c=c(\Omega,n,p)$ *is a suitable positive constant. Moreover,*

$$(\lambda - c\,\|v\|_{2,p})\,\|u\|_{-1,p} \le c_0\,\|f\|_{-1,p}\,. \tag{3.3}$$

Above, u and f are real functions defined in Ω (a similar result holds for systems of equations in spaces $W^{j,p}$, for $j \ge -1$, $p > 1$, as proved in reference [4]). In the sequel, we give just a sketch of the proof of theorem 3.1.

Let $q=p/(p-1)$, and define the operator $B\phi = \lambda\phi - v\cdot\nabla\phi - (\operatorname{div} v)\phi$, whose domain is

$$D(B) = \{\phi \in W_0^{1,q} : v\cdot\nabla\phi \in W_0^{1,q}\}\ .$$

B is a closed operator, and $\mathcal{D}(\Omega) \subset D(B)$. Hence $D(B)$ is dense in $W_0^{1,q}$. Assume, for a while, that B is invertible (i.e. that $B^{-1} \in \mathcal{L}(W_0^{1,q})$), and that

[2] In this section the notation differes from the one used before.

(3.4) $$\|B^{-1}\| \leq 1/(\lambda - c\|v\|_{2,p}) ,$$

if $\lambda > c\|v\|_{2,p}$. Then, the adjoint operator B^* is invertible, moreover $(B^*)^{-1} = (B^{-1})^* \in \mathcal{L}(W^{-1,p})$. Hence, the equation

(3.5) $$B^* u = f$$

has a unique solution $u=(B^{-1})^* f$, for each $f \in W^{-1,p}$.

Equation (3.4) is equivalent to $\langle B\phi,u\rangle = \langle\phi,f\rangle$, $\forall\, \phi \in D(B)$, where $\langle\, , \rangle$ denotes the duality pairing between $W_0^{1,q}$ and $W^{-1,p}$. In particular,

$$\langle \lambda\phi - \mathrm{div}(\phi v),u\rangle = \langle\phi,f\rangle, \qquad \forall\, \phi \in \mathcal{D}(\Omega).$$

Hence $u=(B^{-1})^* f$ is a solution, in the distributional sense, of equation (3.1). Moreover, (3.3) holds, since $\| (B^{-1})^* \| = \| B^{-1} \|$.

Let us show that $B^{-1} \in \mathcal{L}(W_0^{1,q})$, and that (3.4) holds, for a suitable constant c. We will consider the operator B in the slightly more general form

(3.6) $$B\phi = \lambda\phi + v \cdot \nabla\phi + a\phi ,$$

where $a \in W^{1,p}$, and we will prove that

(3.7) $$\|B^{-1}\| \leq 1/[\lambda - c(\|v\|_{2,p} + \|a\|_{1,p})] ,$$

if $\lambda > c(\|v\|_{2,p} + \|a\|_{1,p})$.

Uniqueness. Assume that $u \in W^{1,q}$ is a solution of $Bu=0$. By multiplying both sides of this equation by $(\delta + |u|^2)^{(q-2)/2}\, u$, and by passing to the limit as $\delta \to 0^+$, one easily verifies that $[\lambda - (1/q)\, |\mathrm{div}\, v|_\infty - |a|_\infty]\, |u|_q^q = 0$. Hence $u=0$, if λ is sufficiently large.

Existence in $W_0^{2,q}$. At this stage, a is requested to belong to $W^{2,p}$. Let $\varepsilon > 0$, and let $u_\varepsilon \in W_0^{4,q}$ be the solution of the problem

$$(3.8)\qquad \begin{cases} -\varepsilon\Delta u_\varepsilon + \lambda u_\varepsilon + v\cdot\nabla u_\varepsilon + a u_\varepsilon = f, & \text{in } \Omega, \\ (u_\varepsilon)_{|\Gamma} = 0, \end{cases}$$

where $f \in W_0^{2,q}$. A crucial point in our proof is that the approximating solution u_ε verifies

$$(3.9)\qquad (\Delta u_\varepsilon)_{|\Gamma} = 0,$$

as follows from (3.8). Set $\Lambda = (\delta + |\Delta u_\varepsilon|^2)^{1/2}$, where $\delta > 0$. Since (3.9) holds, one has

$$(3.10)\qquad -\int_\Omega \Delta(\Delta u_\varepsilon)\Lambda^{q-2}\,\Delta u_\varepsilon\,dx = \int_\Omega \nabla(\Delta u_\varepsilon)\cdot\nabla(\Lambda^{q-2}\,\Delta u_\varepsilon)dx.$$

The identity (left to the reader)

$$\nabla(\Delta u_\varepsilon)\cdot\nabla(\Lambda^{q-2}\,\Delta u_\varepsilon) = \Lambda^{q-2}\,|\nabla\Delta u_\varepsilon|^2 + [(q-2)/4]\;\Lambda^{q-4}\,|\nabla(|\Delta u_\varepsilon|^2)|^2,$$

shows that the left hand side of (3.10) is non-negative, if $q \geq 2$; this result holds again if $1<q\leq 2$, since in that case one has the inequality (left to the reader)

$$\nabla(\Delta u_\varepsilon)\cdot\nabla(\Lambda^{q-2}\,\Delta u_\varepsilon) \geq [(q-1)\,|\Delta u_\varepsilon|^2 + \delta]\;\Lambda^{q-4}\,|\nabla(\Delta u_\varepsilon)|^2.$$

Hence the left hand side of (3.10) is non-negative, if ε and δ are positive integers. On the other hand one has the following identity

$$\nabla(\Delta u_\varepsilon)\,\Lambda^{q-2}\,\Delta u_\varepsilon = (1/q)\,\nabla\Lambda^q,$$

which shows that

$$(3.11)\qquad \int_\Omega (v\cdot\nabla\Delta u_\varepsilon)\,\Lambda^{q-2}\,\Delta u_\varepsilon\,dx = -(1/q)\int_\Omega (\operatorname{div} v)\,\Lambda^q\,dx,$$

since $v\cdot\upsilon = 0$ on Γ.

By applying the Laplace operator Δ to both sides of equation $(3.8)_1$, by multiplying both sides by $\Lambda^{q-2}\,\Delta u_\varepsilon$, by integrating in Ω, by using (3.11), and by taking in account that the left hand side of (3.10) is non-negative, one gets

$$(3.12)\quad \begin{aligned}&\lambda \int_\Omega |\Delta u_\varepsilon|^2 \Lambda^{q-2}\,dx - (1/q)\int_\Omega (\operatorname{div} v)\Lambda^q\,dx \le \\ &\le \int_\Omega [\,2\,|\textstyle\sum_{i,j} (D_j v_i)(D^2_{ij}u_\varepsilon)\,| + |\Delta v\cdot\nabla u_\varepsilon| + |\Delta(a u_\varepsilon)| + |\Delta f|\,]\,|\Delta u_\varepsilon|\,\Lambda^{q-2}\,dx\,.\end{aligned}$$

By passing to the limit as $\delta \to 0^+$, one verifies that equation (3.12) holds again if Λ is replaced by $|\Delta u_\varepsilon|$. In particular, it follows that

$$(3.13)\quad [\lambda - c\,(\|v\|_{2,p} + \|a\|_{2,p})]\;|\Delta u_\varepsilon|_q \le |\Delta f|_q\,,$$

where c depends only on Ω, n, p. Note that the quantity $|\Delta u|_q$ is a norm in the space $W_0^{2,q}$.

Assume that

$$(3.14)\quad \lambda > c(\|v\|_{2,p} + \|a\|_{2,p})\,.$$

It follows from (3.13) that there exists a subsequence u_ε weakly convergent to a function $u \in W_0^{2,q}$, as $\varepsilon \to 0$. Moreover, $\varepsilon\,|\Delta u_\varepsilon|_q \to 0$. By passing to the limit in equation $(3.8)_1$, one proves that $Bu = f$.

■

Existence in $W_0^{1,q}$. Let $f \in W_0^{1,q}$, and let $f_m \in W_0^{2,q}$ be a sequence of functions convergent to f in $W_0^{1,q}$. Let λ verify (3.14), and let $u_m \in W_0^{2,q}$ be the solution of the equation

$$(3.15)\quad \lambda u_m + v\cdot\nabla u_m + a u_m = f_m\,.$$

Due to the regularity of u_m, we are allowed to do the following: Aplly the gradient operator ∇ to both sides of (3.15); multiply both sides (scalarly) by $(\delta + |\nabla u_m|^2)^{(q-2)/2}\,\nabla u_m$; and integrate over Ω. One gets, after some calculations, the uniform estimates

$$[\lambda - c(\|v\|_{2,p} + \|a\|_{1,p})]\;|\nabla u_m|_q \le |\nabla f_m|_q\,.$$

Hence (a subsequence) $u_m \to u$, weakly in $W_0^{1,q}$. By passing to the limit in equation (3.15) we show that $Bu = f$. Moreover,

$$[\lambda - c(\|v\|_{2,p} + \|a\|_{1,p})] \, |\nabla u|_q \leq |\nabla f|_q \ ,$$

which proves (3.7). ■

References

[1] **S. Agmon, A. Douglis and L. Nirenberg,** Estimates near the boundary for solution of elliptic partial differential equations satisfying general boundary conditions, II, Comm. Pure Appl. Math 17 (1964), 35-92.

[2] **H. Beirão da Veiga,** Stationary motions and the incompressible limit for compressible viscous fluids, MRC Technical Summary Report #2883, Mathematics Research Center, University of Wisconsin-Madison, (1985); to appear in the Houston Journal of Mathematics.

[3] **H. Beirão da Veiga,** An L^p-theory for the n-dimensional, stationary, compressible Navier-Stokes equations, and the incompressible limit for compressible fluids. The equilibrium solution, Communications in Mathematical Physics 109 (1987), 229-248.

[4] **H. Beirão da Veiga,** Existence results in Sobolev spaces for a stationary transport equation, to appear in Ricerche di Matematica in the volume in honour to Prof. C. Miranda. See also: On a stationary transport equation, Ann. Univ. Ferrara 32 (1986).

[5] **H. Beirão da Veiga,** On a linear non-elliptic system concerning the dynamic of stationary motions, to appear in Rend. Sem. Mat. Fis. Milano.

[6] **A. De Franceschi,** On the stationary, compressible and incompressible Navier-Stokes equations, to appear in Ann. Mat. Pura Appl..

[7] **S. Klainerman and A. Majda,** Singular limits of quasilinear hyperbolic systems with large parameters and the incompressible limit of compressible fluids, Comm. Pure App. Math. 34 (1981), 481-524.

[8] **M. Padula,** Existence and uniqueness for viscous steady compressible motions, to appear.

[9] **J. Serrin,** Mathematical principles of classical fluid mechanics, Handbuch der Physik, Bd. VIII/1, Springer-Verlag, Berlin-Göttingen-Heidelberg (1959).

[10] **A. Valli,** Periodic and stationary solutions for compressible Navier-Stokes equations via a stability method, Ann. Sc. Normale Sup. Pisa (1984), 607-647.

[11] **A. Valli,** On the existence of stationary solutions to compressible Navier-Stokes equations, Ann. Inst. H. Poincaré, Anal. Non Lineaire (1986).

[12] **A. Valli and W.M. Zajaczkowski,** Navier-Stokes equations for compressible fluids: Global existence and qualitative properties of the solutions in the general case, Comm. Math. Physics 103 (1986), 259-296.

ON A CLASS OF HYPERBOLIC OPERATORS WITH DOUBLE CHARACTERISTICS

E.Bernardi
A.Bove
C.Parenti

Dept.of Math.,University of Bologna, Piazza di Porta S.Donato 5, Bologna, Italy.

Dedicated to Hans Lewy

0. Introduction

It is well known that, as a general matter, existence and regularity results for (the Cauchy problem for) hyperbolic operators with characteristics of multiplicity higher than 1 may depend on the lower order terms (see e.g. the pioneering work of O.Olejnik [8]).
Fundamental contribution in this subject have been the papers of Ivrii-Petkov [7], Ivrii [5], [6] and Hörmander [4] which put in evidence the importance of the Hamiltonian matrix, yielding a "classification" of the possible occurring cases, and gave necessary and almost sufficient conditions for the C^∞ well posedness of the Cauchy problem. Furthermore Ivrii [6] gave also some propagation results for the C^∞ wave front set of solutions of the Cauchy problem. Ivrii's proofs require the existence of a suitable factorization of the operator.
Aim of this note is to prove a characterization of the existence of the Ivrii decomposition of the operator in terms of some geometric properties of the principal symbol.

1. Statement of the results.

We shall denote $x \in \mathbb{R}^{n+1}$ by $x = (x_0,x_1,...,x_n) = (x_0,x')$, $D_{x_j} = \frac{1}{i}\frac{\partial}{\partial x_j}$, $j = 0,...,n$; by $T^*\mathbb{R}^{n+1}\backslash 0$ the cotangent bundle minus the zero section, and by (x,ξ) a point of $T^*\mathbb{R}^{n+1}\backslash 0$. $T^*\mathbb{R}^{n+1}$ is naturally equipped with a symplectic structure given by the symplectic 2-form σ, in local coordinates $\sigma = d\xi_0 \wedge dx_0 + \sum_{j=1}^{n} d\xi_j \wedge dx_j$. Without loss of generality we shall consider the (pseudo)-differential operator of order two:

(1) $$P(x,D_x) = - D_{x_0}^2 + Q(x,D_{x'}),$$

where $Q(x,\xi') \sim \sum_{j=0}^{\infty} q_{2-j}(x,\xi')$, are classical symbols in ξ' smoothly dependent on x_0, of order 2-j (this means that q_{2-j} , are C^∞ functions if $\xi' \neq 0$, homogeneous of degree 2-j). Thus the principal symbol of P will be

(2) $$p(x,\xi) = - \xi_0^2 + q_2(x,\xi').$$

We shall suppose that p is hyperbolic:

(3) $$q_2(x,\xi') \geq 0.$$

Let $\Sigma = \{(x,\xi) \in T^*\mathbb{R}^{n+1}\backslash 0 \mid p(x,\xi) = 0\}$ be the characteristic manifold of p, and $\Sigma_2 = \Sigma \cap \{p=0, dp = 0\}$.

Definition 1. Let $\rho \in T^*\mathbb{R}^{n+1}$, then σ defines an isomorphism

$$T_\rho^* T^*\mathbb{R}^{n+1} \ni \gamma \rightarrow H_\gamma \in T_\rho T^*\mathbb{R}^{n+1}, \quad \gamma = \sigma(\cdot, H_\gamma)$$

and, using this, the Hamilton matrix at ρ is defined by:

$$F_p(\rho): T_\rho T^*\mathbb{R}^{n+1} \ni v \rightarrow H_\gamma \in T_\rho T^*\mathbb{R}^{n+1}, \quad \gamma = \text{Hess } p(\rho)(v,\cdot)$$

or, in other words,

$$\langle \text{Hess } p(\rho)v, w \rangle = \sigma(w, F_p(\rho)v).$$

Remark. It is well known that there exist at most a positive real λ, such that $sp(F_p(\rho)) \subset i\mathbb{R} \cup \{-\lambda, \lambda\}$. If such a $\lambda > 0$ exists we say that p is effectively hyperbolic.When sp $F_p(\rho) \subset i\mathbb{R}$, $\forall \rho \in \Sigma_2$ we say that p is non-effectively hyperbolic.

Definition 2 (see also [5]). We say that p is almost factorized (a.f.) if there exist symbols $\lambda, \mu \in S^1$, homogeneous of degree 1, $\tilde{q} \in S^2$, $\tilde{q} \geq 0$, $\tilde{q}$ homogeneous of degree 2, such that, locally near Σ_2,

(4) $$p(x,\xi) = -(\xi_0 - \lambda(x,\xi'))(\xi_0-\mu(x,\xi')) + \tilde{q}(x,\xi'), \text{ where}$$

(5) $$|\{\xi_0-\lambda(x,\xi'),\xi_0-\mu(x,\xi')\}| \le \sqrt{q(x,\xi')} + |\lambda(x,\xi')-\mu(x,\xi')|,$$

(6) $$|\{\xi_0-\lambda(x,\xi'), \tilde{q}(x,\xi')\}| \le \tilde{q}(x,\xi').$$

Remark. i) If p is written in the form (4) then $\Sigma_2 = \{\xi_0 = \lambda = \mu, q= 0\}$, ii) (5) implies that p is non-effectively hyperbolic.

In the symplectic case the following well known result holds (see [5]).

Theorem 3. Let $P(x,D_x)$ be as in (1). Let us assume that

i) Σ_2 is a C^∞ submanifold of $T^*\mathbb{R}^{n+1}$ and $\forall \rho\in \Sigma_2$ rank Hess $p(\rho) = \operatorname{codim} T_\rho\Sigma_2$

ii) $\sigma|_{\Sigma_2}$ has constant rank

iii) $\forall\rho\in\Sigma_2$, $\operatorname{spec}(F_p(\rho))\subset i\mathbb{R}$ and
$\ker F_p^2(\rho) \cap \operatorname{Im} F_p^2(\rho) = \{0\}$.

Then $p(x,\xi)$ can be decomposed according to Definition 2. Our main result deals with the case in which $\ker F_p^2(\rho)$ is not symplectic; it is stated in the following:

Theorem 4. Let $P(x,D_x)$ be as in (1) and assume i), ii) of Theorem 3. Moreover we assume

iv) $\forall\rho\in\Sigma_2$, $\operatorname{spec}(F_p(\rho))\subset i\mathbb{R}$ and
$\ker F_p^2(\rho) \cap \operatorname{Im} F_p^2(\rho) \neq \{0\}$.

Then there exists a C^∞ vector field $z(\rho)$ on Σ_2, z homogeneous of degree 0 such that

1) $z(\rho) \in \ker F_p^2(\rho) \cap \operatorname{Im} F_p^2(\rho)$,

2) let $v(\rho) = -F_p(\rho)\, z(\rho)$; then $v(\rho) \neq 0$ and if $w\in [v(\rho)]^\sigma$ we have
a) $\sigma(w,F_p(\rho)w) \ge 0$
b) $\sigma(w,F_p(\rho)w) = 0$ if and only if $w\in \ker F_p(\rho) \oplus [z(\rho)]$.

3) $p(x,\xi)$ admits a decomposition of the form (4) such that $H_{\xi_0-\lambda}(\rho)$ is colinear with $F_p(\rho)\, z(\rho)$, $\rho\in\Sigma_2$.

Denote by S a C^∞ real function homogeneous of degree 0 such that

(7) $$H_S(\rho) - z(\rho) \in \ker F_p(\rho),\ \forall\rho\in\Sigma_2.$$

Moreover this decomposition verifies (5), (6) if and only if the following equivalent conditions hold:

I) $\sigma(H_{\{S,\wedge\}}(\rho), F_p(\rho)\ H_{\{S,\wedge\}}(\rho)) = 0, \forall \rho \in \Sigma_2$.

II) $(H_S^3\, p)(\rho) = 0$ and $(H_\varphi H_S^2\, p)(\rho) = 0$,

$\forall H_\varphi(\rho) \in \operatorname{Im} F_p^2(\rho) / (\operatorname{Im} F_p^2(\rho) \cap \ker F_p^2(\rho))$.

Remark. 1 - The vector field z in 1), 2) is uniquely determined up to a non-zero factor.
2 - Conditions I) - II) are independent of the choice of S verifying (7).
3. Conditions I) -II) are related to third order terms in the Taylor expansion of p near Σ_2 and also to the behavior of the Hamiltonian flow of p near Σ_2.

2. Proof of Theorem 4.

The proof will be accomplished in several steps.
As a preliminary result we prove 1) and 2).

We explicitly point out that all the arguments below are carried on in a suitable neighborhood of (a point of) Σ_2, but this will be tacitly understood.
Due to hypotheses ii) and iv) we can choose $z(\rho) = F_p^2(\rho)\ \theta(\rho), \rho \in \Sigma_2$, where $\theta(\rho)$ generates the one dimensional gap between $\ker F_p^3(\rho)$ and $\ker F_p^4(\rho)$. It is then easy to see that $z(\rho)$ verifies conditions 1) and 2) a), b).
We turn now to the second step. We recall that $p(x,\xi) = -\xi_0^2 + q_2(x,\xi')$; we may assume that $q_2(x,\xi') = \sum_{j=1}^{r} \psi_j(x,\xi')^2$, $1+r = \operatorname{codim} \Sigma_2$, $\xi_0 = 0 = \psi_1 = \ldots = \psi_r$ being the independent local equations of Σ_2.
Since $z(\rho) \in \ker F_p^2(\rho) \cap \operatorname{Im} F_p^2(\rho)$, $\sigma(z(\rho), F_p(\rho)\, z(\rho)) = 0$ and we may suppose that $\sigma(z(\rho), H_{\xi_0}) = 1$. Let us define $\gamma_j(\rho) = \sigma(z(\rho), H_{\psi_j}(\rho))$, $j = 1,\ldots,r$, and denote by $\gamma(\rho)$ the vector $(\gamma_1(\rho),\ldots,\gamma_r(\rho))$. We have $|\gamma(\rho)| = 1$ and we may choose a C^∞ homogeneous of degree zero real extension of γ in such a way that $|\gamma(\rho)| \le 1$ near Σ_2. Thus we have the decomposition

$$p = -\wedge M + Q, \tag{8}$$

where $\wedge = \xi_0 - \lambda_1(x,\xi')$, and $\lambda_1(x,\xi') = \langle \gamma(x,\xi'), \psi(x,\xi')\rangle$, $M = \xi_0 + \lambda_1(x,\xi')$, $Q(x,\xi')$

$= |\psi(x,\xi')|^2 - <\gamma,\psi>^2(x,\xi')$.

Obviously we may choose γ in such a way that $Q \geq 0$ and vanishes precisely on Σ_2.

We now prove the

Lemma 5.

The following conditions are equivalent:

1 - $\sigma(H_{\{S,\wedge\}}(\rho), F_p(\rho)\, H_{\{S,\wedge\}}(\rho)) = 0, \qquad \rho \in \Sigma_2.$

2- $F_Q(\rho)\, H_{\{S,\wedge\}}(\rho)) = 0, \qquad \rho \in \Sigma_2.$

3 - The vector $(\{S,\{\wedge,\psi_j\}\} - \{\wedge,\{S,\psi_j\}\})_{j=1,\dots,r}(\rho)$ is colinear with $(\{S,\psi_j\}_{j=1,\dots,r}(\rho)$ on Σ_2.

4 - If $\gamma_j(\rho)$, $j = 1,\dots,r$ are real C^∞ functions such that if $\rho \in \Sigma_2$, $\gamma_j(\rho) = \sigma(z(\rho), H_{\psi_j}(\rho))$, $j = 1,\dots,r$, $H_\wedge\psi_j(\rho) = \sum_{k=1}^{r} \alpha_{jk}(\rho)\,\psi_k(\rho)$, $j = 1,\dots,r$, then the vector $(H_\wedge\gamma_j - \sum_{k=1}^{r}\alpha_{jk}\gamma_k)_{j=1,\dots,r}$ is colinear with the vector γ on Σ_2.

Proof.

Since $\{S,\wedge\} = 0$ on Σ_2 we have

$$\sigma(H_{\{S,\wedge\}}, F_p\, H_{\{S,\wedge\}}) = \sigma(H_{\{S,\wedge\}}, F_Q\, H_{\{S,\wedge\}}) \text{ on } \Sigma_2.$$

Thus 1 $\Leftrightarrow$ 2.

We now prove that 2 $\Leftrightarrow$ 3. Since Q vanishes of order two on Σ_2, $Q = \sum_{i,j=1}^{r} q_{ij}\,\psi_i\,\psi_j$.

Therefore $\sigma(H_{\{S,\wedge\}}, F_q\, H_{\{S,\wedge\}}) = \sum_{i,j=1}^{r} q_{ij}\,\theta_i\,\theta_j$ on Σ_2, where $\theta_i = \{S,\{\wedge,\{S,\psi_i\}\} - \{\wedge, \{S,\psi_i\}\}$, $i = 1,\dots,r$. Now 2 is equivalent to saying that $(\theta_1,\dots,\theta_r) \,//\, \gamma = (\{S,\psi_1\},\dots,\{S,\psi_r\})$. This proves the assertion.

The equivalence 3 $\Leftrightarrow$ 4 follows since $\{S,\{\wedge,\psi_j\}\} = \{S, \sum_{k=1}^{r}\alpha_{jk}\,\psi_k\} = \sum_{k=1}^{r}\alpha_{jk}\{S,\psi_k\} = \sum_{k=1}^{r}\alpha_{jk}\,\gamma_k$ on Σ_2, $\forall j = 1,\dots,r$. This proves the Lemma.

We point out that 3 of Lemma 5 is independent from the choice of the function S satisfying (7).

We also remark that condition 2 of Lemma 5 is independent from the choice of a decomposition of p; in fact if $p = -\Lambda' M' + Q'$, where $H_{\Lambda'} \ // \ F_p z$ on Σ_2, then $\Lambda - \Lambda'$ vanishes to the second order on Σ_2. Let us prove that $F_{\Lambda-\Lambda'} z = 0$. If $\Lambda = \xi_0 - <\gamma,\psi>$, $\Lambda' = \xi_0 - <\gamma',\psi>$ we have $\gamma - \gamma' = A\psi + B\psi$, where $A = {}^tA$, $B = -{}^tB$ and we set $\tilde{\gamma} = \gamma' + B\psi$, so that $\Lambda' = \xi_0 - <\tilde{\gamma},\psi>$ and $\gamma - \tilde{\gamma} = A\psi$. Then

$$F_{\Lambda-\Lambda'}(\rho)\, v = \Big(\sum_{k=1}^{r} a_{jk}(\rho)\, \sigma(v, H_{\psi_j}(\rho))\, H_{\psi_k}(\rho)\Big)_{j=1,\dots,r}.$$

Now since $|\gamma|^2 = 1 + O(|\psi|^2)$, $|\tilde{\gamma}|^2 = 1 + O(|\psi|^2)$ it follows that $\Sigma\, a_{jk}\psi_k \tilde{\gamma}_j = O(|\psi|^2)$. Therefore $\sum_{j=1}^{r} a_{jk}(\rho)\, \tilde{\gamma}_j(\rho) = 0 \ \forall\, k = 1,\dots,r$ and finally $\sum_{j=1}^{r} a_{jk}(\rho)\, \sigma(z(\rho), H_{\psi_j}(\rho)) = O$, which proves our claim, and this implies that $H_{\{S,\Lambda'\}} = H_{\{S,\Lambda\}}$ on Σ_2.

Let us now suppose I holds. Let consider p decomposed as in (8). We have

$$\{\Lambda, Q\} = \{\xi_0 - <\gamma,\psi>,\ |\psi|^2 - <\gamma,\psi>^2\}.$$

Let us introduce new local coordinates $(\varphi_1,\dots,\varphi_r)\ (x,\xi')$ such that $<\gamma,\psi> = \varphi_r|\gamma|$, $|\psi|^2 = |\varphi|^2$. We obtain

$$\{\Lambda, Q\} = \{\xi_0 - \varphi_r|\gamma|,\ \sum_1^{r-1}\varphi_j^2 + (1-|\gamma|^2)\,\varphi_r^2\} =$$

$$= 2\sum_{j,k=1}^{r-1}\alpha_{jk}\varphi_j\varphi_k + 2\Big[\sum_{j=1}^{r-1}\alpha_{jr}\varphi_j + (1-|\gamma|^2)\sum_{k=1}^{r-1}\alpha_{rk}\varphi_k\Big]\varphi_r +$$

$$+ [2(1-|\gamma|^2) + \{\Lambda, 1-|\gamma|^2\}]\,\varphi_r^2 \ ,$$

where $\{\Lambda,\varphi_j\} = \sum_{k=1}^{r}\alpha_{jk}\varphi_k$. Now the first and the third terms in the last expression can be estimated by Q. Let us put $1-|\gamma|^2 = <\beta\varphi,\varphi>$ for a suitable C^∞ real positive definite matrix β. Thus the fifth term is also estimated by Q since $\Lambda\varphi = 0$ on Σ_2. It remains to estimate the second term. If $\alpha_{jr} = O(|\varphi|)$ we have

$$\Big|\sum_{j=1}^{r-1}\alpha_{jr}\varphi_j\varphi_r\Big| \lesssim |\varphi|\,|\varphi'|\,|\varphi_r| \lesssim |\varphi|^2\varphi_r^2 + |\varphi'|^2 \le (1-|\gamma|^2)\varphi_r^2 + |\varphi'|^2 = Q.$$

We shall now show that $F_Q H_{\{S,\Lambda\}} = 0$ on Σ_2 is equivalent to saying that $\alpha_{jr} = O(|\varphi|)$, $j=1,...,r-1$.
In fact in the chosen coordinates $\sigma(z,H_{\varphi_j}) = 0$ on Σ_2, $j = 1,...,r-1$. If we now take $\gamma = (0,...,0,|\gamma|)$, condition 4 of Lemma 5 is equivalent to $\alpha_{jr} = 0$ on Σ_2.
This proves that I) ⇒ (6). Condition (5) is then easily proved.
The above argument shows also that the converse implication (6) ⇒ I) holds.
The last step will be to show that II is equivalent to I. We shall in fact prove that II is equivalent to condition 2 of Lemma 5.
First let us suppose that condition 2 of Lemma 5 holds. Then p admits a decomposition, $p = -\Lambda M+Q$, verifying (4) - (6). We have $F_p z = -\sigma(z,H_\Lambda)H_M - \sigma(z,H_M)H_\Lambda + F_Q z$, so $F_p H_S = -H_\Lambda = -\sigma(H_S,H_\Lambda)H_M - \sigma(H_S,H_M)H_\Lambda + F_Q H_S$.
Since $H_S \in \operatorname{Im} F_p$, we have $\sigma(H_S,H_\Lambda) = 0$. Therefore $F_Q H_S = 0$ and $\sigma(H_S,H_M) = 1$. Now our hypothesis yields $F_p H_{\{S,\Lambda\}} = -\sigma(H_{\{S,\Lambda\}},H_\Lambda)H_M - \sigma(H_{\{S,\Lambda\}},H_M)H_\Lambda = -\sigma(H_{\{S,\Lambda\}},H_M)H_\Lambda$. We have $H_S^3 p|_{\Sigma_2} = -3H_S^3\Lambda|_{\Sigma_2} + H_S^3 Q|_{\Sigma_2}$. Let us note that $H_S^2\Lambda|_{\Sigma_2} = 0$ since $H_S^2\Lambda|_{\Sigma_2} = \sigma(H_\Sigma, H_{\{S,\Lambda\}}) = 0$, therefore $H_S^3 p|_{\Sigma_2} = H_S^3 Q|_{\Sigma_2}$.
Let us note that $H_S^3 Q|_{\Sigma_2} = \sigma(H_S, H_{\{S,\{S,Q\}\}})$; now $\{S,\{S,Q\}\} \geq 0$ and vanishes on Σ_2, hence $H_S^3 Q|_{\Sigma_2} = 0$. This proves the first claim. Let $H_\varphi \in \operatorname{Im} F_p^2$: $H_\varphi H_S^2 p|_{\Sigma_2} = -2H_\varphi H_S\Lambda|_{\Sigma_2} H_S M|_{\Sigma_2} + H_\varphi H_S^2 Q|_{\Sigma_2}$. Our next aim is to show that $H_\varphi H_S\Lambda|_{\Sigma_2} = \sigma(H_\varphi, H_{\{S,\Lambda\}}) = 0$. In fact we shall show that

(*) $$H_{\{\Lambda,S\}} \in \ker F_Q \Leftrightarrow H_{\{\Lambda,S\}} - \sigma(H_{\{\Lambda,S\}},H_M)H_S \in \ker F_p.$$

Assume that $H_{\{\Lambda,S\}} \in \ker F_Q$. Therefore $F_p H_{\{\Lambda,S\}} = \sigma(H_{\{\Lambda,S\}},H_M) F_p H_S$. Conversely if $H_{\{\Lambda,S\}} - \sigma(H_{\{\Lambda,S\}},H_M)H_S = \omega \in \ker F_p$ we have

$$F_Q H_{\{\Lambda,S\}} = F_Q\omega = F_p\omega + \sigma(\omega,H_\Lambda)H_M + \sigma(\omega,H_M)H_\Lambda .$$

This is zero since $H_\Lambda \in \ker F_p$ and $\sigma(H_S,H_M) = 1$. On the other hand it is easy to verify that

$$\frac{\operatorname{Im} F_p^2}{\operatorname{Im} F_p^2 \cap \ker F_p^2} = \frac{\operatorname{Im} F_p \cap [H_S]^\sigma}{\operatorname{Im} F_p \cap [H_S]^\sigma \cap (\ker F_p + [H_S])}$$

Thus $\sigma(H_\varphi,H_{\{\Lambda,S\}}) = 0$, and this proves that $H_\varphi H_S^2 p|_{\Sigma_2} = 0$ $\forall\varphi$ such that $H_\varphi \in \operatorname{Im} F_p^2$.

At last we remark that if $H_\varphi \in \ker F_p + [H_S]$, then $H_\varphi H_S^2 p_{|\Sigma_2} = 0$.

This proves condition II.
The converse implication is proved in [2].

References

[1] *E.Bernardi, A.Bove, C.Parenti*, preprint 1986 (in preparation).

[2] *E.Bernardi, A.Bove*, Geometric results for a class of hyperbolic operators with double characteristics, Comm.P.D.E., **13**(1)(1988), 61-86.

[3] *L.Hörmander*, The Analysis of linear partial differential operator III, Berlim 1984.

[4] *L.Hörmander*, The Cauchy problem for differential equations with double characteristics, J.Analyse Math.**32**(1977), 118-196.

[5] *V.Ivrii*, The well posedness ... III, the energy integral, Trans.Moscow Math.Soc.**34**(1978), 149-168.

[6] *V.Ivrii*, WF of solutions of certain hyperbolic pseudodifferential equations, Trans.Moscow Math.Soc.**39**(1981), 87-119.

[7] *V.Ivrii and V.M.Petkov*, Necessary conditions for the Cauchy problem for non strictly hyperblic equations to be well posed, Russian Math. Surveys **29**(1974), 1-70.

[8] *O.A.Olejnik*, On the Cauchy problem for weakly hyperbolic equations, C.P.A.M.**23**(1970), 569-586.

[9] *S.Wakabayashi*, Singularities of solutions of the Cauchy problem for symmetric hyperbolic systems, Comm.P.D.E.**9**(1984), 1147-1177.

RELAXATION PROBLEMS IN CONTROL THEORY

Giuseppe Buttazzo
Scuola Normale Superiore
Piazza dei Cavalieri, 7
56100 Pisa (Italy)

Dedicated to Hans Lewy

1. Introduction

An optimal control problem can be formulated in the following way. Let Y (the space of *state variables*) and U (the space of *controls*) be two sets, let $J : U \times Y \to [0,+\infty]$ be a function (the *cost function*), and let Λ be a subset of $U \times Y$ (the set of *admissible pairs*). The problem of optimal control is then

$$\min\left\{ J(u,y) : (u,y) \in \Lambda \right\} . \tag{1.1}$$

In a large number of situations the space Y is a Sobolev space $W^{1,p}(\Omega)$ (where Ω is a bounded open subset of $\mathbf{R}^n$), the space U is $L^q(\Omega)$, the cost function is an integral functional of the form

$$J(u,y) = \int_A f(x,y(x),u(x))dx ,$$

and the set Λ of admissible pairs is often determined by differential equations or differential inclusions, constraints, etc....

A well-known method to prove the existence of solutions for problem (1.1) is the so-called *Direct Method* which consists in the following: find a topology τ on U and a topology σ on Y such that

i) Λ is (sequentially) $\tau \times \sigma$-closed in $U \times Y$ and J is (sequentially) $\tau \times \sigma$-lower semicontinuous on Λ;

ii) J is (sequentially) $\tau \times \sigma$-coercive on Λ, i.e. for every $k \geq 0$ the set

$$\left\{ (u,y) \in \Lambda : J(u,y) \leq k \right\}$$

is (sequentially) $\tau \times \sigma$-relatively compact in $U \times Y$.

When condition i) is not fulfilled, in general the existence of a solution of problem (1.1) may fail. Nevertheless, it may be interesting to study the asymptotic behaviour (as $h \to +\infty$) of the minimizing sequences (u_h, y_h) which, under hypothesis ii), are $\tau \times \sigma$ compact). To do this, we introduce the so-called *relaxed problem*

(1.2) $$\min\left\{ F(u,y) : (u,y) \in U \times Y \right\}$$

where $F : U \times Y \to [0,+\infty]$ is the greatest (sequentially) $\tau \times \sigma$-lower semicontinuous functional on $U \times Y$ which is less than or equal to J on Λ.

The main features of the functional F can be summarized in the following proposition.

Proposition 1.1. *We have:*

a) F *is (sequentially)* $\tau \times \sigma$*-lower semicontinuous on* $U \times Y$;

b) $\inf\{J(u,y) : (u,y) \in \Lambda\} = \inf\{F(u,y) : (u,y) \in U \times Y\}$;

c) *if* J *is (sequentially)* $\tau \times \sigma$*-coercive on* Λ, *then* F *is (sequentially)* $\tau \times \sigma$*-coercive on* $U \times Y$;

d) *if* (u_h, y_h) *is a minimizing sequence for* (1.1) *with* $u_h \xrightarrow{\tau} u$ *and* $y_h \xrightarrow{\sigma} y$, *then* (u,y) *is a solution of* (1.2);

e) *if* $G : U \times Y \to \mathbf{R}$ *is (sequentially)* $\tau \times \sigma$*-continuous, then the relaxed problem of* $\min\{J(u,y) + G(u,y) : (u,y) \in \Lambda\}$ *is* $\min\{F(u,y) + G(u,y) : (u,y) \in U \times Y\}$.

I shall consider here just some particular cases of control problems governed by ordinary differential equations (more details can be found in [7], [8], [10] and in the references quoted there). Nevertheless, this will show that in many situations the relaxed functional F can be computed explicitely, and we shall see that the relaxed problem (1.2) may have a form rather different from the original problem (1.1).

This phenomenon become more evident when we deal with sequences of optimal control problems: in Section 3 we construct a sequence of optimal control problems governed by linear ordinary differential equations, such that in the "limit problem" the differential state equation disappears (i.e. the set Λ coincides with the whole space $U \times Y$).

2. Some examples of relaxation

In this section we consider control problems governed by ordinary differential equations. The problem we are interested in is

(2.1) $$\min\left\{\int_0^1 f(t,y,u)dt : y' = g(t,y,u), y(0) = y_0\right\}.$$

More precisely, let

$Y = W^{1,1}(0,1)$ endowed with the uniform topology $L^\infty(0,1)$;

$U = L^p(0,1)$ endowed with its weak topology;

$J(u,y) = \int_0^1 f(t,y,u)dt$ where $f : [0,1] \times \mathbf{R} \times \mathbf{R} \to [0,+\infty]$ is a Borel function;

$\Lambda = \{(u,y) \in U \times Y : y' = g(t,y,u)$ a.e. on $[0,1]$, $y(0) = y_0\}$ where $g : [0,1] \times \mathbf{R} \times \mathbf{R} \to \mathbf{R}$ is a given function, and $y_0 \in \mathbf{R}$ is fixed.

Therefore, problem (2.1) can be written in the form (1.1). When the function $g(t,s,z)$ is linear in z, the relaxed problem associated to (2.1) is similar to the original problem (2.1). In fact, the following result holds (see for instance [7], [10]).

Theorem 2.1. *Assume that* $g(t,s,z) = a(t,s) + b(t,s)z$ *and that:*

1) *the functions* $a(t,\cdot)$ *and* $b(t,\cdot)$ *are Lipschitz continuous (uniformly in* t*);*

2) $f(t,0,z) \geq |z|^p$ *for every* $t \in [0,1]$, $z \in \mathbf{R}$;

3) *for every* $t \in [0,1]$, $r, s \in \mathbf{R}$, $z \in \mathbf{R}$ *we have*

$$f(t,r,z) \leq [1 + \omega(|r-s|)]f(t,s,z) + \rho(t,|r-s|)$$

where $\rho(t,\cdot)$ *and* $\omega(\cdot)$ *are increasing and continuous,* ρ *is integrable in* t, *and* $\rho(t,0) = \omega(0) = 0.$

Then the relaxed problem associated to (2.1) is

$$\min\left\{\int_0^1 f^{**}(t,y,u)dt : y' = a(t,y) + b(t,y)u,\ y(0) = y_0\right\},$$

where $f^{**}(t,s,z)$ denotes the greatest function convex and lower semicontinuous in z less than or equal to $f(t,s,z)$.

But, when the function $g(t,s,z)$ is not linear in z, the relaxed problem may drastically change its form, and the differential equation $y' = g(t,y,u)$ may relaxe to a differential inclusion. Moreover, in the cost functional, a new term depending on the derivative y' may appear. This phenomenon is illustrated in the following example (further generalizations and details can be found in [7]).

Example 2.2. Consider the problem ($h \in L^2(0,1)$ and $y_0 \in \mathbf{R}$ are fixed)

$$\min\left\{\int_0^1\left[u^2+u^{-2}+|y-h(t)|^2\right]dt : uy'=1,\ u>0,\ y(0)=y_0\right\}. \tag{2.2}$$

In this case, the function g is $g(t,s,z) = 1/z$, and the function f is given by

$$f(t,s,z) = \begin{cases} z^2+z^{-2}+|s-h(t)|^2 & \text{if } z>0 \\ +\infty & \text{otherwise}. \end{cases}$$

The relaxed problem associated to (2.2) can be explicitely computed (see [7]), and we can write it in the form

$$\min\left\{\int_0^1\left[u^2+y'^2+|y-h(t)|^2+2(uy'-1)\right]dt : uy'\geq 1,\ u>0,\ y(0)=y_0\right\}.$$

We see that the differential equation $uy' = 1$ relaxes to the differential inclusion $y' \in [1/u,+\infty[$, and in the cost functional, the extra term $2(uy'-1)$ appears.

3. An example of limit of a sequence of control problems

When we deal with sequences of control problems

$$(P_\varepsilon) \qquad \min\left\{J_\varepsilon(u,y) : (u,y)\in\Lambda_\varepsilon\right\} \quad (\varepsilon>0)$$

we may be interested in the asymptotic behaviour (as $\varepsilon \to 0$) of the optimal pairs $(u_\varepsilon,y_\varepsilon)$ of (P_ε). This study can be done rigorously by introducing the concept of Γ-convergence (see for instance [10], [20], [21], [28]); but we just show here an example in which the following phenomenon arrives:

the optimal pairs $(u_\varepsilon,y_\varepsilon)$ of (P_ε) tend in $U\times Y$ to a pair (u,y) which is optimal for a "limit problem" (P_0) which has no differential constraint (i.e. the admissible set Λ_0 coincides with the whole space $U\times Y$).

For every $\varepsilon > 0$ consider the following linear-quadratic control problem (we refer to [8] for further details):

$$(P_\varepsilon) \qquad \min \left\{ \int_0^1 \left[u^2 + |y - h(t)|^2 \right] dt : y' = a(t)y + b_\varepsilon(t)u, y(0) = y_0 \right\} .$$

Here the space U is $L^2(0,1)$ endowed with its weak topology, the space Y is $W^{1,1}(0,1)$ endowed with the uniform topology $L^\infty(0,1)$, and $h \in L^2(0,1)$, $a \in L^1(0,1)$, $b_\varepsilon \in L^2(0,1)$. Assume that

$$\begin{cases} b_\varepsilon(t) \text{ tends to } b(t) \text{ weakly in } L^2(0,1) \quad (\text{as } \varepsilon \to 0) \\ b_\varepsilon^2(t) \text{ tends to } \beta^2(t) \text{ weakly in } L^1(0,1) \quad (\text{as } \varepsilon \to 0) . \end{cases}$$

Then, for every $\varepsilon > 0$ there exists one and only one optimal pair $(u_\varepsilon, y_\varepsilon)$ of (P_ε).

It is easy to see the sequences (u_ε) and (y_ε) are compact in U and Y respectively; then the problem is to characterize their limit point as the solutions of a "limit problem" (P_0). In order to compute problem (P_0) we introduce a new control variable $v(t) = b_\varepsilon(t)\, u(t)$, so that problem (P_ε) becomes the minimizing problem for

$$\int_0^1 u^2 + |y - h(t)|^2 dt$$

subjected to the constraints

$$\begin{cases} y' = a(t)y + v \\ y(0) = y_0 . \end{cases} \quad \text{and} \quad v = b_\varepsilon(t)u$$

We may equivalently write

$$(P_\varepsilon) \qquad \min \left\{ \int_0^1 f_\varepsilon(t,y,u,v)dt : y' = a(t)y + v, y(0) = y_0 \right\}$$

where f_ε is the function given by

$$f_\varepsilon(t,y,u,v) = \begin{cases} u^2 + |y - h(t)|^2 & \text{if } v = b_\varepsilon(t)u \\ +\infty & \text{otherwise} . \end{cases}$$

Note that in this new form the control space of (P_ε) is $U \times V = L^2(0,1) \times L^2(0,1)$ which will be endowed with its weak topology.

Since the state equation is now linear and independent of ε, we have only to compute the limit cost functional. This can be done by using a result by Marcellini & Sbordone (see [28], Theorem 3.4), and we get that the limit cost functional is

$$\int_0^1 f(t,y,u,v)dt$$

where, denoting by $f^*(t,y,u,v)$ the polar function (with respect to (u,v) of $f(t,y,u,v)$ it is

$$f^*(\bullet,y,u,v) = \lim_{\varepsilon\to 0} f_\varepsilon^*(\bullet,y,u,v) \quad \text{(in the weak } L^1(0,1) \text{ sense).}$$

In our case, an easy computation gives

$$f(t,y,,u,v) = u^2 + |y - h(t)|^2 + \frac{|v - b(t)u|^2}{\beta^2(t) - b^2(t)}$$

so that, after elimination of the auxiliary control variable v, the limit problem (P_0) can be written in the form

$$(P_0) \qquad \min \left\{ \int_0^1 \left(u^2 + |y - h(t)|^2 + \frac{|y' - a(t)y - b(t)u|^2}{\beta^2(t) - b^2(t)} \right) dt : y(0) = y_0 \right\} .$$

Since (P_0) admits one and only one optimal pair (u_0,y_0), we have that the whole sequence $(u_\varepsilon,y_\varepsilon)$ of the optimal pairs of (P_ε) tends to (u_0,y_0) in $U \times Y$.

Note that, when $b_\varepsilon \to b$ strongly in $L^2(0,1)$ we have $\beta^2 = b^2$, so that Problem (P_0) becomes

$$\min \left\{ \int_0^1 \left[u^2 + |y - h(t)|^2 \right] dt : y' = a(t)y + b(t)u \text{ a.e. on } [0,1], y(0) = y_0 \right\} .$$

But in general, only the inequality $\beta^2 \geq b^2$ a.e. on $[0,1]$ is guaranteed, so that we may have situations such that the "limit cost functional" J_0 is finite everywhere on $U \times Y$, whereas the "limit admissible set" Λ_0 reduces to $\{(u,y) \in U \times Y : y(0) = y_0\}$.

For instance, if $b_\varepsilon(t) = \sin(t/\varepsilon)$, we get at the limit

$$(P_0) \qquad \min \left\{ \int_0^1 \left(u^2 + |y - h(t)|^2 + 2\,|y' - a(t)y|^2\right) dt : y(0) = y_0 \right\}.$$

References

We add to the references quoted in this paper some other references on semicontinuity, relaxation, direct methods in control theory, and on related subjects.

[1] **E. Acerbi and N. Fusco,** Semicontinuity problems in the calculus of variations, Arch. Rational Mech. Anal. 86 (1984), 125-145.

[2] **L. Ambrosio,** Nuovi risultati sulla semicontinuità inferiore di certi funzionali integrali, Atti Accad. Naz. Lincei Rend. Cl. Sci. Fis. Mat. Natur., (to appear).

[3] **H. Attouch,** Variational Convergence of Functionals and Operators, Pitman, Appl. Math. Ser., Boston 1984.

[4] **E.J. Balder,** A general approach to lower semicontinuity and lower closure in optimal control theory, SIAM J. Control Optim. 4 (1984), 570-598.

[5] **L.D. Berkowitz,** Lower semicontinuity of integral functionals, Trans. Amer. Math. Soc. 192 (1974), 51-57.

[6] **G. Buttazzo,** Problemi di semicontinuità e rilassamento in calcolo delle variazioni, Proceedings "Equazioni Differenziali e Calcolo delle Variazioni", Pisa 1985, Edited by L. Modica, ETS, Pisa (1985), 23-36.

[7] **G. Buttazzo,** Some relaxation problems in optimal control theory, J. Math. Anal. Appl., (to appear).

[8] **G. Buttazzo and E. Cavazzuti,** Paper in preparation.

[9] **G. Buttazzo and G. Dal Maso,** Γ-limits of integral functionals, J. Analyse Math. 37 (1980), 145-185.

[10] **G. Buttazzo and G. Dal Maso,** Γ-convergence and optimal control problems, J. Optim. Theory Appl. 38 (1982), 385-407.

[11] **L. Carbone and C. Sbordone,** Some properties of Γ-limits of integral functionals, Ann. Mat. Pura Appl. 122 (1979), 1-60.

[12] **M. Castaing and M. Valadier,** Convex Analysis and Measurable Multifunctions, Springer-Verlag, Lecture Notes in Math. 580, Berlin 1977.

[13] **L. Cesari,** Semicontinuità e convessità nel calcolo delle variazioni, Ann. Scuola Norm. Sup. Pisa Cl. Sci. 18 (1964), 389-423.

[14] **L. Cesari,** Optimization-Theory and Applications, Springer-Verlag, New York 1983.

[15] **F.H. Clarke,** Admissible relaxation in variational and control problems, J. Math. Anal. Appl. 51 (1975), 557-576.

[16] **F.H. Clarke,** Optimization and Nonsmooth Analysis, Wiley Interscience, New York 1983.

[17] **G. Dal Maso and L. Modica,** A general theory of variational integrals, "Topics in Functional Analysis 1980-81", Scuola Normale Superiore, Pisa (1982), 149-221.

[18] **E. De Giorgi,** Convergence problems for functionals and operators, Proceedings "Recent Methods in Nonlinear Analysis", Rome 1978, Edited by E. De Giorgi, E. Magenes and U. Mosco, Pitagora, Bologna (1979), 131-188.

[19] **E. De Giorgi,** Some semicontinuity and relaxation problems, Proceedings "Ennio De Giorgi Colloquium", Paris 1983, Edited by P. Kree, Pitman, Res. Notes in Math. 125, Boston (1985), 1-11.

[20] **E. De Giorgi and G. Dal Maso,** Γ-convergence and calculus of variations, Proceedings "Mathematical Theories of Optimization", S. Margherita Ligure 1981, Edited by J.P. Cecconi and T. Zolezzi, Springer-Verlag, Lecture Notes in Math. 979, Berlin (1983), 121-143.

[21] **E. De Giorgi and T. Franzoni,** Su un tipo di convergenza variazionale, Atti Accad. Naz. Lincei Rend. Cl. Sci. Fis. Mat. Natur. 58 (1975), 842-850.

[22] **I. Ekeland and R. Temam,** Convex Analysis and Variational Problems, North-Holland, Amsterdam 1976.

[23] **C. Goffman and J. Serrin,** Sublinear functions of measures and variational integrals, Duke Math. J. 31 (1964), 159-178.

[24] **A.D. Ioffe,** On lower semicontinuity of integral functionals I, SIAM J. Control Optim. 15 (1977), 521-538.

[25] **A.D. Ioffe,** On lower semicontinuity of integral functionals II, SIAM J. Control Optim. 15 (1977), 991-1000.

[26] **A.D. Ioffe and V.M. Tihomirov,** Theory of Extremal Problems, North-Holland, Amsterdam (1979).

[27] **E.B. Lee and L. Marcus,** Foundations of Optimal Control Theory, John Wiley and Sons, London 1968.

[28] **P. Marcellini and C. Sbordone,** Dualità e perturbazione di funzionali integrali, Ricerche Mat. 26 (1977), 383-421.

[29] **P. Marcellini and C. Sbordone,** Semicontinuity problems in the calculus of variations, Nonlinear Anal. 4 (1980), 241-257.

[30] **E.J. McShane,** Relaxed controls and variational problems, SIAM J. Control Optim. 5 (1967), 438-485.

[31] **C.B. Morrey,** Multiple Integrals in the Calculus of Variations, Springer-Verlag, Berlin 1966.

[32] **F. Murat,** Contre-exemples pour divers problèmes où le controle intervient dans les coefficients, Ann. Mat. Pura Appl. 112 (1977), 49-68.

[33] **C. Olech,** Weak lower semicontinuity of integral functionals, J. Optim. Theory Appl. 19 (1976), 3-16.

[34] **Y. Reshetniak,** General theorems on semicontinuity and on convergence with a functional, Siberian Math. J. 8 (1967), 801-816.

[35] **R.T. Rockafellar,** Convex Analysis, Princeton University Press, Princeton 1972.

[36] **J. Serrin,** On the definition and properties of certain variational integrals, Trans. Amer. Math. Soc. 101 (1961), 139-167.

[37] **L. Tonelli,** Fondamenti di Calcolo delle Variazioni, Vols. 1,2, Zanichelli, Bologna 1921-23.

[38] **J. Warga,** Relaxed variational problems, J. Math. Anal. Appl. 4 (1962), 111-128.

[39] **J. Warga,** Necessary conditions for minimum in relaxed variational problems, J. Math. Anal. Appl. 4 (1962), 129-145.

[40] **L.C. Young,** Lectures on the Calculus of Variations, W.B. Saunders, Philadelphia 1969.

[41] **T. Zolezzi,** Book in preparation.

THE INCLINATION OF AN H-GRAPH

Robert Finn
Department of Mathematics
Stanford University
Stanford, CA 94305 (USA)

Dedicated to Hans Lewy

1.

In the year 1914, the Russian mathematician S. Bernstein published, in a journal with limited circulation, his now celebrated theorem that a solution $u(x,y)$ of the minimal surface equation, defined and regular in the entire (x,y) plane, is itself necessarily a plane [1].

Five years earlier in 1909, Bernstein had already published in a more widely distributed journal the theorem that there is no surface $z = u(x,y)$ of mean curvature $H \geq H_0 > 0$ defined over a region that strictly includes a disk of radius $1/H_0$ [2]. The connection between the two results is evident from the observation that minimal surfaces are characterized as having mean curvature zero.

Despite its more accessible presentation, the earlier result was apparently ignored until 1955, when E. Heinz [3] published a simpler proof of it.

2.

In fact, a stronger result is true [4]:

If $z = u(x,y)$ *has mean curvature* $H \geq H_0 > 0$ *in an open disk of radius* $1/H_0$, *then* $H \equiv H_0$ *and* $u \equiv u_0 + H_0^{-1} - \sqrt{H_0^{-2} - (x^2 + y^2)}$. *That is, any such surface is necessarily a lower hemisphere.*

In proving the result, we may clearly assume $H_0 \equiv 1$, as that normalization can be achieved by a uniform dilation. The function $u(x,y)$ then satisfies the relation

$$\text{(1)} \qquad \operatorname{div} Tu = 2H \geq 2\,, \quad Tu = \frac{\nabla u}{\sqrt{1 + |\nabla u|^2}}\,,$$

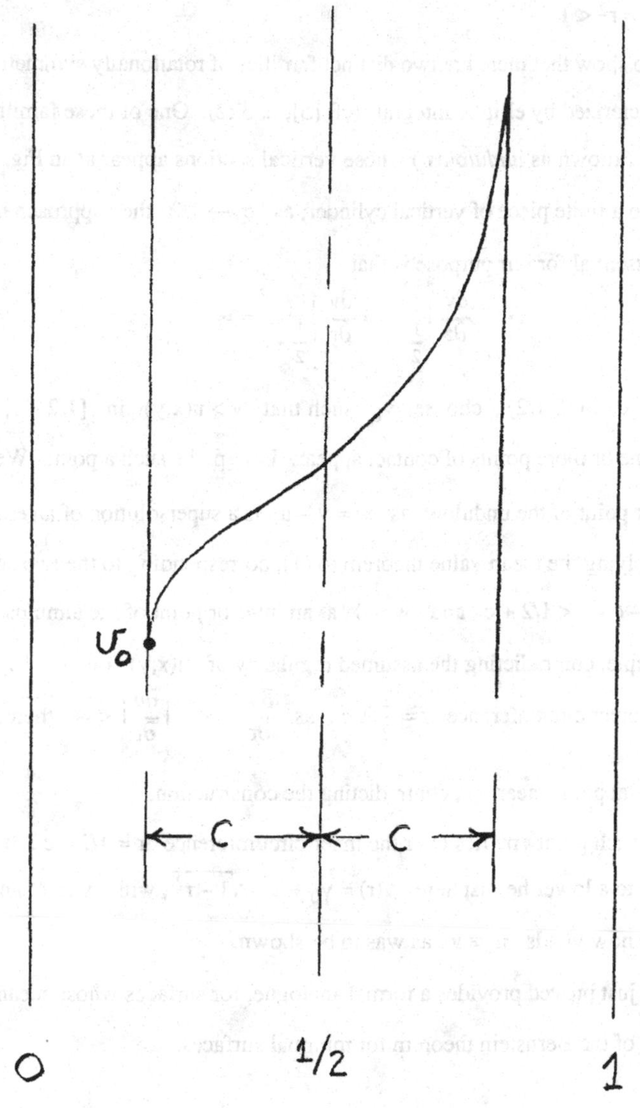

Figure 1: Section of unduloid

in the disk $x^2 + y^2 = r^2 < 1$.

It is easy to show that there are two distinct families of rotationally symmetric solutions of (1) with $H \equiv 1$, characterized by elliptic integrals (cf. [5], p. 322). One of these families consists of surfaces $v(r;c;v_0)$ (known as *unduloids*) whose vertical sections appear as in Fig. 1. As $c \to 0$, these surfaces tend to a finite piece of vertical cylinder; as $c \to 1/2$ they approach uniformly a lower unit hemisphere. Essential for our purpose is that

$$\frac{\partial v}{\partial r}\Big|_{\frac{1}{2}-c} = \frac{\partial v}{\partial r}\Big|_{\frac{1}{2}+c} = \infty .$$

For given c in $(0,1/2)$, choose v_0 such that $v > u(x,y)$ in $[1/2 - c, 1/2 + c]$, then let v_0 decrease until one or more points of contact appear. Let p be such a point. We observe that p cannot be an interior point of the unduloid, as $w = v - u$ is a supersolution of an elliptic equation (this is easily seen by applying the mean value theorem to (1), corresponding to the two solutions). If $w \geq 0$ in the annulus $1/2 - c < r < 1/2 + c$ and $w = 0$ at an interior point of the annulus, then $w \equiv 0$ by the maximum principle, contradicting the assumed regularity of $u(x,y)$ on $r = 1/2 \pm c$. Further, p cannot lie over the outer circumference $r = \frac{1}{2} + c$, as $\frac{\partial v}{\partial r} = \infty$, $\left|\frac{\partial u}{\partial r}\right| < \infty$ there, and thus v would lie below u at points near p, contradicting the construction.

Thus, any such point p lies over the inner circumference $r = 1/2 - c$. If we now let $c \to 1/2$, v tends to a lower hemisphere $v(r) = v_0 + 1 - \sqrt{1 - r^2}$, with $v \geq u$ and $v_0 = u(0)$. The maximum principle now yields $u \equiv v$, as was to be shown.

The result just proved provides a formal analogue, for surfaces whose mean curvature is bounded from zero, of the Bernstein theorem for minimal surfaces.

3.

For surfaces defined over a disk we may obtain further information that does not seem to have a parallel for minimal surfaces; it turns out that surfaces of prescribed $H \neq 0$ continue to be constrained by the domain of definition, provided this domain does not differ too much from an "extremal" one. In [6] the following theorem is proved:

There exists $R_0 = 0{,}5654062332...$ *and a nonincreasing function* $A(R)$ *in* $R_0 < R \leq 1$, *with* $A(1) = 0$, *such that if* $u(x,y)$ *is a solution of*

(2) $$\operatorname{div} Tu = 2$$

in a disk $r < R$, *then* $|\nabla u(0)| < A(R)$. *The value* R_0 *cannot be improved.*

We emphasize here the distinction between this result and the usual gradient estimates for elliptic equations, in that no bound is imposed on $|u|$. Compare for example [7,8,9], where an estimate of the form $|\nabla u(0)| < f(M/R)$ is proved for solutions $u(x,y)$ of (2) that satisfy $|u| < M$ in a disk B_R of radius R. In general, M cannot be deleted from the estimate. That can be seen from the above result with $R \leq R_0$, or more directly by considering the solutions of (2) defined for any $\alpha \in [0, \pi/2)$ by the cylinders

(3) $$u \cos \alpha = x \sin \alpha + \frac{1}{2} - \sqrt{\frac{1}{4} - y^2}$$

all of which cover the disk $r < 1/2$, and for which $u_x(0) \to \infty$ as $\alpha \to \pi/2$.

To prove the theorem, we adopt in place of the unduloids a different comparison function, motivated by physical considerations. We seek a "capillary surface" $v(x,y;R) = v_R$ under zero gravity conditions, in a cylindrical tube with "moon" section, as illustrated in Fig. 2. The surface is to meet the (vertical) bounding walls with zero contact angle on Σ, and with contact angle π on Γ. Then, denoting unit exterior normal by ν, we must have $\nu \cdot T v_R = 1$ on Σ, $\nu \cdot T v_R = -1$ on Γ. (For background information see, e.g. [10], Chapters 1 and 6.) Integration of (2) over the section Ω yields the necessary condition

(4) $$2\,|\Omega| = |\Sigma| - |\Gamma| .$$

In view of the particular geometry, (4) turns out to be sufficient; that follows, e.g. from Theorem 7.9 in [10]. The existence of this surface is crucial in what follows. The condition (4) uniquely determines the position of the center of Γ on any given line through the center of Σ. The value R_0 is the unique value of R for which Γ passes through 0.

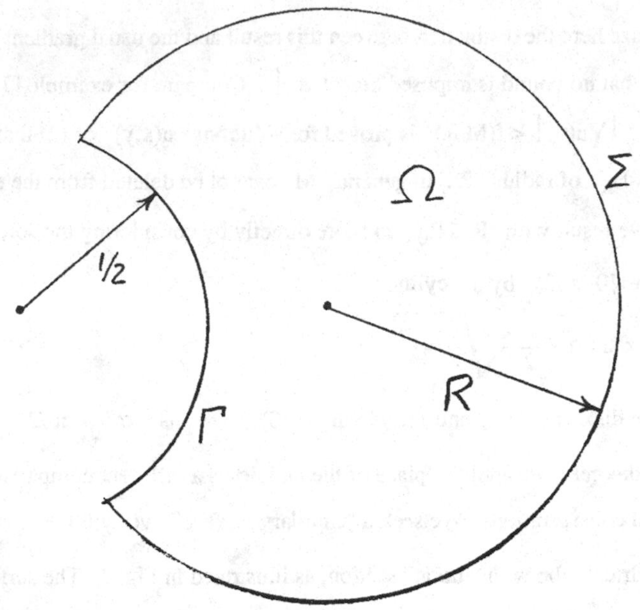

Figure 2: Moon domain

Let $u(x,y)$ be a solution of (2) in the disk $B_R : x^2 + y^2 < R^2$, with $R > R_0$. Then a "moon surface" v_{R0} can be situated over a corresponding Ω_{R0}, such that (i) $\Omega_{R0} \subset\subset B_R$, (ii) the center of B_R (taken as the origin) lies interior to Ω_{R0}, and (iii) $\nabla v_{R0}(0)$ coincides in direction with $\nabla u(0)$, see Fig. 3.

We assert that then $|\nabla u(0)| < |\nabla v_{R0}(0)|$. If not, we observe (see Appendix M) that $\nabla v_{R0} \to \infty$ in a direction normal to Γ_{R0} at almost all points of Γ_{R0}; thus by shifting the position of Ω_{R0} so that Γ_{R0} approaches the origin, we could obtain a configuration with $\Omega_{R0} \subset\subset B_R$ and with $\nabla u(0) = \nabla v_{R0}(0)$. By adding a constant to v_{R0}, we can additionally obtain $u(0) = v_{R0}(0)$. There would then be (at least) four regions distinguished, as in Fig. 3, in which alternatively $v_{R0} > u$ or $v_{R0} < u$ (see, e.g., [6,11,12,19]). A formal integration by parts leads in at least one of these regions to a contradiction with the (extremal) boundary conditions for v_{R0} on $\partial\Omega_{R0}$, see [6] for details.

The above reasoning extends immediately to show the existence of a bound for $|\nabla u|$ *throughout the disk* $B_{R-R_0-\epsilon}$, *for any* $\epsilon > 0$.

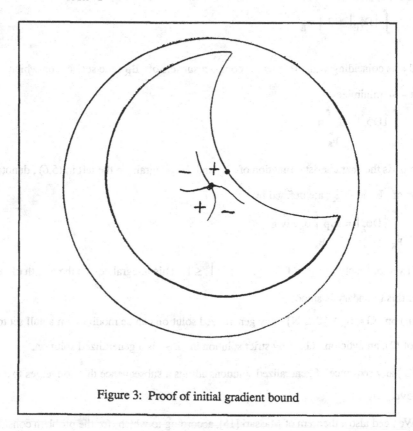

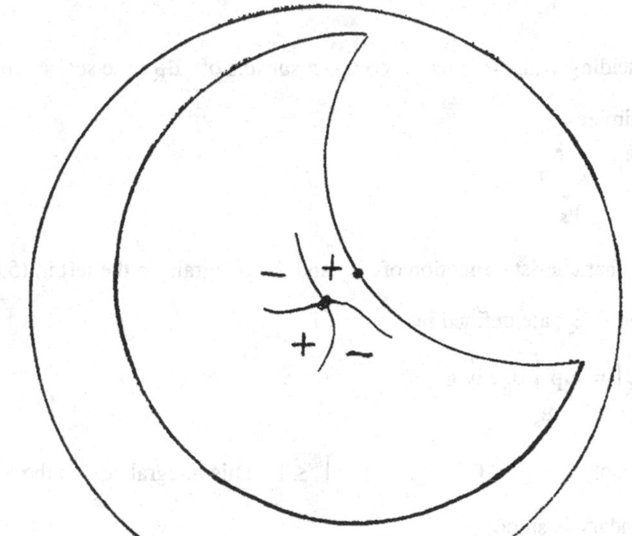

Figure 3: Proof of initial gradient bound

4.

We proceed to use this information to outline the proof of a more precise result [13]:

Let $R > R_0$, *let* $\hat{R}(R)$ *be the radius of the largest disk concentric to* B_R , *such that* $B_{\hat{R}} \subset \Omega_R$. *There exists* $A(R; \epsilon) < \infty$ *such that for any solution* $u(x,y)$ *of* (2) *in* B_R *and any* $\epsilon > 0$ *there holds* $|\nabla u(x,y)| < A(R; \epsilon)$ *in* $B_{\hat{R}-\epsilon}$. *The value* $\hat{R}$ *cannot be improved.*

As before, we emphasize that the estimate in no way depends on a bound for $|u|$. We prove the statement by combining the previous result with a reasoning based on properties of the generalized solutions introduced by Miranda [14]. These solutions are determined by variational conditions, and can be infinite on sets of positive Lebesgue measure. For our purposes, the essential properties we shall need are as follows:

(i) For a generalized solution $u(x,y)$ in B_R, the set P on which a solution $u(x,y) \equiv +\infty$ minimizes the expression

$$(5) \qquad \int_{B_R} |D\phi_E| + 2\int_{B_R} \phi_E$$

among all sets coinciding with P outside compact subsets of B_R; the set N on which $u(x,y) \equiv -\infty$ minimizes

$$(6) \qquad \int_{B_R} |D\phi_E| - 2\int_{B_R} \phi_E \ .$$

Here ϕ_E is the characteristic function of E, and the integrals on the left in (5,6), denoted as the *perimeter* of E in B_R, are defined by

$$(7) \qquad \int_{B_R} |D\phi_E| = \sup \int_{B_R} \phi_E \operatorname{div} g$$

among all vector functions $g \in C^1{}_0(B_R)$, $|g| \leq 1$. This integral equals the length of ∂E in B_R whenever this boundary is smooth.

(ii) on $G = B_R \setminus \{P \cup N\}$ any generalized solution can be modified on a null set to be a strict solution of (2), analytic on G. Any strict solution in B_R is a generalized solution.

(iii) any sequence of generalized solutions admits a subsequence that converges to a generalized solution over B_R.

We need also a theorem of Massari [15], according to which (for the problem considered here) if P (or N) is neither the null set $\emptyset$ nor B_R, then it is bounded in B_R by circular arcs of radius $1/2$; P must lie on the side of these arcs opposite to that into which the curvature vector points, N on the side into which that vector points.

We return to the proof of the stated theorem. Given $R > R_0$ and any family of solutions $\{u(x,y)\}$ of (2) in B_R, we introduce the normalization $u(0) = 0$, which can be achieved by additive constants. In view of the above properties, we see that it suffices to prove that neither P nor N can coincide with B_R, and that no arcs of ∂P or ∂N can enter $B_{\tilde{R}}$.

It is clear from the above gradient bound in $B_{R-R0-\varepsilon}$ that neither P nor N can coincide with B_R. The following lemma is proved in [13].

If, for some domain $\mathcal{D}$, E_0 *minimizes*

$$\int_{\mathcal{D}} |D\phi_E| - 2\int_{\mathcal{D}} \phi_E$$

then

(8) $$\int |D\phi_{E_0}| - 2\int \phi_{E_0} \le \int |D\phi_{\mathcal{D}}| - 2\int \phi_{\mathcal{D}}$$

the integrals here being taken over the entire space $\mathbf{R}^2$.

In our case, we take $\mathcal{D} = B_R$, $E_0 = N$ and obtain

(9) $$2\ |B_R \setminus N| \le |\Sigma^*| - |\Gamma| \ .$$

Suppose an arc Γ' of ∂N were to enter $B_{\hat{R}}$, as in Fig. 4. We introduce an extremal "moon surface" v_R corresponding to $B_{\hat{R}}$, as indicated. Then v_R satisfies (2) in $B_R \setminus N$, has finite derivatives on $\Gamma = \partial(B_R \setminus N) \cap B_R$, and $\nu \cdot Tv_R = 1$ almost everywhere on $\Sigma^* = \partial(B_R \setminus N) \cap \partial B_R$. Integrating over $B_R \setminus N$, we obtain

(10) $$2|B_R \setminus N| = |\Sigma^*| + \int_{\Gamma} \nu \cdot Tv_R \ .$$

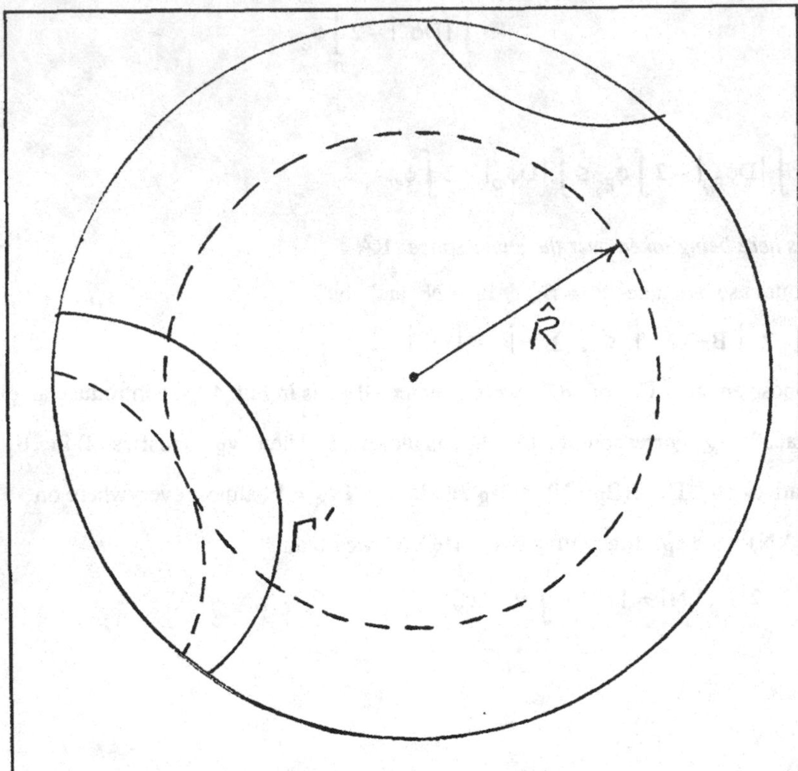

Figure 4: Proof that $N \cap B_{\hat{R}} = \emptyset$

Since however

(11) $$\nu \cdot Tv_R = \frac{\nu \cdot \nabla v_R}{\sqrt{1 + |\nabla v_R|^2}} > -1$$

we find

(12) $$2|B_R \setminus N| > |\Sigma^*| - |\Gamma|$$

contradicting (9) and thus completing the proof.

We consider now the set P. Since $|\nabla u|$ is bounded in $B_{R-R0-\varepsilon}$, the origin cannot be interior to P. Since $R > 1/2$ and P is bounded by circular arcs of radius 1/2, the only possibility is a set Ω^* of the form indicated in Fig. 5. We compare the functional (5) for P with the value obtained when Γ is replaced by the dotted curve in Fig. 5, with $\Omega^* = \phi$. The change in value of the functional is exactly

$$-\left\{|\Gamma| - |\Sigma^*| - |\Sigma'| + 2|\Omega^*| + 2|\Omega'|\right\} - \left\{|\Gamma| - 2|\Omega'|\right\} = -\Phi[\Gamma] - \Psi[\Gamma'] .$$

The configuration of the "moon domains" is determined by the criterion $\Phi = 0$ (see Appendix M). It is proved in [13] that if $R > R_0$ then the moon domain Ω_R contains the origin, also that for fixed R, Φ increases as Γ moves into B_R contains the origin. Thus $\Phi > 0$ in the configuration of Fig. 5. But also $\Psi > 0$ (cf. [17]), and it follows that Ω^* does not minimize. We conclude that not only do the boundary arcs of P not enter B_R, but in fact $P = \emptyset$.

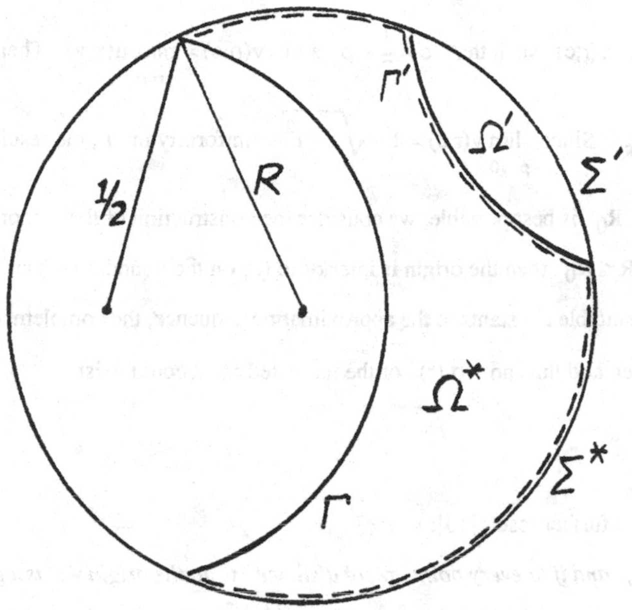

Figure 5: Proof that $P = \emptyset$

5.

In view of what has just been shown, we may now state:

There exists a nonincreasing $\sigma(r) < \infty$ *in* $R_0 < r \le 1$, *with* $\sigma(1) = 0$, *such that if* $R > R_0$ *and* $u(x,y)$ *satisfies* (2) *in* B_R, *then*

$$u(x,y) - u(0) \le 1 - \sqrt{1 - r^2} + \sigma(R) \quad . \tag{13}$$

The value R_0 *cannot be improved, and is not achieved.*

That is, after a normalization all solutions are bounded above, by a function asymptotic to a lower hemisphere. Since the "moon surfaces" can be approximated by solutions defined in all of B_R (Appendix M), no such bound can be found from below.

To prove the statement, we observe that since $P = \emptyset$, all (normalized) solutions are equibounded above in any compact subdomain of B_{R_0}, hence in $B_{1/2}$. Thus, there exists $M(\rho) < \infty$, with $\lim_{\rho \to 0} M(\rho) = 0$, such that $u - u(0) < M(\rho)$ in $0 < \rho < 1/2$. Setting $R' = 1 - \rho$, we choose an unduloid $v(r;c)$ such that $c = \frac{1}{2} - \rho$ and $v(\rho;c) = \max_{r=\rho} u(x,y)$. Then $u(x,y) \le u(0) + M(\rho) + v$ in $B_{R'}$. Since $\lim_{\rho \to 0} v(r;c) = 1 - \sqrt{1 - r^2}$ uniformly in r, the result follows.

To see that R_0 is best possible, we consider the construction of the "moon surfaces" (Appendix M). If $R \le R_0$ then the origin is interior to (or on the boundary of) an "N" domain for B_R. By addition of suitable constants to the approximating sequence, the complementary domain can be made into a "P" set, and thus no $\sigma(r)$ of the indicated form could exist.

6.

We indicate a further result [13]:

If $R > 1/2$, *and if to every point* p *of distance* r *to the origin we associate the function*

$$\mathcal{P}(r;R) = \frac{1 + 2(r^2 - R^2)}{2r} \quad ,$$

then there corresponds to each such p *an outward radial sector* S *of opening*

$$\psi(r) = \begin{cases} 2\cos^{-1}\mathcal{P} & \text{if } |\mathcal{P}(r;R)| < 1 \\ 0 & \text{if } \mathcal{P}(r;R) \ge 1 \\ 2\pi & \text{if } \mathcal{P}(r;R) \le -1 \end{cases} \tag{14}$$

(see Fig. 6), *and a finite* $M(r;R)$, *such that for any solution* $u(x,y)$ *of* (2) *in* B_R *for which* $\nabla u(p)$ *is directed into* S, *there holds* $|\nabla u(p)| < M$.

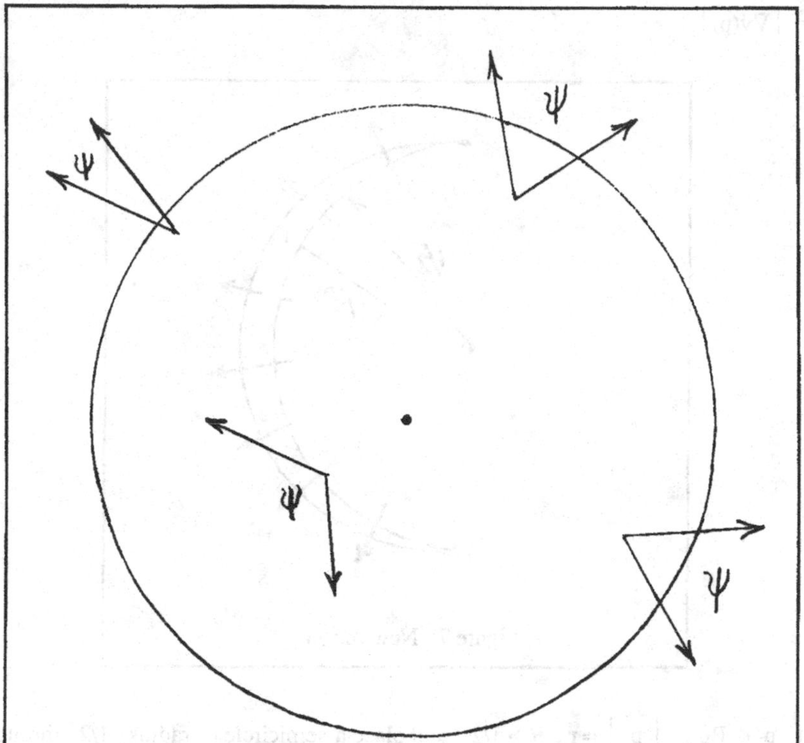

Figure 6: Excluded sectors

The interest in this result is that a gradient bound is provided up to the boundary of B_R, in the sense that if $|\nabla u|$ is large then ∇u must be directed outside the radial sector S. In directions outside S, $|\nabla u|$ can in fact become arbitrarily large (cf. Appendix M).

We note that if $R = 1/2^+$ then a bound is provided only near ∂B_R. A bound is obtained at the origin (and then for all directions) whenever $R > \sqrt{2}/2$. However, for this purpose more information is provided by the previously indicated results. If a bound is obtained for any given r, then bounds continue to hold for all larger r, the sector opening decreasing with r. This behaviour is consistent with that of the lower hemisphere for the case $R = 1$.

We prove the statement by using as comparison surfaces the "moon surfaces" v_R for the case $R = 1/2^+$. We then obtain a "new moon" as indicated in Fig. 7 (for details see [13]) for which ∇v is directed (essentially) as indicated in the figure. *Whenever such a region can be placed interior to the region of definition of* $u(x,y)$ *so that the gradient directions coincide at a point* p, *then*

$|\nabla u(p)| < |\nabla v(p)|$.

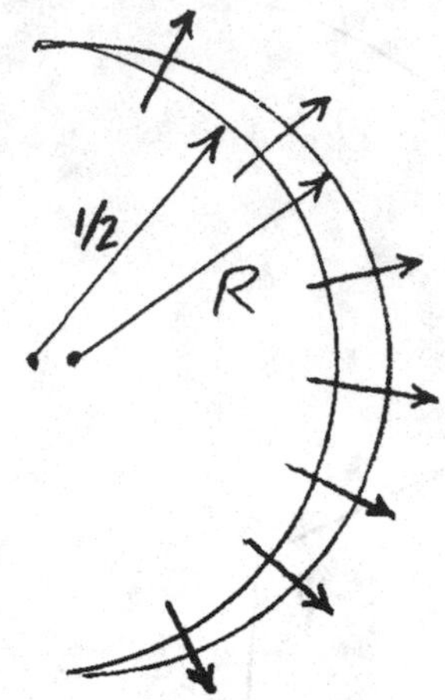

Figure 7: New moon

Let $p \in B_R$, $|p| = r$, $R > 1/2$, and place a semicircle of radius 1/2 through p with its vertices on ∂B_R, as indicated in Fig. 8. We assert that the angle between the two radius vectors determines half the sector S. To show that, it suffices to prove that a "new moon" can be placed interior to B_R with its outer radius directed along any ray of the sector.

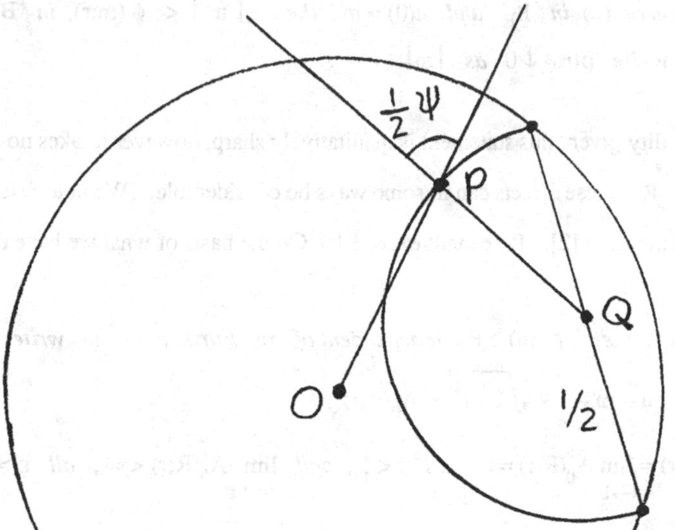

Figure 8: Proof of excluded sector property

Move the semicircle away from p with the two vertices sliding along ∂B_R, then move it orthogonal to its diameter until p is once more contacted. In this procedure the orientation of the diameter changes continuously. Finally, p will be contacted again by the reflection, across the radial direction through p, of the original semicircle; in this configuration the line Qp achieves the other extremal direction of S. Thus, all radial directions in S are achieved. The relation (14) follows by applying the law of cosines to an extremal configuration.

7.

We now wish to show that the methods described above yield as special case a conceptually simpler proof of a Harnack inequality due to Serrin [7] and give in some ways more explicit information on behavior.

We recall Serrin's theorem, which in our normalization can be stated as follows:

There exists $\rho(m) > 0$ *and a function* $\phi(m;r) < \infty$ *in* $r < \rho$, *such that if* $u(x,y)$ *is a nonvanishing solution of* (2) *in* B_R *and* $u(0) = m$, *then* $|u| < \phi(m;r)$ *in* B_R. *There holds* $\lim_{r\to\rho} \phi(m;r) = \infty$, *while* $\rho(m) \downarrow 0$ *as* $|m| \to \infty$.

In the generality given, this statement is qualitatively sharp, however it takes no account of the effect of changes in R; these effects can in some ways be considerable. (We note that in view of Bernstein's original theorem [2], R cannot exceed 1.) On the basis of what we have done thus far, we may state:

1. *If* $R > R_0$, *then* $\rho(m) \geq \hat{R}$, *independent of* m. *Further, we may write*

$$\text{(15)} \qquad A_0^-(R;r) < u - m - 1 + \sqrt{1-r^2} < A_0^+(R;r) \quad ,$$

where $\lim_{R\to 1} A_0^-(R;r) = \lim_{R\to 1} A_0^+(R;r) = 0$, *all* $r < 1$, *and* $\lim_{r\to R-\varepsilon} A_0^+(R;r) < \infty$, *all* $\varepsilon > 0$.

2. If $R > R_0$, $m > 0$, *then* $\rho(m) = \infty$, *independent of* m. *We have*

$$\text{(16)} \qquad u - m - 1 + \sqrt{1-r^2} < A_1(R)$$

independent of r. *Here* $A_1(R) \downarrow 0$ *as* $R \to 1$.

Both statements 1. and 2. follow easily from the earlier material of this paper. It remains to provide the promised new proof of the original Serrin theorem.

Were the theorem false, there would exist, for fixed m, R, a sequence u_n of non vanishing solutions of (2) in B_R, such that $\lim_{n\to\infty} |u_n(p_n)| \to \infty$ at a sequence $p_n \to 0$, and such that $u_n(0) = m$, all n. By Miranda's theorem, there would be a subsequence (which we again label u_n) convergent to a generalized solution u, for which either N or P must be null. Suppose first $m > 0$, so that $N = \emptyset$. Since the sequence converges uniformly on compact subsets of the regularity set G of u, it is clear from the construction that $P \neq \emptyset$. The convergence is also uniform on compact subsets of $B_R \setminus \bar{G}$, hence the origin cannot be an interior point of P. Since P is bounded by circular arcs Γ of radius 1/2, one such arc, Γ_0, must pass through the origin.

Using the results of [10], Chapter 7, it is easy to show that in the configuration $\mathcal{D}$ of Figure 9, in which all four arcs have radius 1/2, the relative dimensions can be adjusted so that the diameter of $\mathcal{D}$ will be as small as desired, and so that a solution v of (2) in $\mathcal{D}$ will exist, with $\nu \cdot Tv = 1$ on $\mathcal{C}_1^+$, $\mathcal{C}_2^+$; $\nu \cdot Tv = -1$ on $\mathcal{C}_1^-$, $\mathcal{C}_2^-$. (An alternative proof could be obtained from results of Spruck [16]; these results are however derived using methods of [7], and our reasoning would thus become circular). By Theorem 1 of [18], $v \to +\infty$, a.e. on C_1^+, C_2^+; $v \to -\infty$ a.e. on C_1^-, C_2^-.

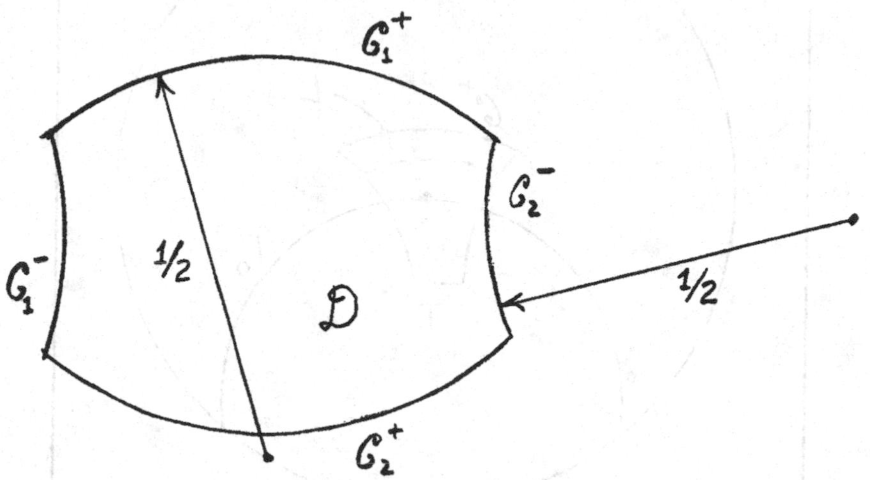

Figure 9: Extremal configuration

Let us position the upper half $\mathcal{D}^+$ of $\mathcal{D}$ as indicated in Fig. 10, and adjust v by an additive constant to that $v < 0$ on the symmetry segment T. On the dotted curve L, $u_n \to \infty$ uniformly [14], hence $u_n > v$ on L for large enough n. By the general comparison principle (Theorem 5.1 in [10]) we find $u_n(0) > v(0)$. But by shifting $\mathcal{D}^+$ so that C^+_1 tends toward Γ_0, we can make $v(0)$ as large as desired. This contradicts the choice $u(0) = m$.

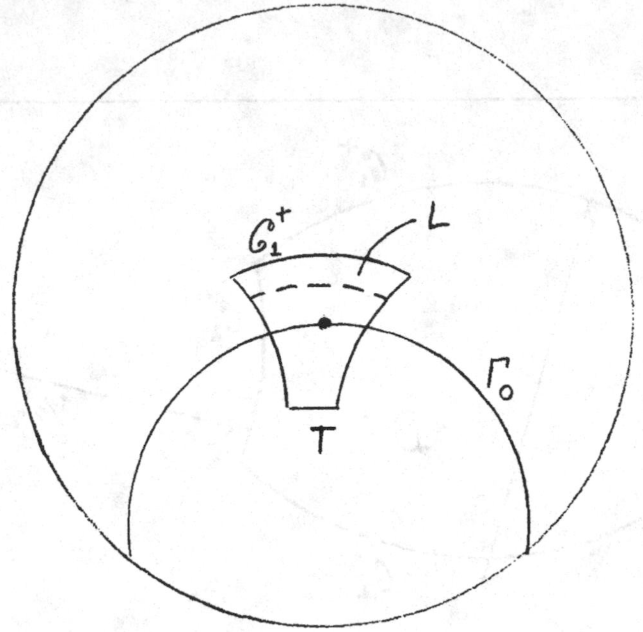

Figure 10: Proof of Harnack inequality

If $m < 0$ we repeat (essentially) the same reasoning, using a 90° rotation of $\mathcal{D}$.

Remarks:

1. The statements 1 and 2 above yield considerably sharper (and qualitatively different) information than that encompassed in the original Serrin formulation. However, for small values (especially) of R the Serrin procedure provides suggestive information as to the structure of ϕ tha does not follow from the present (indirect) procedures, see [7], p. 378.

2. From Statement 2, we see that for $R > R_0$ all solutions are bounded above, uniformly throughout B_R. There can be no such bound from below, as can be seen from the construction of the "moon surfaces" in Appendix M. We have already remarked that the value R_0 cannot be improved.

Appendix M: Moon landing

In order to establish sufficiency for the bound in $B_{\hat{R}}$ it is necessary to know the existence of the moon surfaces $v_R = v(x,y;R)$; in order to know that $\hat{R}$ is sharp we must be able to approximate v_R by solutions that are regular over the entire disk B_R.

The existence of v_R follows, for example, from Theorem 7.9 in [10], and is obtained by prescribing the data $\gamma = 0$ on Σ, $\gamma = \pi$ on Γ. It is clear that there are no extremal arcs in Ω that meet Σ in the prescribed angles, and thus the conditions of the theorem are satisfied.

For given $R > 1/2$, a necessary condition for existence of a solution $v(x,y)$ over the entire B_R, with data γ_1 on Σ_1, γ_2 on Σ_2, is obtained by integrating (2) over B_R. We find (see Fig. 11)

(17) $$|\Sigma_1| \cos \gamma_1 + |\Sigma_2| \cos \gamma_2 = 2\,|B_R| \;.$$

For v_R the corresponding values are $\cos \gamma_1 = 1$ on Σ_1, $\cos \gamma_2 = -1$ on Γ, and hence

(18) $$\Phi[\Omega] \equiv |\Sigma_1| - |\Gamma| - 2|\Omega| = 0 \;.$$

From this relation it is clear that there is no solution of (2) in all of B_R, with $\cos \gamma_1 \equiv 1$, as then the derivatives could not be bounded on Γ. We therefore choose $\varepsilon > 0$ and seek, as approximant to v_R, a solution v in B_R with data $\cos \gamma_1 = 1 - \varepsilon$. Then $\cos \gamma_2 = \sigma(\varepsilon)$ is determined by the relation

(19) $$(1-\varepsilon)\,|\Sigma_1| + \sigma(\varepsilon)\,|\Sigma_2| = 2\,|B_R|$$

and we seek a solution v^{ε} in B_R with data $\gamma_1 = \cos^{-1}(1-\varepsilon)$, $\gamma_2 = \cos^{-1} \sigma(\varepsilon)$.

In order for the procedure to make sense, it must be shown that $-1 \le \sigma(\varepsilon) \le 1$. It is a technical exercise to do so, and the details are carried out in [17], where it is shown that the stronger result $-1 < \sigma(\varepsilon) < 0$ holds for all sufficiently small $\varepsilon > 0$.

Thus the basic necessary condition can be satisfied. This condition ensures that the value $H \equiv 1$ is the value determined by the conditions of Theorem 7.1 in [10]. In order to apply that theorem to obtain the existence of a solution, it suffices to show that there is no subarc of a semicircle of radius $1/2$ that meets Σ at two points with angles determined by the data and for which $\Phi \le 0$ (cf. the discussion in §§ 6.8, 7.7 of [10]). Since $R > 1/2$, it is clear that no such arc can meet Σ_1. By Theorem 6.10 in [10], the arc Γ meets Σ in an angle β, with $0 < \beta < \gamma_2 = \cos^{-1} \sigma$, see Fig. 11. Thus, there is a unique symmetrically situated arc $\hat{\Gamma}$ of radius $1/2$, meeting Σ_2 in the angle γ_2. It suffices to consider this symmetric case, as every other such arc meeting Σ_2 at both end points yields the same value for Φ.

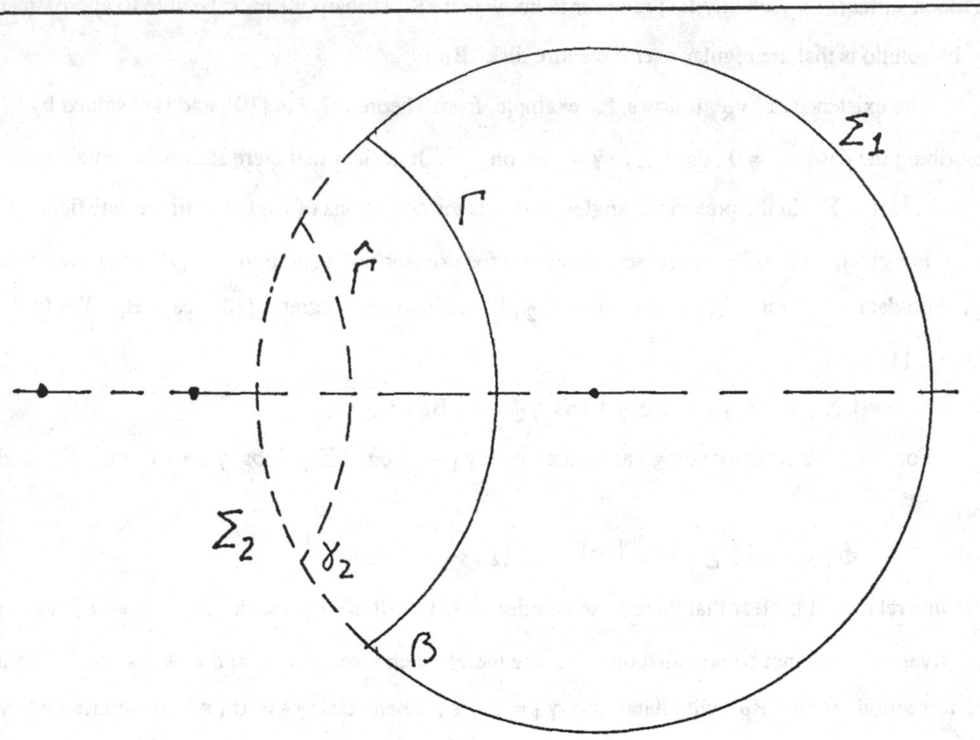

Figure 11: Existence criterion for moon surfaces

Again it is a technical exercise to show that the corresponding $\Phi[\hat{\Omega}]$ is positive; the details are given in [17]. We conclude from Theorem 7.1 of [10] that there exists a solution $v(x,y;\varepsilon)$ to (2) in B_R, with data as above, corresponding to any sufficiently small $\varepsilon > 0$.

We now let $\varepsilon \to 0$. Since $\Phi[\Omega] = 0$, we find at once that

$$\int_\Gamma \nu \cdot Tv \, ds \to |\Gamma| \quad .$$

Since $|\nu \cdot Tv| < 1$, and since

$$\int_\Gamma (1 + \nu \cdot Tv) \, ds \to 0 \quad ,$$

we conclude that $v \cdot Tv \to -1$ almost everywhere on Γ, and hence $|\nabla v| \to \infty$ a.e. on Γ. Thus the disk $B_{\hat{R}}$, in which we have shown that a gradient bound can be found, cannot be replaced by any disk of larger radius.

The solutions $v(x,y;\varepsilon)$ in fact converge to v_R in the sense of Miranda [14]. To see that, we first observe that by Miranda's theorem [14] any sequence of them admits a subsequence that converges to a generalized solution $w(x,y)$. In [17] it is shown that the regularity domain of w coincides with Ω. Since the boundary data converge on Γ_0 and on Σ_1 to those of v_R, a uniqueness reasoning shows the identity of the two functions in Ω. Thus v_R is in fact the restriction to Ω of a generalized solution determined by the v^ε. Since v_R is uniquely determined up to an additive constant (Theorem 7.6 in [10]) we conclude that every sequence v^ε converges (after normalization) to the same generalized solution in B_R.

In [17] it is shown that $v_R \to -\infty$, $Tv_R \to -\nu$ at all points of Γ_0, while $Tv_R \to \nu$ at all points of Σ_1, and v_R is bounded at Σ_1. This information could be of use toward finding explicit gradient estimates for solutions of (2).

This research was supported in part by the National Science Foundation, and in part by the National Aeronautics and Space Administration. The paper was prepared while the author was visiting at Universität Bonn under the auspices of Sonderforschungsbereich 72.

References

[1] **S.N. Bernstein,** Sur un théorème de géometrie et ses applications aux équations aux dérivees partielles du type elliptique, Comm. Soc. Math. Kharkov (2) 15 (1915-1917), 38-45; see also the German translation in Math. Zeit. 26 (1927), 551-558.

[2] **S.N. Bernstein,** Sur les surfaces définies au moyen de leur courbure moyenne et totale, Ann. Sci. Ecole Norm. Sup. 27 (1910), 233-256.

[3] **E. Heinz,** Über Flächen mit eineindeutiger Projektion auf einer Ebene, deren Krümmungen durch Ungleichungen eingeschränkt sind, Math. A. 129 (1955), 451-454.

[4] **R. Finn,** Remarks relevant to minimal surfaces and to surfaces of prescribed mean curvature, J. Analyse Math. 14 (1965), 139-160.

[5] **P. Concus and R. Finn,** The shape of a pendent liquid drop, Phil. Trans. Roy. Soc. London, Ser. A 292 (1979), 307-340.

[6] **R. Finn and E. Giusti,** On nonparametric surfaces of constant mean curvature, Ann. Scuola Norm. Sup. Pisa 4 (1977), 13-31.

[7] **J.B. Serrin,** The Dirichlet problem for surfaces of constant mean curvature, Proc. Lon. Math. Soc. (3) 21 (1970), 361-384.

[8] **O.A. Ladyzhenskaia and N.N. Ural'tseva,** Local estimates for gradients of solutions of non-uniformly elliptic and parabolic equations, Comm. Pure Appl. Math. 23 (1970), 677-703.

[9] **E. Heinz,** Interior gradient estimates for surfaces $z = f(x,y)$ of prescribed mean curvature, J. Diff. Geom. 5 (1971), 149-157.

[10] **R. Finn,** Equilibrium capillary surfaces, Springer-Verlag, New York, 1986. (Grundlehren der Mathem. Wiss. #284).

[11] **R. Finn,** New estimates for equations of minimal surface type, Arch. Rat. Mech. Anal. 14 (1963), 337-375.

[12] **P. Hartman and A. Wintner,** On the local behavior of solutions of non-parabolic partial differential equations, Amer. J. Math. 75 (1953), 449-476.

[13] **Fei-tsen Liang,** On nonparametric surfaces of constant mean curvature, Dissertation, Stanford University, 1986; to appear.

[14] **M. Miranda,** Superfici minime illimitate, Ann. Scuola Norm. Sup. Pisa 4 (1977), 313-322.

[15] **U. Massari,** Esistenza e regolarità delle ipersuperfici di curvatura media assegnata in $\mathbf{R}^n$, Arch. Rat. Mech. Anal. 55 (1974), 257-282.

[16] **J. Spruck,** Infinite boundary value problems for surfaces of constant mean curvature, Arch. Rat. Mech. Anal. 49 (1972/3), 1-31.

[17] **R. Finn,** Moon Surfaces, Preprint 816, Universität Bonn, Sonderforschungsbereich 72; to appear.

[18] **R. Finn,** A note on the capillary problem, Acta Math. 132 (1974), 199-206.

[19] **R. Finn,** On equations of minimal surface type. Annals of Math. 60 (1954), 397-416, esp. p. 397.

ON THE MATHEMATICAL THEORY OF VORTEX SHEETS

Paul R. Garabedian
Courant Institute of Mathematical Sciences
New York University, 251 Mercer Street
New York, N.Y. 10012 (U.S.A.)

Dedicated to Hans Lewy

1. A variational principle

In this paper we discuss mathematical and computational methods to study vortex sheets in fluid dynamics. These are slip streams across which the pressure p and speed q of the flow remain continuous. In light fog they can be observed rolling up behind the wings of an airplane. Also when a helicopter hovers before landing, vortex sheets that trail from one rotor blade to impinge on the next are largely responsible for the noise that is heard. We shall be concerned here primarily with questions of existence, regularity, and numerical approximation for the simplest models of a vortex sheet in potential flow.

Let

$$\phi = x + \frac{\alpha x}{r^3} + \ldots$$

represent the velocity potential at infinity of the steady irrotational flow of an incompressible fluid past a fixed ring-shaped body R that surrounds and includes a thin film or free surface S. The function ϕ is harmonic throughout the flow and has a normal derivative $\partial\phi/\partial\nu$ that vanishes at the boundary. If the speed $q = |\nabla\phi|$ remains continuous across S, then the pressure p becomes continuous, too, and S is called a vortex sheet.

The virtual mass of the flow is defined by the Dirichlet integral

$$E = \iiint \left(q^2 - 1\right) dx\, dy\, dz \quad .$$

It is related to the volume V of R and the coefficient α in the expansion of ϕ at infinity by the identity

$$4\pi\alpha = E + V ,$$

which follows from an application of the divergence theorem. When the free surface S is subjected to an infinitesimal shift δv along its normal, the virtual mass is found to satisfy the Hadamard variational formula

$$\delta\alpha = \frac{1}{4\pi}\iint_S q^2\,\delta v\,d\sigma \quad ,$$

where the integral is to be evaluated over both sides of the surface, but with opposite signs for the normal shift [6]. The formula is easy to verify because ϕ satisfies a natural boundary condition.

If q is discontinuous across S then in general $\delta\alpha \neq 0$. On the other hand, when S is a vortex sheet, so that q has no jump, the two integrals from the opposite sides of S cancel and $\delta\alpha = 0$. This result gives rise to a variational principle [8] asserting that vortex sheets are characterized by the extremal property

$$\alpha = \text{minimum}$$

within a suitable class of admissible surfaces S.

The variational principle can be extended to the case of a compressible fluid. Then the nonlinear partial differential equation

$$\left(\rho\phi_x\right)_x + \left(\rho\phi_y\right)_y + \left(\rho\phi_z\right)_z = 0$$

for the velocity potential is known to be the Euler equation for a problem

$$-\iiint p\left(q^2\right)\,dx\,dy\,dz = \text{minimum}$$

in the calculus of variations in which p is defined as a function of $q^2 = (\nabla\phi)^2$ by Bernoulli's law

$$\frac{1}{2}q^2 + \int \frac{dp}{\rho} = \text{const.}$$

and the usual equation of state $p = A\rho^\gamma$. We can introduce an appropriate finite part of the energy integral occurring here and minimize it with respect to the shape of S to arrive at the requirement for a vortex sheet that q have no jump. Finite element versions of this extremal problem provide a numerical method to locate the vortex sheet trailing behind a wing or rotor blade in transonic flow [1].

A minimum energy principle of the kind we have described has been used to prove the existence of plane or axially symmetric flows with free streamlines [7]. In two dimensions the regularity of the free boundary is established by a technique of analytic continuation due to Lewy [13]. However, for truly three-dimensional flows with no symmetry the question of existence and regularity of solutions becomes more subtle. We shall deal with it briefly in the next section.

2. Existence and regularity

It is possible to prove analyticity of free boundaries for simple Dirichlet problems in three dimensions [11]. However, this is not true of a Neumann problem like the one for the velocity potential ϕ. In fact, for three-dimensional flow any surface generated by a one-parameter family of streamlines may be viewed as a vortex sheet, and clearly it need have very little regularity. Such examples are familiar in magnetohydrodynamics, where it has also been established that no smooth solution of the problem of vortex sheets should exist in certain cases with periodic geometry [9]. The idea of the proof is to consider flux surfaces $\psi = \text{const.}$ satisfying the first order partial differential equation

$$\nabla\phi \bullet \nabla\psi = 0$$

inside a torus where ϕ represents a potential flow with zero normal derivative at the boundary. The Kolmogoroff-Arnold-Moser theory of dynamical systems shows there is a difficulty in finding periodic solutions ψ with toroidal level surfaces that might be interpreted as vortex sheets. Vortices of this kind are relevant to the theory of coherent structures in turbulence.

In connection with the issue of regularity, it is interesting to consider the time-dependent motion of a vortex sheet in the plane. Let the sheet be represented in complex notation as a curve $x + iy = z(s,t)$ defined for any time $t \geq 0$ on the interval $-1 \leq s \leq 1$, where the parameter s stands in a fixed relationship to the distribution of lift $\Gamma = \Gamma(s)$. Birkhoff [3] has derived an integral equation of the form

$$\overline{z_t(s,t)} = \frac{1}{2\pi i} \int_{-1}^{1} \frac{d\Gamma(\sigma)}{z(s,t) - z(\sigma,t)}$$

to determine z when Γ is given. Here a Cauchy principal value is used for the integral on the right. Initial value problems have been studied in which the lift is prescribed more or less arbitrarily along a slit on the real axis. For an elliptical distribution of lift the sheet is known to roll up at the ends, and similarity solutions have been found that describe its behavior there.

To investigate regularity properties of the vortex sheet it is natural, as in Lewy's work on free streamlines, to perform an analytic continuation of the unknown function $z(s,t)$ into a domain of the complex s-plane. Assuming that Γ is an analytic function, one can replace the Cauchy principal value of the integral in Birkhoff's equation by a contour integral in the σ-plane that circumvents the pole at $\sigma = s$. The residue at that pole contributes an additional term on the right, so that we have

$$\overline{z_t(s,t)} = \frac{1}{2}\frac{\Gamma_s(s)}{z_s(s,t)} + \frac{1}{2\pi i}\int_{-1}^{1}\frac{d\Gamma(\sigma)}{z(s,t) - z(\sigma,t)} \quad ,$$

where the contour of integration joining -1 to $+1$ has been deformed into the lower half-plane. Conjugates have been introduced on the left in a fashion making it apparent from the principle of analytic continuation that the equation becomes an identity valid in the complex domain. In particular, both sides are seen to be analytic functions when s is located in the upper half-plane and σ is located in the lower half-plane. On the other hand, if the path of integration is moved to the upper half-plane then there is a change in the sign of the term found from the residue theorem.

The functional equation we have obtained for z in the complex domain has the form of a reflection rule with arguments s and $\bar{s}$ that lie in opposite half-planes. A single iteration of the rule brings these arguments together and simultaneously transforms the leading terms into a quasilinear partial differential operator

$$L[z] = z_{tt}(s,t) + \frac{1}{4}\,\frac{\Gamma_s(s)\,\overline{\Gamma_s(\bar{s})}}{z_s(s,t)^2\,\overline{z_s(\bar{s},t)}^2}\,z_{ss}(s,t) + \ldots$$

which becomes of the hyperbolic type when the complex differentiations with respect to s are evaluated in suitable characteristic directions. This suggests that the equation can be used as the constructive element in a proof of the existence of solutions in cases with initial data that are analytic. A number of authors have dealt with such problems, and for the model of Kelvin-Helmholtz instability estimates have been made of the time an analytic disturbance takes to curl up and become singular [12,14,17].

We wish to apply our functional equation to a special example considered by Schwartz [16]. He imposes the cusped lift distribution

$$\Gamma(s) = \left(1 - s^2\right)^{3/2} / 3$$

along a segment of the real axis on which $x = s$ initially. Numerical evidence of the analyticity of the solution of this problem is provided by computing it as a formal power series and studying convergence. We observe that in these circumstances the integral in our functional equation can be evaluated over a closed contour surrounding the vortex sheet because z is supposed to be single-valued, whereas a square root occurs in the specification of Γ. More precisely, by substituting $s = \sin\theta$ we find that

$$\overline{z_t(\overline{\sin\theta},t)} = \frac{1}{2}\,\frac{\sin\theta\cos\theta}{z_s(\sin\theta,t)} + \frac{1}{4\pi i}\int \frac{\sin\theta'\cos^2\theta'\,d\theta'}{z(\sin\theta,t)-z(\sin\theta',t)} ,$$

where the integration is performed over the interval $-\pi \le \mathrm{Re}(\theta') \le \pi$ on a horizontal line $\mathrm{Im}\{\theta'\} = \mathrm{const}$. The zeros of $\cos\theta$ at $\theta = \pm\pi/2$ serve to offset difficulties with denominators there.

This form of the functional equation lends itself to analysis by a method of Caflisch and Orellana [4]. It would appear that their decomposition into upper and lower functions should be adequate to prove the conjecture of Schwartz that an analytic solution exists for finite time.

3. Computational methods

By now codes to calculate transonic flow in three dimensions have received wide acceptance. Especially remarkable is the time-dependent analysis by Chang [5] of potential flow over rotor blades. Although the codes handle shock waves adequately, their success in modeling the vortex sheets that appear behind wings or rotor blades is less satisfactory. An attractive proposal is to use the Euler equations and attempt to capture vortex sheets in a fashion similar to the treatment of shocks [10]. However, that may be objectionable because numerical viscosity tends to generate substantial errors in the vorticity, and the solution is not unique when there are closed streamlines. One might therefore also wish to make potential flow calculations in which vortex sheets are located by means of the variational principle developed in Section 1. In particular, this would be desirable for the problem of the tip vortex. We cannot discuss such matters in more detail here, but we do present some recent numerical results concerning the interaction of vortices with a wing tip that were obtained from codes based on a less satisfactory linearized theory of the vortex sheet [2].

In his Ph.D. thesis Ross [15] has developed a computer code called NACROSS to analyze the transonic flow over a supercritical wing and to redesign the wing after an engine nacelle has been included in the configuration. The nacelle is modeled by singularities in the velocity potential ϕ analogous to sources and sinks. The numerical method is based on adding carefully selected artificial time and artificial viscosity terms to the partial differential equation for ϕ that involve derivatives such as ϕ_t, ϕ_{xt} and ϕ_{xxx}. Convergence is achieved by freezing the coefficients of this equation in a natural way at the singularities representing the nacelle. Good results have been obtained showing that the problem of redesign to eliminate shocks close to an engine can largely be solved by twisting wing sections nearby.

We have modified the NACROSS code to treat a similar problem concerned with the effect of the vortex sheet from one rotor blade of a helicopter impinging on the succeeding blade. The aim is to redesign the blade tips so as to reduce the shocks and noise caused when they collide with the vortex sheets. The mathematical model we consider consists of a straight wing in transonic flow with a rectilinear vortex filament parallel to the motion at infinity included in the representation of the velocity potential ϕ. Thus we express ϕ in the form

$$\phi = Ux + \Gamma \tan^{-1} \frac{y - y_0}{z - z_0} + \ldots ,$$

where U is the speed at infinity, Γ is the strength of the vortex filament, y_0 and z_0 are coordinates specifying the location of the filament, and the dots stand for regular terms. The singularity of ϕ at $y = y_0$, $z = z_0$ is handled in much the same way as the sources and sinks that represented an engine nacelle.

The vortex filament code works well in realistic cases simulating the interaction of a vortex sheet with the tip of a supercritical wing designed to be shockless in a free stream. Figs. 1 and 2 display the pressure profiles, vortex filament, wing sections and shocks occurring in a typical calculation. The Mach number at infinity, lift coefficient, wave drag coefficient, aspect ratio, and normalized strength of the vortex filament are shown. The shocks that are seen as black marks in Fig. 1 are caused by vorticity from the filament, which alters the effective angle of attack of the flow locally. The shocks more or less disappear in Fig. 2 because the tip of the wing has been twisted up through 10° to compensate for a resulting loss of lift. Although the code has a capability to change the shape of the wing so as to reduce wave drag even further, we have found that it is usually not necessary to pursue that option. In the future it would be desirable to upgrade the code so as to incorporate more of the ideas we have suggested to resolve the problem of the tip vortex. A more self-consistent, periodic model could be analyzed in which the vortex sheet trailing from one blade evolves into a pair of vortex filaments impinging on the next one.

Acknowledgements

Work supported by NASA Grant NAG2-345 and NSF Grant DMS-8320430. I.C. Chang and R. Krasny made several helpful suggestions, and D. Ross modified his code to provide the data for Figs. 1 and 2.

References

[1] **F. Bauer, O. Betancourt and P. Garabedian,** Magnetohydrodynamic Equilibrium and Stability of Stellarators, Springer-Verlag, New York, 1984.

[2] **F. Bauer, P. Garabedian, A. Jameson and D. Korn,** Supercritical Wing Sections II, a Handbook, Lecture Notes in Economics and Mathematical Systems 108, Springer-Verlag, New York, 1975.

[3] **G. Birkhoff,** Helmholtz and Taylor instability, Amer. Math. Soc., Proc. Symp. Appl. Math. 12 (1962), 55-76.

[4] **R. Caflisch and O. Orellana,** Long time existence for a slightly perturbed vortex sheet, to be published.

[5] **I.C. Chang,** Transonic flow analysis for rotors, NASA TP-2375, 1984.

[6] **P. Garabedian,** Partial Differential Equations, Chelsea, New York, 1986.

[7] **P. Garabedian, H. Lewy and M. Schiffer,** Axially symmetric cavitational flow, Ann. Math. 56 (1952), 560-602.

[8] **P. Garabedian and M. Schiffer,** Convexity of domain functionals, J. Anal. Math. 2 (1953), 281-368.

[9] **H. Grad,** Toroidal containment of a plasma, Phys. Fluids 10 (1967), 137-154.

[10] **A. Jameson, T. Baker and N. Weatherill,** Calculation of inviscid transonic flow over a complete aircraft, AIAA Paper 86-0103, 1986.

[11] **D. Kinderlehrer and L. Nirenberg,** Regularity in free boundary problems, Ann. Scuola Norm. Sup. Pisa, Ser. IV, 4 (1977), 373-391.

[12] **R. Krasny,** A study of singularity formation in a vortex sheet by the point vortex approximation, J. Fluid Mech. 186 (1986).

[13] **H. Lewy,** Free surface flow in a gravity field, Comm. Pure Appl. Math. 5 (1952), 413-414.

[14] **D. Moore,** The spontaneous appearance of a singularity in the shape of an evolving vortex sheet, Proc. Roy. Soc. London A 365 (1979), 105-119.

[15] **D. Ross,** Computation of the transonic flow about a swept wing in the presence of an engine nacelle, Research and Development Report DOE/ER/03077-267, Courant Inst. Math. Sci., N.Y.U., 1985.

[16] **L. Schwartz,** A semi-analytic approach to the self-induced motion of vortex sheets, J. Fluid Mech. 111 (1981), 475-490.

[17] **C. Sulem, P. Sulem, C. Bardos and U. Frisch,** Finite time analyticity for the two and three dimensional Kelvin-Helmholtz instability, Comm. Math. Phys. 80 (1981), 485-516.

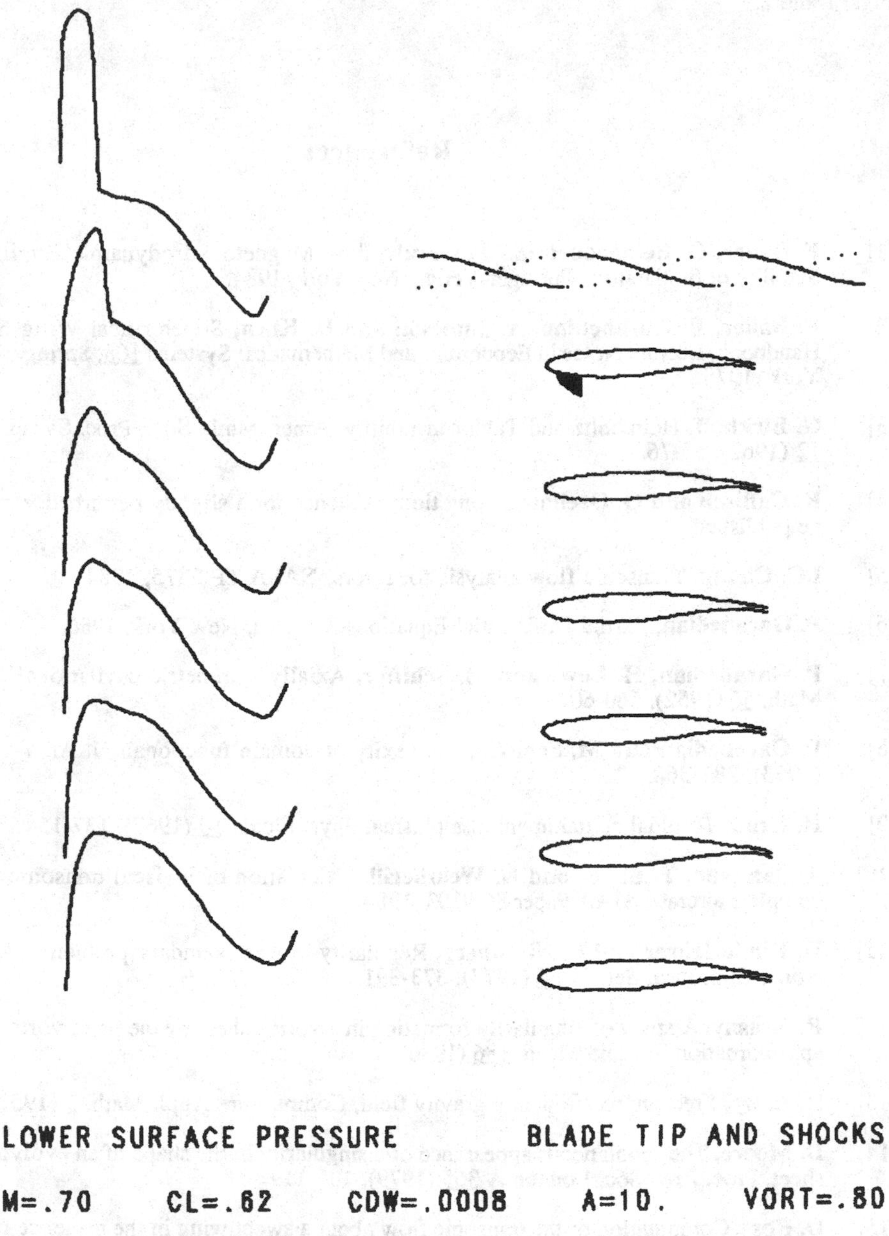

Fig. 1. Vortex sheet incident on a supercritical blade.

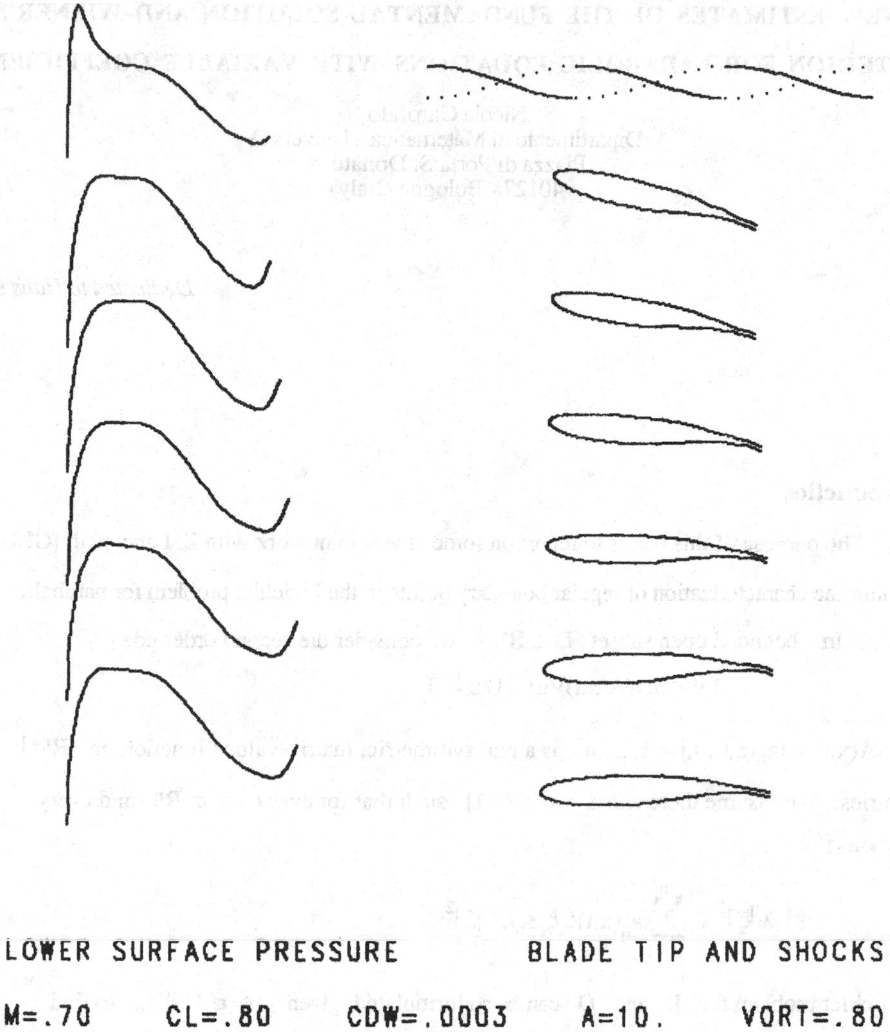

Fig. 2. Redesigned blade twisted to suppress shock.

NEW ESTIMATES OF THE FUNDAMENTAL SOLUTION AND WIENER'S CRITERION FOR PARABOLIC EQUATIONS WITH VARIABLE COEFFICIENTS

Nicola Garofalo
Dipartimento di Matematica - Università
Piazza di Porta S. Donato, 5
40127 - Bologna (Italy)

Dedicated to Hans Lewy

1. Introduction

The purpose of this talk is to report on some recent joint work with E. Lanconelli [GL] concerning the characterization of regular boundary points in the Dirichlet problem for parabolic equations. In a bounded open subset $D \subset \mathbf{R}^{n+1}$ we consider the second order pde

$$Lu = \mathrm{div}(A(x,t)\nabla u) - D_t u = 0 \quad , \tag{1.1}$$

where $A(x,t) = (a_{ij}(x,t), i,j = 1,\ldots\ldots,n,$ is a real symmetric, matrix-valued function on $\mathbf{R}^{n+1}$ with C^∞ entries. We assume there exists $\lambda \in (0,1]$ such that for every $\xi \in \mathbf{R}^n$ and every $(x,t) \in \mathbf{R}^{n+1}$

$$\lambda|\xi|^2 \le \sum_{i,j=1}^{n} a_{ij}(x,t)\xi_i\xi_j \le \lambda^{-1}|\xi|^2 \quad . \tag{1.2}$$

The Dirichlet problem for L and D can be so formulated: given $\phi \in C(\partial D)$, to find $u \in C^\infty(D) \cap C(\overline{D})$ such that

$$\begin{cases} Lu = 0 \quad \text{in } D \\ u|_{\partial D} = \phi \quad . \end{cases} \tag{1.3}$$

Such a problem always has a generalized solution provided by the Perron-Wiener-Brelot-Bauer method. An open set $U \subset \mathbf{R}^{n+1}$ is said L-*regular* if for any $\phi \in C(\partial U)$ the Dirichlet problem (1.3) admits a unique (classical) solution H_ϕ^U such that $H_\phi^U \ge 0$ if $\phi \ge 0$.

A function $w : D \to \mathbf{R}$ is said L-*superparabolic* in D if: (i) $-\infty < w \leq +\infty$ in a dense subset of D; (ii) w is lower semi-continuous; (iii) for every regular subset $U \subset \overline{U} \subset D$, and every $\phi \in (\partial U)$, if $w|_{\partial U} \geq \phi$, then $w \geq H_\phi^U$ in U. $w : D \to \overline{\mathbf{R}}$ is said L-*subparabolic* if $-w$ is L-superparabolic. Given $\phi \in C(\partial D)$ the Perron-Wiener-Brelot-Bauer solution to (1.3) is defined for $(x,t) \in D$ by

(1.4) $$u(x,t) = \inf \Big\{ w(x,t) \,|\, w \text{ is L-superparabolic in } D \text{ and } \lim_{\substack{(x,t)\to(x_0,t_0) \\ (x,t)\in D}} w(x,t) \geq \phi(x_0,t_0), \text{ for every } (x_0,t_0) \in \partial D \Big\} .$$

Classical potential theory assures that $u \in C^\infty(D)$ and $Lu = 0$ in D in the classical sense. However, it is not true in general that u attains continuously the boundary value ϕ. A point $(x_0,t_0) \in \partial D$ is said to be L-*regular* if

(1.5) $$\lim_{\substack{(x,t)\to(x_0,t_0) \\ (x,t)\in D}} u(x,t) = \phi(x_0,t_0) .$$

The primary motivation of [GL] has been to characterize those points $(x_0,t_0) \in \partial D$ which are L-regular for the Dirichlet problem. In order to state the results we need a few more definitions. For L as in (1.1) we let $\Gamma(x,t;y,s)$ be its fundamental solution. We recall that, under the assumptions made on the matrix $A(x,t)$, $\Gamma(x,t;y,s)$ is a C^∞ function for $(x,t) \neq (y,s)$. We indicate with $M^+(\mathbf{R}^{n+1})$ the cone of all nonnegative Radon measures on $\mathbf{R}^{n+1}$, and, for a closed set $F \subset \mathbf{R}^{n+1}$, we let $M^+(F) = \{ \mu \in M^+(\mathbf{R}^{n+1}) \mid \operatorname{supp} \mu \subset F \}$. If $\mu \in M^+(\mathbf{R}^{n+1})$ we set

(1.6) $$\Gamma_\mu(x,t) = \int_{\mathbf{R}^{n+1}} \Gamma(x,t;y,s)\, d\mu(y,s) .$$

Γ_μ is called the L-*potential* of μ. For a closed set $F \subset \mathbf{R}^{n+1}$ the L-*capacity* of F is defined as

(1.7) $$\operatorname{cap}_L(F) = \sup \left\{ \mu(\mathbf{R}^{n+1}) \,|\, \mu \in M^+(F), \Gamma_\mu \leq 1 \text{ on } \mathbf{R}^{n+1} \right\} .$$

Finally, given $(x_0,t_0) \in \mathbf{R}^{n+1}$ and a $k \in \mathbf{N}$ we define

(1.8) $$A(x_0,t_0;2^{-k}) = \left\{ (x,t) \in \mathbf{R}^{n+1} \,|\, \left(4\pi 2^{-(k+1)}\right)^{-n/2} \geq \Gamma(x_0,t_0;x,t) \geq \left(4\pi 2^{-k}\right)^{-n/2} \right\} .$$

$A(x_0,t_0;2^{-k})$ is the "annular" region included between the two level sets of

$$\Gamma: \left\{ \Gamma(x_0,t_0;x,t) = \left(4\pi 2^{-(k+1)}\right)^{-n/2} \right\} \text{ and } \left\{ \Gamma(x_0,t_0;x,t) = \left(4\pi 2^{-k}\right)^{-n/2} \right\} .$$

The main results in [GL] are

Theorem 1.1. *Given a bounded open subset* $D \subset \mathbf{R}^{n+1}$ *, a point* $(x_0,t_0) \in \partial D$ *is* L-*regular iff*

$$\sum_{k=1}^{n} 2^{kn/2} \mathrm{cap}_L\left(D^c \cap A\left(x_0,t_0;2^{-k}\right)\right) = +\infty , \tag{1.9}$$

where $D^c = \mathbf{R}^{n+1} \setminus D$.

Theorem 1.2. *Let* $A(x,t)$, $B(x,t)$ *be two real, symmetric, matrix-valued functions on* $\mathbf{R}^{n+1}$ *with* C^∞ *entries and for which there exist* $\lambda, \mu \in (0,1]$ *such that*

$$\lambda |\xi|^2 \le \sum_{i,j=1}^{n} a_{ij}(x,t)\xi_i\xi_j \le \lambda^{-1} |\xi|^2$$

$$\mu |\xi|^2 \le \sum_{i,j=1}^{n} b_{ij}(x,t)\xi_i\xi_j \le \mu^{-1} |\xi|^2 ,$$

for every $\xi \in \mathbf{R}^n$, $(x,t) \in \mathbf{R}^{n+1}$. *Consider the two parabolic operators*

$$L = \mathrm{div}(A(x,t)\,\nabla) - D_t, \quad M = \mathrm{div}(B(x,t)\,\nabla) - D_t . \tag{1.10}$$

Let D *be a bounded open subset of* $\mathbf{R}^{n+1}$, *and let* $(x_0,t_0) \in \partial D$ *be such that*

$$A(x_0,t_0) = B(x_0,t_0) . \tag{1.11}$$

Then (x_0,t_0) *is* L-regular *iff it is* M-regular.

For the proofs of Theorems 1.1 and 1.2 we refer to [GL]. It would be impossible to present them here. In the next section we discuss the main ideas behind the results and give a description of the main ingredients (or lemmas) which constitute the building bricks of the proofs of Theorems 1.1 and 1.2.

2. Discussion of the results

Theorem 1.1 above is modelled on Wiener's well known criterion for the regularity of a boundary point for Laplace's operator. In 1924 Wiener [W] proved that: *given a bounded domain* $\Omega \subset \mathbf{R}^n, n \ge 3$, *a point* $x_0 \in \partial\Omega$ *is* Δ-regular *for the Dirichlet problem iff*

$$\sum_{k=1}^{\infty} 2^{k(n-2)} \mathrm{cap}_\Delta\left(\Omega^c \cap A\left(x_0;2^{-k}\right)\right) = +\infty . \tag{2.1}$$

In (2.1) $\Omega^c = \mathbf{R}^n \setminus \Omega$, for $k \in \mathbf{N}$ $A(x_0;2^{-k}) = \{x \in \mathbf{R}^n \mid 2^{-(k+1)} \le |x - x_0| \le 2^{-k}\}$,

and $\mathrm{cap}_\Delta(\bullet)$ denotes the Newtonian capacity.

In 1982 Evans and Gariepy [EG] succeeded in extending Wiener's result to the heat operator $H = \Delta - D_t$ in $\mathbf{R}^{n+1}$. They proved

Theorem 2.1 (Evans and Gariepy). *Given a bounded domain* $D \subset \mathbf{R}^{n+1}$, *a point* $(x_0,t_0) \in \partial D$ *is* H-regular *iff*

$$\sum_{k=1}^{\infty} 2^{kn/2}\, \mathrm{cap}_H\left(D^c \cap A(x_0,t_0;2^{-k})\right) = +\infty . \tag{2.2}$$

Here, for a closed set $F \subset \mathbf{R}^{n+1}$ $\mathrm{cap}_H(F)$ is the H-capacity (see (1.7)) defined via the Gauss-Weierstrass kernel

$$K(x,t;y,s) = K(x-y;t-s) = \begin{cases} (4\pi(t-s))^{-n/2} \exp\left(- \dfrac{|x-y|^2}{4(t-s)}\right), & t > s , \\ 0, \quad t \le s , \end{cases} \tag{2.3}$$

whereas $A(x_0,t_0;2^{-k})$ is defined as in (1.8) by replacing Γ with K. Theorem 1.1 above extends Evans and Gariepy's result to parabolic operators with smooth variable coefficients.

The difficulties one encounters in passing from the elliptic to the parabolic case account for the time-lag between Wiener's and Evans and Gariepy's results. We mention that the "only if" part of Theorem 2.1 had already been proven, together with some sufficient conditions different from (2.2), by Lanconelli in [L1]. Such difficulties are mainly due to the evolutive nature of the operator H, a fact which is reflected in the form of the fundamental solution K in (2.3). In particular, the presence of the exponential factor in (2.3) makes it difficult on one hand to estimate the capacity of the "annular" regions $A(x_0,t_0;2^{-k})$ appearing in (2.2) from above, on the other hand to connect via such an estimate the divergence of the series in (2.2) to information effective in concluding the regularity of the boundary point under consideration. The ingenious strategy in [EG] is based on two main ingredients: 1) Representation formulas for solutions of $Hu = 0$ as averages on the level sets of the fundamental solution K, and their consequences; 2) A strong form of Harnack inequality allowing control of a positive solution of $Hu = 0$ in suitable regions exterior to the parabolic ones in which the usual Harnack inequality holds. A careful analysis of the proof of Evans and Gariepy's result reveals the new order of difficulty one has to face in passing from the Laplace to the heat operator. The situation becomes more complicated as one tries to extend Theorem 2.1 to variable coefficients operators, a fact mainly due to lack of an explicit knowledge of the fundamental solution. This point is crucial. To by-pass it we had to prove the following

Theorem 2.2. *Let* K *be a sufficiently small compact neighborhood of the origin in* $\mathbf{R}^n$, *and let* $\Gamma(x,y,t) = \Gamma(x,t;y,0)$ *be the fundamental solution of* L *in* (1.1) *with pole at* $(y,0)$. *Then as* $t \to 0^+$ *we have the asymptotic expansion*

$$\Gamma(x,y,t) \sim G(x,y,t) \sum_{j=0}^{\infty} t^j u_j(x,y,t) \ , \tag{2.4}$$

Here $G(x,y,t) = (4\pi t)^{-n/2} \exp\left(-\frac{d^2(x,y,t)}{4t} \right)$, $d(x,y,t)$ *being the distance from* x *to* y *in the Riemannian metric* $ds_t^2 = g_{ij}(t)dx_i \otimes dx_j$, *where*

$$(g_{ij}(t)(x)) = \left(a^{ij}(x,t)\right) = A(x,t)^{-1} .$$

In (2.4) $u_j \in C^\infty(K \times K \times [0,T])$, $T > 0$ *and sufficiently small. An expansion similar to* (2.4) *holds for the derivatives of* Γ.

What we mean by (2.4) is that for every $k \in N \cup \{0\}$ there exists $\omega_k : K \times K \times (0,T) \to \mathbf{R}$ such that

$$\Gamma(x,y,t) - G(x,y,t) \sum_{j=0}^{k} t^j u_j(x,y,t) = \omega_k(x,y,t) \ , \tag{2.5}$$

with

$$\omega_k(x,y,t) = o\left(t^{k-n/2} \exp\left(-\frac{\delta |x-y|^2}{4t} \right) \right) \quad \text{as } t \to 0^+ \tag{2.6}$$

uniformly for $x,y \in K$. $\delta > 0$ is a suitable number which depends on λ in (1.2) and n.

Remark. Kannai proved in [K] an asymptotic estimate for operators with time-independent coefficients in which the error is estimated as follows:

$$\omega_k(x,y,t) = o\left(t^{k-n/2} \exp\left(-\frac{d^2(x,y)}{4t} \right) \right), \quad \text{as } t \to 0^+ \ . \tag{2.7}$$

In (2.7) $d(x,y)$ is the distance in the metric $ds^2 = g_{ij}dx_i \otimes dx_j$, with

$g_{ij}(x) = a^{ij}(x) = A(x)^{-1}$. (2.7) gives of course a better evaluation than (2.6), which is, at any rate, sufficient for our purposes. We conjecture that a more careful analysis should lead to an estimate of the error in (2.5) like the one in (2.7) where $d^2(x,y)$ is replaced by $d^2(x,y,t)$.

The proof of Theorem 2.2 uses some facts from geometry. It is modelled on Minakshisundaram and Pleijel's asymptotic evaluation of the fundamental solution of the heat operator on a compact manifold, see, e.g., [BGM]. Some complications arise, because of the time dependence of the coefficients of L in (1.1), in the solution of the "transport" equations which define the u_k's in (2.4) inductively, but the situation turns out to be favorable in the end. In the applications that concern us the value of the first function, u_0, at $x = y$ is of great importance. In the proof of Theorem 2.2 u_0 can be chosen such that

(2.8) $$u_0(x,x,t) \equiv 1, \quad \text{for } t \text{ small}.$$

With Theorem 2.2 in hand we could prove a result which plays a key role in the proof of the sufficiency of (1.9). This is a strong form of Harnack inequality which gives quantitative information on a positive solution of (1.1) in regions where the usual Harnack inequality fails to hold. Its motivation lies in the fact that one would want to infer from the divergence of the series in (1.9) convergence to one of certain n-dimensional averages on sets which are clustering to the boundary point in consideration. Once one knows this the regularity of the point is almost proven. The proof of the strong Harnack inequality, Theorem 2.3 below, relies on a Lemma which is rather interesting in itself. This lemma is more easily understood by looking at the constant coefficients case. Let K be the fundamental solution of $H = \Delta - D_t$ and set for $t > 0$

(2.9) $$E(x,t) = \ln K(x-0, t-0).$$

We have

(2.10) $$\nabla_x E(x,t) = -\frac{x}{2t}, \quad D_t E(x,t) = \frac{|x|^2}{4t^2} - \frac{n}{2t}.$$

It is immediate to recognize from (2.10) that for any $\theta > 1$ there exists a

$\delta = \delta(\theta) = \dfrac{\theta - 1}{2n\theta}$, such that if $t \le \delta |x|^2$ we have

(2.11) $$|\nabla_x E(x,t)|^2 \le \theta D_t E(x,t).$$

In other words (2.11) holds outside a paraboloid with vertex at $(0,0)$ and aperture depending on θ. Now let us denote by Ω_r^* the set

$$\Omega_r^* = \left\{ z \in \mathbf{R}^{n+1} \,|\, \Gamma(z,0) > (4\pi r)^{-n/2} \right\} . \tag{2.12}$$

For $\delta > 0$ we define

$$W_{\delta,r}^* = \left\{ z \in \mathbf{R}^{n+1} \,|\, z \in \Omega_r^*,\, t < \delta |x|^2 \right\} . \tag{2.13}$$

Then we have

Lemma 2.1 (see [GL]). *There exist* $r_0 > 0$ *and for every* $\theta > 1$ *a* $\delta = \delta(\theta) > 0$ *such that for every* $z \in W_{\delta,r}^*$

$$\frac{A(z)\big(\nabla_x \Gamma(z;0)\big) \bullet \nabla_x \Gamma(z;0)}{\Gamma^2(z;0)} \leq \theta \, \frac{D_t \Gamma(z;0)}{\Gamma(z;0)} , \tag{2.14}$$

for all $r \leq r_0$.

In the intrinsic notation of Theorem 2.2, if M_t denotes the manifold $(K, g_{ij}(t))$, K being the compact neighborhood of the origin in $\mathbf{R}^n$ in Theorem 2.2, we let ∇_{M_t} denote the intrinsic gradient in M_t and $|\bullet|_t$ the Riemannian length. Setting $E(z) = \ln \Gamma(z;0)$ we may rewrite (2.14) as

$$|\nabla_{M_t} E(z)|_t^2 \leq \theta D_t E(z) \quad \text{in } W_{\delta,r}^* , \tag{2.15}$$

a formulation that recalls (2.11). The proof of Lemma 2.1 uses the full strength of Theorem 2.2. Lemma 2.1 contains the key information needed to prove the next result. Before stating it we recall some notation from [GL]: in the sequel $\alpha_1, \eta, \sigma, \gamma_1$ will be numbers fixed throughout the discussion which solely depend on the eigenvalue λ in (1.2). For $t < 0$ and $r < 0$ we set

$$R_r^1(t) = 2n\,\alpha_1(-t)\ln\left(-\frac{r\gamma_1}{t}\right) \tag{2.16}$$

$$\Omega_r = \left\{ z \in \mathbf{R}^{n+1} \,|\, \Gamma(0,z) > (4\pi r)^{-n/2} \right\}$$

$$Q_r = \left\{ z \in \Omega_r \,|\, t > -\eta \frac{r}{2} \right\} . \tag{2.17}$$

In the sequel one should keep in mind that σ and η have been choosen in such a way that $\Omega_{\sigma r} \subset Q_{2r}$. Moreover, we have

$$-\eta r < \inf\left\{ t \,|\, (x,t) \in \Omega_{\sigma r} \right\} . \tag{2.18}$$

The picture below illustrates the geometry.

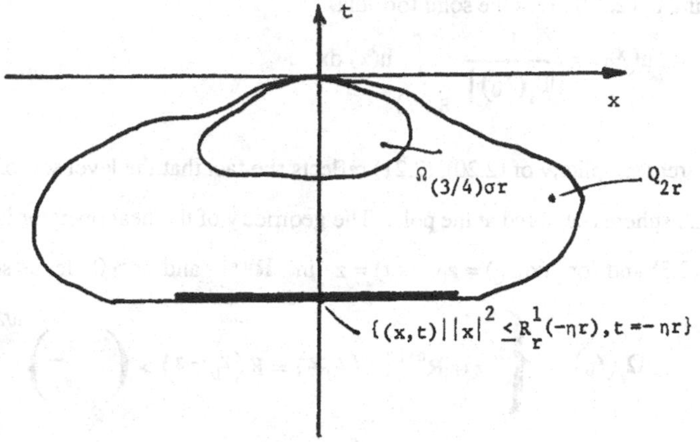

We can now state the following

Theorem 2.3 (Strong Harnack inequality, see [GL]). *There exist* $r_0 > 0$ *and a* $\Lambda > 0$ *such that if* $u > 0$ *is a solution of* $Lu = 0$ *in* Q_{2r}, *with* $u \in C(\partial Q_{2r} \setminus \{(0,0)\})$, *then*

$$(2.19) \qquad \fint_{|x|^2 \leq R_r^1(-\eta r)} u(x,-\eta r)dx \leq \Lambda \inf_{\Omega_{(3/4)\sigma r}} u$$

for every $r \leq r_0$. Λ *does not depend on* r.

In (2.19) $\fint$ denotes the n-dimensional average of u on the n-ball $\{|x|^2 \leq R_r^1(-\eta r)\}$ placed on the hyperplane $\{t = -\eta r\}$. We remark that in virtue of (2.18) the sets $\{z \in \mathbf{R}^{n+1} \mid |x|^2 \leq R_r^1(-\eta r), t = -\eta r\}$ and $\Omega_{(3/4)\sigma r}$ are detached and there is a time-lag between them, see the picture above. Moreover, (2.19) cannot be obtained just from the usual Harnack inequality since the set $\Omega_{(3/4)\sigma r}$ includes a region which is exterior to any paraboloid $\{t = -\delta r|x|^2\}$ with vertex at $(0,0)$ and opening in the negative time-direction.

We now pass to discuss another important ingredient in the proof of the sufficiency of (1.9); the mean-value formulas for solutions of $Lu = f$, alluded to before. If u is a solution of $\Delta u = 0$ in $\mathbf{R}^n$ then for every $x_0 \in \mathbf{R}^n$ and every $r > 0$ we have

(2.20) $$u(x_0) = \frac{1}{|\partial B_r(x_0)|} \int_{\partial B_r(x_0)} u(x)\, d\sigma(x) .$$

Integrating in (2.20) gives the solid formula

(2.21) $$u(x_0) = \frac{1}{|B_r(x_0)|} \int_{B_r(x_0)} u(x)\, dx .$$

The apparent simplicity of (2.20), (2.21) reflects the fact that the level sets of the fundamental solution of Δ are spheres centered at the pole. The geometry of the heat operator is not as simple. Let K be as in (2.3) and for $(x_0,t_0) = z_0$, $(x,t) = z$ in $\mathbf{R}^{n+1}$ and $r > 0$ let us set

(2.22) $$\Omega_r(z_0) = \left\{ z \in \mathbf{R}^{n+1} \mid K(z_0;z) = K(z_0 - z) > \left(\frac{1}{4\pi r}\right)^{n/2} \right\} .$$

By analogy with the Euclidean case we call the set $\Omega_r(z_0)$ the *heat-ball* with "center" at z_0 and radius r. The corresponding *heat-sphere* is the set

(2.23) $$\psi_r(z_0) = \left\{ z \in \mathbf{R}^{n+1} \mid K(z_0;z) = \left(\frac{1}{4\pi r}\right)^{n/2} \right\} .$$

From (2.3) it is clear that an alternative description of (2.22) is

(2.24) $$\Omega_r(z_0) = \left\{ z \in \mathbf{R}^{n+1} \mid |x_0 - x|^2 < 2n(t_0 - t)\ln\left(\frac{r}{t_0 - t}\right), t_0 - r < t < t_0 \right\} .$$

The heat-sphere $\psi_r(z_0)$ is a football shaped surface in $\mathbf{R}^{n+1}$ with the "center" z_0 laying on the surface itself. *For a solution* u *of* $Hu = 0$ *in* $\mathbf{R}^{n+1}$ *the following mean-value formula holds for every* $r > 0$

(2.25) $$u(z_0) = (4\pi r)^{-n/2} \int_{\Omega_r(z_0)} u(z) \frac{|x_0 - x|^2}{4(t_0 - t)^2}\, dx\, dt .$$

(2.25) is obtained by first establishing a surface mean-value formula with respect to a probability measure supported on the heat-sphere, and then integrating it with respect to the "radial" variable. Pini was the first one to discover and use (2.25) for the heat equation in $\mathbf{R}^2$, see [P1], [P2], [P3]. Subsequently, Fulks [F] obtained the $(n + 1)$-dimensional surface formula, mentioned above, and Watson [Wa] the $(n+1)$-dimensional solid one, i.e., (2.25). Evans and Gariepy made the key observation that (2.25) could be used to estimate from above the capacity of the annular regions $A(x_0,t_0;2^{-k})$ in (2.2). This is done through the following Lemma 2.1 and its corollary.

Lemma 2.1. (see [EG]). *For* $u \in C^\infty(\mathbf{R}^{n+1})$ *let*

$$\phi(r) = r^{-n/2} \int_{\Omega_r} u(z) \frac{|x|^2}{t^2}\, dx\, dt \tag{2.26}$$

where $\Omega_r = \Omega_r(0)$. *Then*

$$\phi'(r) = \frac{n}{2} r^{-(n+2)/2} \int_{\Omega_r} Hu(z) \left(\frac{R_r(t) - |x|^2}{t} \right) dx\, dt \ , \tag{2.27}$$

where $R_r(t) = 2n(-t) \ln\left(-\frac{r}{t}\right)$.

Corollary 2.1 (see [EG]). *Suppose that* $Hu \geq 0$ *in* Ω_{2r}. *Then there exists* $c_n > 0$ *such that*

$$\phi(2r) - \phi(r) \geq c_n r^{-n/2} \int_{\Omega_{r/2}} Hu(z)\, dx\, dt \ . \tag{2.28}$$

The direct computations in [EG] are made possible by the explicit knowledge of the fundamental solution K in (2.3). In the variable coefficients case this approach does not work, but a formula like (2.25) still holds. The result is due to E.B. Fabes and the author, see [FG]. In order to state it, let Γ denote the fundamental solution of L in (1.1). For z_0, $z \in \mathbf{R}^{n+1}$, and $r > 0$ we set

$$\Omega_r(z_0) = \left\{ z \in \mathbf{R}^{n+1} \ | \Gamma(z_0;z) > (4\pi r)^{-n/2} \right\} \ , \tag{2.29}$$

$$\psi_r(z_0) = \left\{ z \in \mathbf{R}^{n+1} \ | \Gamma(z_0;z) = (4\pi r)^{-n/2} \right\} \ . \tag{2.30}$$

Theorem 2.4 (see [FG]). *Let* u *be a solution to* $Lu = 0$ *in* $\mathbf{R}^{n+1}$, *and let* $z_0 = (x_0,t_0) \in \mathbf{R}^{n+1}$ *be fixed. Then for* a.e. $r > 0$

$$u(z_0) = \int_{\psi_r(z_0)} u(z) \frac{A(z)\left(\nabla_x \Gamma(z_0;z)\right) \cdot \nabla_x \Gamma(z_0;z)}{|\left(\nabla_x \Gamma(z_0;z), D_t \Gamma(z_0;z)\right)|}\, dH_n(z) \ . \tag{2.31}$$

For every $r > 0$

$$(2.32)\qquad u(z_0)=(4\pi r)^{-n/2}\int_{\Omega_r(z_0)} u(z)\frac{A(z)\left(\nabla_x\Gamma(z_0;z)\right)\cdot\nabla_x\Gamma(z_0;z)}{\Gamma^2(z_0;z)}\,dz\ .$$

We remark that if $A(z)=$ Identity, and hence $L=H$, and $\Gamma=K$, the kernel

$$\frac{A\nabla_x\Gamma\cdot\nabla_x\Gamma}{\Gamma^2}\ \text{in (2.32) becomes}\ \frac{|\nabla_x K|^2}{K^2}=\frac{|x_0-x|^2}{4(t_0-t)^2}\ ,$$

which is the one appearing in (2.25). The main ingredient in obtaining (2.32) from (2.31) is Federer's *co-area formula* of geometric measure theory, see [Fe]: *given* $f\in L^1(\mathbf{R}^{n+1})$ *and* $g\in \mathrm{Lip}(\mathbf{R}^{n+1})$

$$(2.33)\qquad \int_{\mathbf{R}^{n+1}} f(z)\,|Dg(z)|\,dz=\int_{-\infty}^{\infty} d\alpha\left(\int_{g=\alpha} f(z)\,dH_n(z)\right)\ .$$

(2.33) allows to by-pass Watson's and Evans and Gariepy's approach to (2.25) and (2.26) based on the explicit knowledge of K. In [GL] we extend Theorem 2.4 to solutions of the inhomogeneous equation $Lu=f$. We also prove

Theorem 2.5 (see [GL]). *Let* $u\in C^\infty(\mathbf{R}^{n+1})$. *For* $z_0\in\mathbf{R}^{n+1}$ *and* $r>0$ *we set*

$$(2.34)\qquad \phi(r)=(4\pi r)^{-n/2}\int_{\Omega_r(z_0)} u(z)\frac{A(z)\left(\nabla_x\Gamma(z_0;z)\right)\cdot\nabla_x\Gamma(z_0;z)}{\Gamma^2(z_0;z)}\,dz\ .$$

Then

$$(2.35)\qquad \phi'(r)=\frac{n}{2r}\left(\frac{1}{4\pi r}\right)^{n/2}\int_{\Omega_r(z_0)} Lu(z)\ln\left[(4\pi r)^{n/2}\,\Gamma(z_0;z)\right]dz\ .$$

Corollary 2.2. *For* $z_0\in\mathbf{R}^{n+1}$ *and* $\alpha>1$ *let* $u\in C^\infty\left(\Omega_{\alpha^2 r}(z_0)\right)$, *with* $Lu\geq 0$ *in* $\Omega_{\alpha^2 r}(z_0)$. *Then there exists a positive constant* $C=C(n,\alpha)$ *such that*

$$(2.36)\qquad \phi(\alpha^2 r)-\phi(\alpha r)\geq C\,r^{-n/2}\int_{\Omega_r(z_0)} Lu(z)\,dz\ .$$

Remark. If in Theorem 2.5 we take $L=H$, and hence $\Gamma=K$, we obtain

$$(2.37)\qquad \ln\left[(4\pi r)^{n/2}\,K(z_0-z)\right]=\frac{R_r(t_0-t)^2-|x_0-x|^2}{4(t_0-t)}\ .$$

Therefore, Lemma 2.1 is a special case of Theorem 2.5.

Theorem 2.3 and Corollary 2.2 are the primary ingredients in the proof of Theorem 1.1. However, there is still a piece of information missing. In fact, in the course of the proof of the sufficiency of (1.9) it is crucial at one point to be able to deal with regularizations of equilibrium potentials, rather than with the potentials themselves. Since usual mollification does not work in that context our idea has been to regularize using the averaging operator $u \to u_r$, where

$$u_r(z) = \int_{\Omega_r(z)} u(\zeta)\, K_r(z,\zeta)d\zeta \ . \tag{2.38}$$

In (2.38) we have set

$$K_r(z;\zeta) = \frac{1}{(4\pi r)^{n/2}} \frac{A(\zeta)(\nabla_x\Gamma(z;\zeta)) \cdot \nabla_x\Gamma(z;\zeta)}{\Gamma^2(z;\zeta)} , \tag{2.39}$$

see (2.32). With the aid of Theorem 2.2 we have been able to prove the following

Theorem 2.6 (see [GL]). *Let* $u \in L^\infty_{loc}(\mathbf{R}^{n+1})$, *then* $u_r \in C(\mathbf{R}^{n+1})$. *Moreover, if* n *is large enough and* $u \in C(\mathbf{R}^{n+1})$, *then* $u_r \in C^1(\mathbf{R}^{n+1})$ *and we have*

$$D_{x_j} u_r(z) = \int_{\psi_r(z)} u(\zeta) \frac{D_{x_j}\Gamma(z;\zeta)K_r(z;\zeta)}{|(D_x\Gamma(z;\zeta), D_t\Gamma(z;\zeta))|} dH_n(\zeta) + \int_{\Omega_r(z)} u(\zeta) D_{x_j} K_r(z;\zeta)d\zeta \ . \tag{2.40}$$

A similar formula holds for $D_t u_r$. *In general, if* $u \in C^k(\mathbf{R}^{n+1})$, *and* $n \geq n(k)$, *with* $n(k)$ *suitable, then* $u_r \in C^{k+1}(\mathbf{R}^{n+1})$.

Using Theorem 2.6 and a device based on climbing up in the dimension we can regularize solutions of the equation $Lu = -\mu$, μ a positive measure, as much as we wish. Moreover, the regularized u_r has the property that outside the support of μ, where $Lu = 0$, we also have $Lu_r = 0$ for small enough r. This allows to complete the proof of Theorem 1.1. We refer to section 6 in [GL] for complete details.

We now wish to discuss Theorem 1.2. We begin by recalling the following beautiful result of Littman, Stampacchia and Weinberger

Theorem 2.7 (see [LSW]). *Let* $L_0 = \mathrm{div}(A(x)\nabla)$ *be a uniformly elliptic operator with bounded measurable coefficients in a bounded open subset* $\Omega \subset \mathbf{R}^n$. *Then a point* $x_0 \in \partial\Omega$ *is* L_0-regular *iff* x_0 *is* Δ-regular.

Let $\Gamma_0(x,y)$ be the fundamental solution of L_0. The proof of Theorem 2.7 is based on the discovery that there exists a costant $C > 0$ such that for every $x \neq y$

(2.41) $$\frac{C^{-1}}{|x-y|^{n-2}} \le \Gamma_0(x,y) \le \frac{C}{|x-y|^{n-2}} \quad .$$

An estimate like (2.41) is impossible for a parabolic fundamental solution. In fact, if we let K_{α_i} be the fundamental solution of the operator $\alpha_i\Delta - D_t$, $\alpha_i > 0$, $i = 1,2$, then

(2.42) $$\frac{K_{\alpha_1}(x,t)}{K_{\alpha_2}(x,t)} = \left(\frac{\alpha_2}{\alpha_1}\right)^{n/2} \exp\left(-\left(\frac{1}{\alpha_1} - \frac{1}{\alpha_2}\right)\frac{|x|^2}{4t}\right), \quad x \in \mathbf{R}^n,\ t > 0 \ .$$

This fact has a geometrical counterpart, which is enlightened by the following example due to Petrovskii, see [Pe]. Let us consider the function $p(t) = 4|t|\,|\ln|t||$, $t < 0$, and let Ω be the region in $\mathbf{R}^2$ which lies below the curve $x^2 = 4p(t)$ and above the line $\{t = -1\}$. Then the point $(0,0) \in \partial\Omega$ is regular for $H_2 = 2D_{xx} - D_t$, but it is not regular for $H = D_{xx} - D_t$. More in general we can prove, see [GL]

Theorem 2.8. *Let* L *be the operator in* (1.1) *and* $M = \mathrm{div}(B(x,t)\nabla) - D_t$ *be another uniformly parabolic operator with smooth coefficients. If for a given* $(x_0,t_0) \in \mathbf{R}^{n+1}$

$$A(x_0,t_0) \neq B(x_0,t_0) \quad ,$$

then there exists a bounded domain $D \subset \mathbf{R}^{n+1}$ *for which* (x_0,t_0) *is a point of* ∂D L-regular *but not* M-regular (or vice-versa).

In light of the above discussion we might say that Theorem 1.2 constitutes the appropriate substitute for parabolic equations of Theorem 2.7. The proof of Theorem 1.2 relies on Theorem 1.1 and Theorem 2.2. By the former the question of the regularity of $z_0 \in \partial D$ is reduced to study the behavior of the series in (1.9). The latter implies that under the assumption (1.11) there exists a constant $C > 0$ (independent of $k \in N$) such that:

(2.43) $$z \in D^c \cap A_L(z_0;2^{-k}) \text{ implies } C^{-1}\left(4\pi 2^{-(k+1)}\right)^{-n/2} \ge \Gamma_M(z_0;z) \ge C(4\pi 2^{-k})^{-n/2} \quad .$$

In (2.43) we have denoted by $A_L(z_0;2^{-k})$ the set (1.8) relative to the fundamental solution Γ_L of the operator L, whereas Γ_M = fundamental solution of the operator M. Now in spite of the fact that an estimate like (2.41) cannot hold for Γ_L and Γ_M, the following holds: *there is a constant* $B > 0$ *such that*

(2.44) $$B^{-1}\,\mathrm{cap}_L(\bullet) \le \mathrm{cap}_M(\bullet) \le B\,\mathrm{cap}_L(\bullet), \quad \text{(see [L2])}$$

cap_L, cap_M being the parabolic capacities defined as in (1.7). Using (2.43) and (2.44) we conclude that if z_0 is L-regular, then it is also M-regular. A similar argument shows that M-regularity implies L-regularity.

References

[BGM] **M. Berger, P. Gauduchon et E. Mazet,** Le Spectre d'une Variété Riemanniene, Lecture Notes in Math., Springer-Verlag 194, 1971.

[EG] **L.C. Evans and R.F. Gariepy,** Wiener's Criterion for the Heat Equation, Arch. Rat. Mech. Anal. 78 (1982), 293-314.

[FG] **E.B. Fabes and N. Garofalo,** Mean Value Properties of Solutions to Parabolic Equations with Variable Coefficients, Journal of Math. Analysis and Applications, J. of Math. Analysis and Appl. 121 (1987), 305-316.

[Fe] **H. Federer,** Geometric Measure Theory, Die Grundlehren der mathematischen Wissenschaften, Springer-Verlag 1969.

[F] **W. Fulks,** A Mean Value Theorem for the Heat Equation, Proc. Amer. Math. Soc. 17 (1966), 6-11.

[GL] **N. Garofalo and E. Lanconelli,** New Estimates of the Fundamental Solution and Wiener's Criterion for Parabolic Equations with Variable Coefficients, preprint.

[K] **Y. Kannai,** Off Diagonal Short Time Asymptotics for Fundamental Solutions of Diffusion Equations, Comm. in P.D.E. 2 (8) (1977), 781-830.

[L1] **E. Lanconelli,** Sul problema di Dirichlet per l'equazione del calore, Ann. di Mat.Pura e Appl. 97 (1973), 83-114.

[L2] **E. Lanconelli,** Sul problema di Dirichlet per equazioni paraboliche del secondo ordine a coefficienti discontinui, Ann. di Mat. Pura e Appl. 106 (1975), 11-38.

[LSW] **W. Littman, G. Stampacchia and H.F. Weinberger,** Regular Points for Elliptic Equations with Discontinuous Coefficients, Ann. Scuola Norm. Sup. Pisa 17 (1963).

[Pe] **I. Petrovskii,** Zur ersten Randwertaufgabe der Wärmeleitungsgleichung, Compositio Mathematica 1 (1935), 383-419.

[P1] **B. Pini,** Sulle equazioni a derivate parziali, lineari del secondo ordine in due variabili, di tipo parabolico, Ann. di Mat. Pura e Appl. 32 (1951), 179-204.

[P2] **B. Pini,** Maggioranti e minoranti delle soluzioni delle equazioni paraboliche, Ann. di Mat. Pura e Appl. 37 (1954), 249-264.

[P3] **B. Pini,** Sulla soluzione generalizzata di Wiener per il primo problema di valori al contorno nel caso parabolico, Rend. Sem. Mat. Univ. Padova 23 (1954), 422-434.

[Wa] **N.A. Watson,** A Theory of Temperatures in Several Variables, Proc. London Math. Soc. 26 (3) (1973), 385-417.

[W] **N. Wiener,** The Dirichlet Problem, J. Math. and Phys. 3 (1924), 127-146.

GREEN FUNCTION AND INVARIANT DENSITY FOR AN INTEGRO-DIFFERENTIAL OPERATOR

Maria Giovanna Garroni
Dipartimento di Matematica - Università "La Sapienza"
P.zzale A. Moro, 2
00185 Roma (Italy)

Dedicated to Hans Lewy

The purpose of this lecture is to present some new results for the oblique derivative elliptic or parabolic problems with an integro-differential operator. More precisely these results concern the Green function, the invariant probability density and its connection with the asymptotic behavior of the solutions of some equations and inequations. These results were obtained in collaboration with J.L. Menaldi [7,8,9].

1. Hypotheses and main properties

Let Ω be a bounded open subset of $\mathbf{R}^d$, $d \geq 2$, with the boundary $\partial\Omega$ of class C^2. Set $Q_T = \Omega \times]0, T[, 0 < T < +\infty$ and $\Sigma_T = \partial\Omega \times [0,T]$. Consider an integro-differential operator of this kind:

(1) $A = A_0 + D - I$

where

(2) $A_0\phi = -\sum_{ij=1}^{d} a_{ij}(x,t) \frac{\partial^2\phi}{\partial x_i \partial x_j}$; $D\phi = \sum_{i=1}^{d} a_i(x,t) \frac{\partial\phi}{\partial x_i} + a_0(x,t)\phi$;

(3) $I\phi = \int_E \Big[\phi\big(x + \gamma(x,t,\xi),t\big) - \phi(x,t)\Big]\beta(x,t,\xi)\pi(d\xi)$

and the boundary operator

(4) $By = \sum_{i=1}^{d} b_i(x,t) \frac{\partial\phi}{\partial x_i}$.

On (3) $\pi(\bullet)$ is a σ-finite measure on the measurable space (E,F). The assumption on the coefficients are the following:

$$(5)\quad \begin{cases} \text{(i)} & a_{ij}, a_i, b_i \in C^{\alpha,\frac{\alpha}{2}}(\overline{Q}_T), \text{ for some } 0<\alpha<1 , \\ \text{(ii)} & \sum_{i,j=1}^{d} a_{ij}\, \xi_i\, \zeta_j \geq \mu\, |\xi|^2,\ \mu > \infty,\ \forall x,t \in Q_T \\ \text{(iii)} & |\sum_{i=1}^{d} b_i(x,t) n_i(x)\,| \geq \nu > 0,\ \forall x, t \in \Sigma_T \end{cases}$$

where $n = (n_1,\ldots,n_d)$ is the outward normal vector to $\partial\Omega$. The functions $\gamma(x,t,\zeta)$, $\beta(x,t,\xi)$ are continuous for (x,t) in Q_T and F-measurable in ξ, and there exist a F-measurable function $\gamma_0(\xi)$ and a constant $c_0 > 0$ satisfying for every x,x' in Ω, t, t' in $[0,T]$

(i) $0 < |\gamma(x,t,\xi)| \leq \gamma_0(\xi),\ 0 \leq \beta(x,t,\xi) \leq 1,\ \int_E \gamma_0(\xi)\pi(d\xi) = c_0$

(ii) $|\gamma(x,t,\xi) - \gamma(x',t';\xi)| \leq \gamma_0(\xi)\left(|x-x'|^{\alpha} + |t-t'|^{\alpha/2}\right)$

(iii) $|\beta(x,t,\xi) - \beta(x',t',\xi)| \leq |x-x'|^{\alpha} + |t-t'|^{\alpha/2}$.

(iv) For any (x,t,ξ) in $\overline{Q}_T \times E$, s.t. $\beta(x,t,\xi) \neq 0$, the segment $[x, x+\gamma(x,t,\xi)]$ is included in $\overline{\Omega}$.

(6) (v) The function $\gamma(x,t,\xi)$ is continuously differentiable in x and there is a constant $c_0 > 0$ s.t.

$$|x-x'| \leq c_0\, |(x-x') + t'[\gamma(x,t,\xi) - \gamma(x',t,\xi)]\,| ,$$

$$|\gamma(x,t,\xi) - \gamma(x',t',\xi)| \leq c_0\left(|x-x'| + |t-t'|^{1/2}\right) ,$$

$\forall\, x,x' \in \mathbf{R}^d,\ t,t' \in [0,1],\ \xi$ in E .

Integral operators of this kind arise for instance in the study of a diffusion process with interior jumps reflected at the boundary of a bounded domain. The fundamental studies in this subject are those by Bensoussan and Lions [1], Gikhiman and Skorokhod [11], Komatsu [13], Lepeltier and Marchal [14], Stroock [17]. In Bensoussan and Lions [1] the integral operator (3) is given through a Radon measure on $E = \mathbf{R}^d - \{0\} \equiv \mathbf{R}^d_*$, with $\gamma(x,t,\xi) = \xi$, and in Gikhiman and Skorokhod [11] $E = \mathbf{R}^d_*$ with $\beta(x,t,\xi) = 1$ is used, both situations are covered here.

2. Green function and parabolic problems

We are interested in constructing the Green function associated with the integro-differential operator A and the differential boundary operator B, that is a function $G(x,y,t,\tau)$ s.t. for every τ and y fixed, verifies:

$$\text{(7)} \qquad \begin{cases} \dfrac{\partial G}{\partial t} + A_x G = \delta(x-y)\delta(t-\tau) \\ B_x G = 0 \quad \text{on } \Sigma_T \\ G(x,y,t,\tau) = 0, \; t \le \tau \; . \end{cases}$$

(δ is the Dirac function).

G is constructed as a solution of an integral equation

$$\text{(8)} \qquad G(x,y,t,\tau) = G_0(x,y,t,\tau) + \int_\tau^t ds \int_\Omega G_0(x,z,t,s)(I - D)G(z,y,s,\tau)dz$$

$\forall\, x,y \in \overline{\Omega}, \; 0 \le \tau < t \le T$,

by the method of successive approximations and is expressed in the form of a series

$$\text{(9)} \qquad G(x,y,t,\tau) = G_0(x,y,t,\tau) + \sum_{K=1}^{\infty} G^{(K)}(x,y,t,\tau) \quad ,$$

where G_0 is the Green function for the initial-boundary value problem associated with A_0 and B as constructed in Garroni and Solonnikov [10] and $G^{(K+1)}$ is an iterate kernel

$$\text{(10)} \qquad G^{(K+1)}(x,y,t,\tau) = \int_t^\tau ds \int_\Omega G_0(x,z,t,s)(I - D)G^{(K)}(z,y,s,\tau)dz \; .$$

The most delicate point of this construction is, as always, the exact evaluation of punctual or integral estimates for the iterates $G^{(K)}$. Usually in these estimates (cfr. [12] and [10]) the exponential of the heat-kernel type $e^{-\frac{|x-y|^2}{t-\tau}}$, or an adequate transformation of it, play an essential role.

Let us point out in this respect that the heat-kernel $e^{-\frac{|x-y|^2}{t-\tau}}$ in G_0 disappears during the iterations. This imposes the use of convenient functions spaces in which the integral operator works and to which the iterates $G^{(K)}$ belong. Let us denote by $C^K_\alpha(R^d)$, $0<\alpha<1$, $K \geq -1$, the space of all continuous functions $\phi(x,y,t,\tau)$ defined for x,y in R^d and $0<\tau<t\leq T$, with values in R^d and s.t. the following infima (11) are finite:

$$C(\phi,K) = \inf\left\{ C \geq 0 : |\phi(x,y,t,\tau)| \leq C(t-\tau)^{\left(\frac{k-d}{2}\right)}, \forall x,y,t,\tau \right\}$$

$$K(\phi,k) = \inf\left\{ K \geq 0 : \left[\int_\Omega |\phi(x,z,t,\tau)|dz\right] \vee \left[\int_\Omega |\phi(z,y,t,\tau)|dz\right] \leq K(t-\tau)^{\frac{k}{2}}, \forall x,y,t,\tau \right\}.$$

$$M(\phi,k) = \inf\{M \geq 0 : |\phi(x,y,t,\tau) - \phi(x',y',t',\tau')| \leq M(|x-x'|^\alpha + |t-t'|^{\frac{\alpha}{2}} + |y-y'|^\alpha +$$

$$+ |\tau-\tau'|^{\frac{\alpha}{2}}) \times \left[(t-\tau)^{\frac{k-d-\alpha}{2}} \vee (t'-\tau')^{\frac{k-d-\alpha}{2}}\right], \forall x,y,t,\tau,x',y',t',\tau', \text{ with } \tau \vee \tau' < t \wedge t' \Big\}.$$

$$(11) \quad N(\phi,k) = \inf\{N \geq 0 : [\int_\Omega |\phi(x,z,t,\tau) - \phi(x',z,t',\tau')|dz \leq N(|x-x'|^\alpha + (t-t')^{\alpha/2} +$$

$$+ |\tau-\tau'|^{\alpha/2}) \times \left[(t-\tau)^{(k-\alpha)/2} \vee (t'-\tau')^{(k-\alpha)/2}\right], \forall x,y,t,\tau,x',t',\tau', \text{ with } \tau \vee \tau' < t \wedge t' ;$$

$$R(\phi,k) = \inf\left\{ R \geq 0 : \left[\int_\Omega |\phi(x,t,Z,\tau) - \phi(x,t,Z',\tau)|J_\eta(Z,Z')dz\right] \vee \right.$$

$$\left. \vee \left[\int_\Omega |\phi(Z,t,y,\tau) - \phi(Z',t,y,\tau)|J_\eta(Z,Z')dz\right] \leq R\eta^\alpha(t-\tau)^{(k-\alpha)/2}, \forall x,t,y,\tau \text{ and } \eta > 0 \right\},$$

where $\wedge$ and $\vee$ denote the minimum and the maximum between two real numbers, the change of variables $Z(z)$ and $Z'(z)$ are diffeomorphism of class one in R^d and the Jacobian

$$J_\eta(Z,Z') = \begin{cases} |\det(\nabla Z)| \wedge |\det(\nabla Z')| & \text{if } |Z-Z'| \leq \eta \\ & \text{and } Z, Z' \in \overline{\Omega}, \\ 0 \quad \text{otherwise}. \end{cases}$$

In the above formulae, $\det(\bullet)$ stands for the determinant of a $d \times d$ matrix $\nabla Z, \nabla Z'$ mean the matrices of the first partial derivatives of $Z(z)$, $Z'(z)$ w.r. to variable z.

Definition. *Let* G^k_α *be the space of all functions* $\phi(x,y,t,\tau)$ *such that* ϕ *belongs to* C^k_α *and its gradient* $D_x\phi$ *belongs to* $C_\alpha^{k-1}, k \geq 0$.

The operators I and D map the space G_α^{k+1} into the space C^k_α $\forall\, k=-1,0,1,\ldots$.

Observe that the Green function G_0 belongs to G^0_α.

We prove

Theorem 1. *There exists a unique solution of* (7) *and we have:*

i) $G(x,y,t,\tau) \in G^0_\alpha$

ii) $G(x,y,t,\tau) \geq 0, \ \forall\, x,y \in \overline{\Omega}, \ 0 \leq \tau < t \leq T$,

iii) $\begin{cases} \textit{there exist two constant, } 0<\eta,\ \delta<T \text{ s.t.} \\ G(x,y,t,\tau) \geq \eta, \ \forall x,y \in \overline{\Omega};\ t,\tau \in [0,T],\ t-\tau=\delta \end{cases}$

iv) $\begin{cases} G(x,y,t,\tau) = \displaystyle\int_\Omega G(x,z,t,s)G(z,y,s,\tau)dz \\ \forall x,y \in \overline{\Omega},\ 0 \leq \tau < s < t \leq T \end{cases}$

(12) v) $\begin{cases} \phi(x) = \displaystyle\lim_{t\to\tau} \int_\Omega G(x,y,t,\tau)\phi(y)dy,\ 0 \leq \tau < t \leq T \\ \textit{uniformly in } x \textit{ belonging to } \overline{\Omega} \textit{ and for any continuous function } \phi \textit{ in } \overline{\Omega}. \end{cases}$

Moreover if we take $a_0(x,t) \equiv 0$, *then*

vi) $\displaystyle\int_\Omega G(x,y,t,\tau)dy = 1, \ \forall x \text{ in } \overline{\Omega},\ 0 \leq \tau < t \leq T$.

From (12) i the estimates for $D^s_x G(x,y,t,\tau)$, $s=0,1$, follow. To obtain the estimates for $D^r_t D^s_x G(x,y,t,\tau)$, $2r+s=2$ we use (8); due to the lack of regularity of the coefficients $(b_i(x,t),\ i=1,\ldots,d)$ given on the boundary $\partial\Omega$ the Green function $G_0(x,y,t,\tau)$ possesses a term $F_0(x,y,t,\tau)$ whose second derivatives have an "explosion" at the boundary, that is

$$|D^r_t D^s_x F_0(x,y,t,\tau)| \leq c(t-\tau)^{\frac{-d+1+\alpha}{2}} \left[\rho(x)^{\alpha-1} \vee (t-\tau)^{\frac{\alpha-1}{2}}\right] \exp\left(-c\frac{|x-y|^2}{t-\tau}\right),$$

$$2r+s=2,\ \forall x,y \in \overline{\Omega},\ 0 \leq \tau < t \leq T,\ \rho(x) = \inf\left\{ |z-x| : z \text{ in } \partial\Omega \right\} .$$

One of the difficulties in obtaining the second derivative estimates for G is due to this "explosion".

Consider now the parabolic problem:

$$(13)\qquad \begin{cases} \partial_t u + Au = f & \text{on } Q_T \\ u(x,0) = 0 & \text{on } \overline{\Omega} \\ B\,u = 0 & \text{in } \Sigma_T \end{cases},$$

we can easily prove that the function

$$u(x,t) = \int_0^t d\tau \int_\Omega G(x,y,t,\tau) f(y,\tau) dy$$

is the unique classic or weak solution of problem (13) according to the data, and we have the following estimates:

$$(14)\qquad \begin{cases} \text{i)}\ T^{-2}[u]_{\alpha,T} + T^{-1}[\nabla u]_{\alpha,T} \le c[f]_{0;T} \\ \text{ii)}\ [\nabla^2 u]^*_{\gamma,T} + [\partial_t u]^*_{\gamma,T} \le c[f]_{\gamma,T} \end{cases}$$

where

$$[\phi]_{\alpha,T} = \inf\Big\{ c \ge 0 : |\phi(x,t)| \le c, \text{ and } |\phi(x,t) - \phi(x',t')| \le c\big[\, |x - x'|^\alpha + |t - t'|^{\alpha/2} \big],$$
$$\forall x, x' \in \overline{\Omega},\ t,t' \in [0,T] \Big\},$$

$$[\phi]^*_{\gamma,T} = \inf\Big\{ c \ge 0 : |\phi(x,t)| \le c\, \rho^{\alpha-1}(x), \text{ and } |\phi(x,t) - \phi(x',t')| \le$$
$$\le c\big(\rho^{\varepsilon-1}(x) + \rho^{\varepsilon-1}(x')\big) \big[\, |x - x'|^\gamma + |t - t'|^{\gamma/2} \big],\ \forall x,x' \in \Omega,\ t,t' \in [0,T],\ \gamma,\varepsilon > 0,\ \alpha = \gamma + \varepsilon,$$

and

$$(15)\qquad \begin{cases} \text{i)}\ t^{-2} \|u(\cdot,t)\|_{L^p(\Omega)} + t^{-1} \|Du(\cdot,t)\|_{L^p(\Omega)} \le c_p \|f\|_{L^p(Q_T)} & 1 \le p \le \infty,\ 0 < t \le T\ . \\ \text{ii)}\ \|D^2 u\|_{L^p(Q_T)} + \|\partial_t u\|_{L^p(Q_T)} \le M_{p,\alpha} \|f\|_{L^p(Q_T)}, & 1 < p < (1-\alpha)^{-1}\ . \end{cases}$$

In (15) ii) the condition $p < (1-\alpha)^{-1}$ is optimal, in fact Bu belongs to $W_p^{1-1/p}(\Sigma_T)$. This in particular implies that we cannot use generalized maximum principles to prove the uniqueness of the weak solution.

3. Invariant measure and asymptotic behaviour

Through this section we suppose that the integro-differential operator A and the differential boundary operator B are independent of the variable t, and $a_0 \equiv 0$. We want to consider the following three problems:

$$(16) \qquad \begin{cases} \partial_t u + Au = g & \text{in } \Omega x(0,\infty) \\ u(x,0) = 0 & \text{on } \Omega \\ Bu = 0 & \text{on } \partial\Omega x[0,\infty) \end{cases}$$

$$(17) \qquad \begin{cases} Au_\lambda + \lambda u_\lambda = g & \text{in } \Omega \\ Bu_\lambda = 0 & \text{on } \partial\Omega \end{cases}$$

$$(18) \qquad \begin{cases} Au_0 = g & \text{in } \Omega \\ Bu_0 = 0 & \text{on } \partial\Omega \end{cases}$$

λ positive constant. We first prove the existence and uniqueness for problem (16) and (17), (in fact all results we know are obtained in the case of more regular coefficients (cfr. e.g. [1]), then we want to know under which compatibility conditions for g, the problem (18) possess one or more solutions and when we have:

$$(19) \qquad \begin{cases} \text{i)} \; u(\bullet,t) \to u_0 & \text{as } t \to +\infty \\ \text{ii)} \; u_\lambda \to u_0 & \text{as } \lambda \to 0 \;. \end{cases}$$

These questions lead to the study of the invariant measures of the homogeneous semigroup generated by the integro-differential operator A and the differential boundary operator B. We now proceed similarly to I. Capuzzo-Dolcetta and M.G. Garroni [4,5].

Denote by R_λ the resolvent operator of the stationary problem (17) i.e.

$$R_\lambda g(x) = \int_0^{+\infty} e^{-\lambda t} \int_\Omega G(x,y,t,0)g(y)dy, \quad \forall\, x \in \overline{\Omega} \;.$$

It is clear that R_λ can be considered as a linear compact operator in $L^p, 1 < p < \infty$, $0 < \lambda < 1$, and according to the assumptions on the function g, $u_\lambda = R_\lambda g$ is the solution of (17). Using this operator and its ajoint we prove:

Theorem 2. *There exists a unique Hölder continuous function* $m(\cdot)$ *in* Ω, *with exponent* α, *s.t.*

$$(20)\qquad \begin{cases} m(y) = \int_0^\infty e^{-\lambda t}\,dt \int_\Omega G(x,y,t,0)m(x)dx, \quad \forall y \in \overline{\Omega},\ \lambda > 0\ , \\ \int_\Omega m(x)dx = 1\ . \end{cases}$$

Moreover, there exists a positive constant ν *s.t.:*

$$(21)\qquad m(y) \geq \nu > 0, \quad \forall y \in \overline{\Omega}$$

and for every $t \geq 1$, $g \in L^1(\Omega)$, $x \in \Omega$

$$(22)\qquad \left| \int_\Omega G(x,y,t,0)g(y)dy - \int_\Omega g(y)m(y)dy \right| \leq c \exp(-\nu t)\, ||g||_{L^1(\Omega)}\ ,$$

c *independent of* t,g,p.

Furthermore:

$$(23)\qquad m(y) = \int_\Omega G(x,y,t,0)m(x)dx, \quad \forall y \in \overline{\Omega},\ t > 0\ .$$

Remarks. If a_{ij} and b_i are regular (f.i. C^1) and $I \equiv 0$, then m belongs to C^2 and coincides with the solution of differential ajoint homogeneous problem:

$$\begin{cases} A^* m = 0 \ \text{ on } \Omega \\ B^* m = 0 \ \text{ in } \partial\Omega \end{cases}$$

(see f.i. [3]).

If g is regular enough, (f.i. g belongs to $C^0(\Omega)$ or belongs to $W_p^{2-2/p}(\Omega)$, $1 < p < \dfrac{1}{1-\alpha}$), and satisfies convenient compatibility conditions, then $\int_\Omega G(x,y,t,0)g(y)dy$ is the classic or weak solution $v(x,t)$ of the problem

$$\begin{cases} \partial_t v + Av = 0 \ \text{ on } Q_T\ , \\ v(x,0) = g(x) \ \text{ on } \Omega \\ Bv = 0 \qquad \text{ in } \Sigma_T\ ; \end{cases}$$

since the property (23) implies

$$\int_\Omega g(y)m(y)dy = \int_\Omega \left[\int_\Omega G(x,y,t,0)g(y)dy \right] m(x)dx, \quad \forall t > 0 \ ,$$

we have that the average of the solution $v(x,t)$ is constant for every positive t; the property (22) gives the asymptotic behaviour of the solution $v(x,t)$ when t goes to infinity.

Theorem 2 has a probabilistic interpretation. Actually, we can define probability transition functions P by setting

$$P(x,t,E) = \int_E G(x,y,t,0)dy, \ \forall \text{ Borel subset } E \text{ of } \overline{\Omega} \ .$$

These are related to a diffusion process reflected at the boundary $\partial\Omega$ according to the vector field B and with interior jumps. The property (20) says that $d\mu = mdx$ is the invariant measure of this process (see [8] for details). The condition (22) is clearly an ergodic property of this measure, cfr. [2].

At this point it is clear that m is a "good" candidate for the compatibility condition. Actually by means of the Fredholm alternative and Theor. 2 we establish the following:

Theorem 3. *The limit problem* (18) *possesses a solution unique (up to an additive constant) if an only if*

$$\int_\Omega g(x)m(x)dx = 0 \tag{24}$$

where m *is the function defined in Theorem 2. Moreover, the solution of problem* (18), *satisfying* (24) *with* u_0 *instead of* g, *admits the representation*

$$u_0(x) = \int_0^\infty dt \int_\Omega G(x,y,t,0)g(y)dy \equiv \lim_{t\to+\infty} u(x,t), \ \forall x \in \overline{\Omega} \tag{25}$$

and

$$u_0(x) = \lim_{\lambda\to 0} u_\lambda(x) = \lim_{\lambda\to 0} \int_0^\infty e^{-\lambda t}dt \int_\Omega G(x,y,t,0)\ (y)dy \ . \tag{26}$$

The class of functions to which u_0 *belongs is determined by the following estimates according to how smooth the data* g *is*

$$[u_0]_\alpha + [Du_0]_\alpha \le c[g]_0$$

$$[\nabla^2 u_0]^*_\gamma \le c_\gamma [g]_\alpha \ , \ 0 < \gamma < \alpha$$

$$\|u_0\|_{L^p(\Omega)} + \|\nabla u_0\|_{L^p(\Omega)} \le c_p \|g\|_{L^p(\Omega)}, \quad 1 < p < \infty ,$$

$$\|D^2 u_0\|_{L^p(\Omega)} \le M_{p,\alpha} \|g\|_{L^p(\Omega)}, \quad 1 < p < (1-\alpha)^{-1} ,$$

where the constants $c, c_\gamma, c_{p,\alpha}$ *are independent of* g.

Remark. The invariant measure is the key to solving other non coercive stationary problems motivated by the control theory of stochastic processes with jumps, for instance the study of the limiting behaviour of solutions of unilateral problems [9].

References

[1] **A. Bensoussan and J.L. Lions,** Contrôle impulsionnel et inequations quasi-variationnelles, Dunod, Paris (1982).

[2] **A. Bensoussan, J.L. Lions and G. Papanicolau,** Asymptotic analysis for periodic structures, North-Holland, Amsterdam (1978).

[3] **A. Bensoussan and J.L. Lions,** On the asymptotic behaviour of the solution of variational inequalities in summer School on the theory, Akademic Verlag, Berlin (1978), 25-40.

[4] **I. Capuzzo-Dolcetta and M.G. Garroni,** Comportement asymptotic de la solution de problèmes non sous forme divergence, C.R.A.S. V. 299 S.I. 17 (1984), 843-846.

[5] **I. Capuzzo-Dolcetta and M.G. Garroni,** Oblique derivative problemes and invariant measures, to appear in Annali della Scuola Normale Sup. Pisa.

[6] **M. Chaleyat-Maurel, N. El Kavom and B. Marchal,** Réflextion discontinue et systémes stochastiques, Annals Prob. 8 (1980), 1049-1067.

[7] **M.G. Garroni and J.L. Menaldi,** Fonction de Green et densité invariante pour des problèmes intégro-différentiels de deuxième ordre, C.R.A.S. V. 303, S.I. (1986), 787-790.

[8] **M.G. Garroni and J.L. Menaldi,** Green function and invariant density for an integro-differential operator of second order, to appear.

[9] **M.G. Garroni and J.L. Menaldi,** On the asymptotic behaviour of solutions of inequalities for an integro-differential operator, to appear.

[10] **M.G. Garroni and V.A. Solonnikov,** On parabolic oblique derivative problem with Hölder continuous coefficients, Comm. in Partial Diff. Equations 9 (14) (1984), 1323-1372.

[11] **I.S. Gikhiman and A.V. Skorokhod,** The theory of stochastic processes, vol. I, II, III Springer Verlag, Berlin (1975).

[12] **O.A. Ladyzhenskaya, V.A. Solonnikov and N.N. Uraltzeva,** Linear and quasilinear equations of parabolic type, Nauka, Moscow (1967).

[13] **T. Komatsu,** Markov processes associated with certain integro-differential operators, Osaka J. Math. 10 (1973), 271-303.

[14] **J.P. Lepeltier and B. Marchal,** Problème des martingales et équations diff. stochatiques associées à un operateur integro-différentiel, Ann. Inst. Henri Poincaré 12 n.1 (1976), 43-103.

[15] **J.L. Menaldi and M. Robin,** An ergodic control problem for reflected diffusion with jumps, I.M.A. J. Math. Control Inf. 1 (1984), 309-322.

[16] **J.L. Menaldi and M. Robin,** Reflected diffusion processes with jumps, Annals Probab. 13 (1985), 319-341.

[17] **D. Stroock,** Diffusion processes associated with Levy generators, Z. Wahrscheinlichkerts Theorie 32 (1975), 209-244.

SOME REMARKS ON THE REGULARITY OF MINIMIZERS

Mariano Giaquinta
Istituto di Matematica Applicata
Facoltà di Ingegneria - Università
Via S. Marta, 3
50139 - Firenze (Italy)

Dedicated to Hans Lewy

Consider a variational integral

$$F[u;\Omega] := \int_\Omega f(x,u,Du)dx \tag{1}$$

where the integrand $f(x,u,p)$ grows polynomially like $|p|^m$. More precisely assume that

H.1- *For a positive constant* Λ *we have*

$$|p|^m \leq f(x,u,p) \leq \Lambda(1+|p|^2)^{m/2}$$

where $m \geq 2$.

A *minimizer* for (1) is a function $u \in W^{1,m}(\Omega)$ such that

$$F[u;\operatorname{spt}\phi] \leq F[u+\phi;\operatorname{spt}\phi] \quad \forall \phi \in W^{1,m}(\Omega), \operatorname{spt}\phi \subset\subset \Omega .$$

In this talk I want to describe some steps leading to the regularity of minimizers, especially in connection with possible degeneracy of the integrand. For the sake of simplicity I shall restrict myself to consideration of mainly the case of scalar minimizers.

Basic regularity.

In 1982 E. Giusti and I, [4] [6], proved that assumption H.1 enables us to prove the following two results:

(a) *If* u *is a minimizer of (1), then* u *is Hölder-continuous with some exponent* α *and for any ball* $B_R(x_0) \subset\subset \Omega$ *we have*

$$\int_{B_R(x_0)} |Du|^m dx \leq c R^{n-m+2\alpha} .$$

(b) *There is a positive constant* ε *such that*

$$Du \in L^{m+\varepsilon}_{loc}(\Omega, R^n) .$$

Moreover the following reverse Hölder inequality (with increasing support) holds:

$$\left(\fint_{B_R(x_0)} H(Du)^{\frac{m+\varepsilon}{m}} dx \right)^{\frac{m}{m+\varepsilon}} \leq c \fint_{B_{2R}(x_0)} H(Du) dx$$

where

$$H(Du) := (1 + |Du|^2)^{m/2}$$

and

$$\fint_{B_R} g\, dx = |B_R|^{-1} \int_{B_R} g\, dx .$$

We notice that $f(x,u,p)$ need not be convex with respect to p. In proving (a) one uses De Giorgi's results on what we now call De Giorgi classes, while in proving (b) one uses a result of M. Giaquinta and G. Modica on reverse Hölder inequalities related to a previous result of F.W. Gehring.

Besides the interest they have by themselves, results (a) and (b) play an important role in studying further regularity of minimizers and in particular in proving Hölder-continuity of the first derivatives.

Hölder continuity of the derivatives of a minimizer.

Assuming convexity or ellipticity of f with respect to p, it is reasonable to expect that the first derivatives of a minimizer will be Hölder-continuous. This is actually true, and, first, I want to describe the main ideas in case $m = 2$. But let me state the hypotheses for $m \geq 2$.

H.2 - *We shall assume that* $f(x,u,p)$ *be if class* C^2 *with respect to* p,

$$(2) \qquad |f_{pp}(x,u,p)| \leq c_1 (\mu^2 + |p|^2)^{\frac{m-2}{2}}$$

and, in the case $\mu > 0$, or $m = 2$, that

$$(3) \qquad |f_{pp}(x,u,p) - f_{pp}(x,u,q)| \leq c_2 (\mu^2 + |p|^2 + |q|^2)^{\frac{m-2}{2}} \omega(|p-q|)$$

where $\omega(t)$ *is a concave function going to zero as* t *goes to zero, while if* $\mu = 0$ *we shall assume that*

$$(3)' \qquad |f_{pp}(x,u,p) - f_{pp}(x,u,q)| \le c_2(\mu^2 + |p|^2 + |q|^2)^{\frac{m-2}{2} - \frac{\alpha}{2}} |p-q|^{\alpha}$$

for some small positive α .

H.3 - *The integrand* f *is elliptic in the sense that*

$$f_{p_\alpha p_\beta}(x,u,p)\xi^\alpha \xi^\beta \ge (\mu^2 + |p|^2)^{\frac{m-2}{2}} |\xi|^2 \;\; \forall \xi \in \mathbf{R}^n$$

so for $m > 2$ and $\mu = 0$, the integrand degenerates at the points where the gradient is zero; this instead, is not the case if $\mu > 0$.

H.4 - *The function*

$$(1 + |p|^2)^{-\frac{m}{2}} f(x,u,p)$$

is Hölder-continuous in (x,u) *uniformly with respect to* p.

In 1983 E. Giusti and I [5] proved that *if* $m = 2$ *and* H.1 ... H.4 *hold, then any minimizer of (1) has first order Hölder-continuous derivatives.*

I would like to describe the main ingredients of the proof.

The idea consists in comparing the minimizer u in a ball $B_R(x_0)$ with the function v minimizing the functional

$$F^\circ[v;B_R(x_0)] := \int_{B_R(x_0)} f(x_0, u_{x_0,R}, Dv)dx$$

among all functions taking value u on ∂B_R.

One easily sees, using H.3, that

$$(4) \qquad \int_{B_R(x_0)} |D(u-v)|^2 dx \le \text{const} \left\{ F^\circ[u;B_R(x_0)] - F^\circ[v;B_R(x_0)] \right\} .$$

Estimates for the comparison function v.

Writing the Euler-Lagrange equation for v and differentiating, we deduce that $|Du|^2$ is a subsolution for an elliptic operator with L^∞ coefficients, so we get

$$(I) \qquad \int_{B_\rho(x_0)} |Dv|^2 dx \le c\left(\frac{\rho}{R}\right)^n \int_{B_R(x_0)} |Dv|^2 dx \quad \forall \rho < R .$$

While from De Giorgi-Nash theorem, we get

$$\text{(II)} \qquad \int_{B_\rho(x_0)} |Dv - (Dv)_{x_0,\rho}|^2\, dx \le c \left(\frac{\rho}{R}\right)^{n+2\mu} \int_{B_R(x_0)} |Dv - (Dv)_{x_0,R}|^2\, dx$$

for some positive μ.

The comparison argument.

1st step: We first show that *the minimizer* u *is Hölder-continuous with any exponent* $\gamma < 1$ *and*

$$\int_{B_R(x_\rho)} |Du|^2\, dx \le c\, R^{n-2+2\gamma}.$$

In fact from (I) we deduce for any $\rho < R$

$$\text{(5)} \qquad \int_{B_\rho} |Du|^2\, dx \le c \left\{ \left(\frac{\rho}{R}\right)^2 \int_{B_R} |Du|^2 dx + \int_{B_R} |D(u-v)|^2\, dx \right\}.$$

The last term can be estimated by (4) as follows

$$\text{(6)} \quad \int_{B_R(x_0)} |D(u-v)|^2\, dx \le c \left\{ \int_{B_R} \left[f(x_0,u_{x_0,R},Du) - f(x,u,Du) \right] dx + \right.$$

$$\left. + \int_{B_R} \left[f(x,u,Du) - f(x,v,Dv) \right] dx + \int_{B_R} \left[f(x,v,Dv) - f(x_0,u_{x_0,R},Dv) \right] dx \right\}.$$

So, using the continuity of u (and v) and H.4, we get

$$\text{(7)} \qquad \int_{B_R(x_0)} |D(u-v)|^2 dx \le c\,\eta(R) \int_{B_R(x_0)} |Du|^2 dx$$

i.e.

$$\text{(8)} \qquad \int_{B_\rho(x_0)} |Du|^2 dx \le c \left[\left(\frac{\rho}{R}\right)^n + \eta(R) \right] \int_{B_R(x_0)} |Du|^2 dx$$

with $\eta(R) \downarrow 0$ for $R \downarrow 0$. From (8) and a well-known result of Morrey one easily gets the result.

2nd step: The first derivatives if a minimizer u *are Hölder-continuous.*

Using (II), (5) can be replaced by

$$\text{(9)} \qquad \int_{B_\rho} |Du - (Du)_\rho|^2 dx \le \left[\left(\frac{\rho}{R}\right)^{n+2\mu} \int_{B_R} |Du - (Du)_R|^2 dx + \int_{B_R} |D(u-v)|^2 dx \right].$$

Using step 1, i.e. the Hölder continuity of u and of f in (x,u), we deduce from (6)

$$\int_{B_R} |D(u-v)|^2 dx \le c\, R^{2\alpha\gamma} \int_{B_R} |Du|^2 dx \le c\, R^{n-2+2\gamma+2\alpha\gamma} \qquad \text{(for some } \alpha > 0\text{)}$$

If γ is close to 1, clearly

$$n-2+2\gamma+2\alpha\gamma = n+2\sigma \qquad \sigma > 0$$

So that we conclude for any ball $B_R(x_0) \subset \Omega$, and for any $\rho < R$

$$(10) \qquad \int_{B_\rho(x_0)} |Du-(Du)_\rho|^2 dx \le c \left(\frac{\rho}{R}\right)^{n+2\mu} \int_{B_R(x_0)} |Du-(Du)_R|^2 dx + c\, R^{n+2\sigma} .$$

This implies, by iteration and a result of Campanato, the Hölder continuity of Du.

The previous argument will work in the general case of polynomial growth, and actually works, if we find out the right substitute of $|Du|$ and $|Du-(Du)_\rho|$ in the estimates (I), (II).

It happens that the right quantities to be considered are, see [8], [1], [9],

$$H(Du) := \left(\mu^2 + |Du|^2\right)^{m/2}$$

and

$$V(Du) := \left(\mu^2 + |Du|^2\right)^{\frac{m-2}{4}} Du .$$

And the point is to study minimizers u of the functional in (1) in case the integrand f does not depend explicitely on x and u, and H.1, H.2, H.3 hold.

Estimate (I). If u is such a minimizer, then one has that H(Du) is a subsolution for an elliptic operator with L^∞ coefficients. More precisely one gets: *for any nonnegative* ϕ *with compact support*

$$(11) \qquad \int_\Omega |DV(Du)|^2 dx + \int_\Omega a^{\alpha\beta} D_\beta H\, D_\alpha \phi\, dx \le 0$$

where $a^{\alpha\beta} \in L^\infty$ *and* $a^{\alpha\beta}\xi_\alpha \xi_\beta \ge \nu |\xi|^2, \forall \xi \in \mathbb{R}^n, \ \nu > 0.$

So estimate (I) follows at once, with $|Du|^2$ replaced by H(Du), from the sup estimate

$$\sup_{B_{R/2}} H(Du) \le c \int_{B_R} H(Du)dx .$$

Estimate (II). Here the situation is much more delicate. In fact as we remarked, our assumption H.3 does not exclude that the equation in variation be degenerate, and actually it might be degenerate. Moreover we know that minimizers for example of the simple integral

$$(12) \qquad \int_\Omega |Du|^m \, dx$$

in general have not continuous second derivatives.

In 1977 K. Uhlenbeck [14] in a very interesting paper proved the following result: *if* $u \in W^{1,m}(\Omega)$ *is a minimizer of*

$$\int_\Omega f(Du)dx$$

and H.1, H.2, H.3 *hold, then the first derivatives of* u *are Hölder-continuous with some exponent* α *and*

$$(13) \qquad \sup_{x,y \in B_R} |Du(x) - Du(y)| \le c \max_{B_{2R}} |Du| \frac{|x-y|^\alpha}{R^\alpha} .$$

Moreover the same result holds for vector valued minimizers u *provided*

$$(14) \qquad f(Du) = g\left(|Du|^2\right)$$

or more generally

$$(15) \qquad f(Du) = g\left(A^{\alpha\beta} a_{ij} D_\alpha u^i D_\beta u^j\right)$$

where A *and* a *are symmetric positive definite matrices.*

Estimate (13) does not play the same role as (II); instead, one would like to have (see [8],[9],[1]) the following estimate

$$(16) \qquad \int_{B_\rho(x_0)} |V(Du) - V(Du)_{x_0,\rho}|^2 dx \le c\left(\frac{\rho}{R}\right)^{n+2\sigma} \int_{B_R(x_0)} |V(Du) - V(Du)_{x_0,R}|^2 dx$$

for any $B_\rho(x_0)$ $B_R(x_0)$ and some positive σ.

Recently N. Fusco and J. Hutchinson [3] have shown for minimizers of the functional (12) that (16) follows, using again the equation, from (13). Independently, G. Modica and I [10] deduced (16), going more carefully through the original proof of K. Uhlenbeck.

I would like to sketch the main steps of the proof in [14], [10].

1st step: It is estimate (11)

2nd step: Set

$$M(x_0,r) := \sup_{B_r(x_0)} H(Du)$$

$$\Phi(x_0,r) := \int_{B_r(x_0)} |V(Du) - V(Du)_{x_0,r}|^2 dx \ .$$

From (11) we see that $M(x_0,R) - H(Du)$ is a supersolution in $B_R(x_0)$ for an elliptic operator, so using Harnack inequality in an appropriate way one gets

$$\Phi(x_0,r) \le c[M(x_0,R) - M(x_0,R/2)] \ .$$

3rd step: By a comparison argument, where the comparison function is given by minimizing the integral obtained by taking the second order Taylor expansion if f, one shows: *there exist* $\tau \in (0,1/4)$ *and* $\varepsilon_0 << 1$, *such that*

$$\Phi\left(x_0,\tau R\right) \le \tau\,\Phi(x_0,R)$$

provided

$$\Phi(x_0,R) \le \varepsilon_0 \, M(x_0,R/2) \ .$$

4th step: Finally from step 2 and 3 one gets: *there exist* $\tau \in (0,1/4)$, $0 < \varepsilon_0 << 1$, δ, $0 < \delta < 1$, *so that either*

$$\Phi\left(x_0,\tau R\right) \le \tau\,\Phi(x_0,R)$$

or

$$\begin{cases} \Phi(x_0,R) \ge \varepsilon_0 \, M(x_0,R/2) \\ \text{and} \\ M(x_0,R/4) \le \delta M(x_0,R/2) \ . \end{cases}$$

Now a kind of algebraic iteration implies estimates (16).

Again steps 1,....,4 carry on in the vector valued case, provided (15) holds.

This permits to show the following, see [10]:

Let $u \in W^{1,m}(\Omega,\mathbb{R}^N)$ *be a minimizer of the integral (1) and let* f *have the special structurure*

$$f(x,u,Du) = g\left(x,u,A^{\alpha\beta}(x,u)a_{ij}(x,u)D_\alpha u^i D_\beta u^j\right) \tag{17}$$

then the first derivatives of u *are Hölder-continuous in an open set* Ω_0 *and the Hausdorff dimension if the singular set* $\Omega - \Omega_0$ *is strictly less than* $n-m$.

This result is in a sense optimal, since it applies for example to energy minimizing maps between Riemannian manifolds, in case the image of the minimizer u lies in a coordinate chart; and we know that even bounded minimizers may be singular. For example it applies to bounded minimizers of

$$\int_{B_1(0)} \frac{|Du|^2}{\left(1+|u|^2\right)^2}\,dx \tag{18}$$

which represents in local coordinates (choosing stereographic coordinates on S^n-{point}) the energy of a map from the disk D_n into S^n-{point}. As it is known [12], [13], [11], even bounded minimizers of (18) may be singular (in dimension $n \geq 7$).

On the contrary, if we do not assume (17), no result is known if $\mu = 0$, i.e. if degeneracy may occur. The situation, instead, is quite satisfactory if $\mu > 0$, see [8], [1], [9], [2].

References

[1] **L.C. Evans,** Quasi convexity and partial regularity in the calculus of variations, pre-print (1984).

[2] **N. Fusco and J. Hutchinson,** $C^{1,\alpha}$ partial regularity of functions minimizing quasiconvex integrals, Manuscripta math. 54 (1985), 121-143.

[3] **N. Fusco and J. Hutchinson,** Partial regularity of minimizers of certain functionals having non quadratic growth, pre-print.

[4] **M. Giaquinta,** Multiple integrals in the Calculus of Variations and nonlinear elliptic systems, Princeton Univ. Press, Princeton 1983.

[5] **M. Giaquinta and E. Giusti,** On the regularity of minima of variational integrals, Acta Math. 148 (1982), 31-46.

[6] **M. Giaquinta and E. Giusti,** Differentiability of minima of nondifferentiable functionals, Inventiones Math. 72 (1983), 285-298.

[7] **M. Giaquinta and E. Giusti,** Quasi-minima, Ann. Inst. H. Poincaré, Analyse non linéaire 1 (1984), 79-107.

[8] **M. Giaquinta and P.A. Ivert,** Partial regularity of minimizers of variational integrals, pre-print 1983.

[9] **M. Giaquinta and G. Modica,** Partial regularity of minimizers of quasiconvex integrals, Ann. Inst. H. Poincaré, Analyse non linéaire 3 (1986).

[10] **M. Giaquinta and G. Modica,** Remarks on the regularity of the minimizers of certain degenerate functionals, Manuscripta math., to appear.

[11] **M. Giaquinta and J. Soucek,** Harmonic maps into a hemisphere, Ann. Sc. Norm. Sup. Pisa 12 (1985), 81-90.

[12] **W. Jäger and H. Kaul,** Rotationally symmetric harmonic maps from a ball into a sphere and the regularity problem for weak solutions of elliptic systems, J. reine u. angew. Math. 343 (1983), 146-161.

[13] **R. Schoen and K. Uhlenbeck,** Regularity of minimizing harmonic maps into the sphere, Inventiones Math. 78 (1984), 89-100.

[14] **K. Uhlenbeck,** Regularity for a class of nonlinear elliptic systems, Acta Math. 138 (1977), 219-240.

QUADRATIC FUNCTIONALS WITH SPLITTING COEFFICIENTS

Enrico Giusti
Istituto Matematico - Università
Viale Morgagni, 67/A
50134 Firenze (Italy)

Dedicated to Hans Lewy

1. Introduction

In this lecture I shall be concerned with the problem of regularity of functions $u : \Omega \to R^n$ minimizing the variational integral:

$$E(v,\Omega) = \int_\Omega \gamma^{\alpha\beta}(x)\, g_{ij}(x,u) D_\alpha u^i D_\beta u^j\, dx \tag{1.1}$$

where γ and g are bounded continuous functions satisfying:

$$\gamma^{\alpha\beta} = \gamma^{\beta\alpha}\,; \quad g_{ij} = g_{ji}$$

$$\gamma^{\alpha\beta}(x) k_\alpha k_\beta \geq |k|^2 \tag{1.2}$$

$$g_{ij}(x,u) h^i h^j \geq |h|^2 \quad .$$

It is well known that such functionals enter in the theory of harmonic mappings between Riemannian manifolds, in which they represent the energy in local coordinates. This explains why they have been studied by various authors. Here however we shall not adopt the point of view of geometry, but rather that of the calculus of variations; although overlapping in many respects, they are not completely coincident. In particular, we shall always consider the coordinate system as fixed.

The reader interested in harmonic maps can refer to the papers by Schoen and Uhlenbeck [11], [12], where similar results have been proved in a geometric context.

The plan of the paper is the following: in section 2 and 3 we deal with interior regularity and with boundary regularity of minimizers taking prescribed boundary values. Section 4 will be devoted to the discussion of the special but interesting case of energy minimizing harmonic maps from a disk into a sphere or an ellipsoid. Finally, in section 5 we discuss the behaviour of minimizers in the neighborhood of a singular point.

2. Regularity of minimizers of quadratic functionals

One of the main tools in the proof of regularity is the Caccioppoli's Inequality:

$$\int_{B(x^0,r)} |Du|^2 dx \le c\,(R-r)^{-2} \int_{B(x^0,R)} |u - u_{x^0,R}|^2 dx \tag{2.1}$$

which holds for every minimizer of the functional (1.1), for every point x^0 and for sufficiently small radii r and R, with $0 < r < R < \mathrm{dist}(x^0,\Omega)$. By $u_{x^0,R}$ we have indicated the average of u on $B(x^0,R)$:

$$u_{x^0,R} = \fint_{B(x^0,R)} u\,dx = |B_R|^{-1} \int_{B(x^0,R)} u\,dx \ .$$

The proof of (2.1) under conditions (1.2) is not easily found in the literature, and therefore we shall give it here in detail. Due to its local character, we may suppose without loss of generality that the coefficients $\gamma^{\alpha\beta}(x)$ are uniformly continuous.

The first step is a lemma of Lax-Milgram type, for which we shall assume that g_{ij} are bounded measurable functions of x alone, satisfying conditions (1.2).

Lemma 2.1. *There exists a positive number* R^0 *such that if* $w \in W^{1,0}(\Omega)$ *and* diam supp(w) $< R^0$, *then for some* $\alpha^0 > 0$ *we have*

$$E^0(w,\Omega) = \int_\Omega \gamma^{\alpha\beta}(x) g_{ij}(x) D_\alpha w^i D_\beta w^j\,dx \ge \alpha^0 \int_\Omega |Dw|^2 dx \ . \tag{2.2}$$

Proof. Let $x^0 \in$ supp(w). After a rotation about x^0 followed by a change of scale for each coordinate, we may suppose that

$$\gamma^{\alpha\beta}(x^0) = \delta^{\alpha\beta} \ . \tag{2.3}$$

Under this change of coordinates, the diameter of supp(w) changes by a factor that can be estimated from above and below independently of x^0. We have:

$$\gamma^{\alpha\beta}(x) = \delta^{\alpha\beta} + \sigma^{\alpha\beta}(x)$$

and from the uniform continuity of γ:

$$|\sigma^{\alpha\beta}(x)| \le \sigma\left(|x - x^0|\right)$$

with $\sigma(0) = 0$.

Denoting again by w and g the transformed functions, we have:

$$E^0(w,\Omega) = \int_\Omega g_{ij}(x) D_\alpha w^i D_\alpha w^j \, dx + \int_\Omega \sigma^{\alpha\beta}(x) g_{ij}(x) D_\alpha w^i D_\beta w^j \, dx \ge$$

$$\ge \int_\Omega |Dw|^2 dx - L\sigma\left(R^0\right) \int_\Omega |Dw|^2 dx \quad .$$

If we choose R^0 so small that $L\sigma(R^0) < 1/2$, we obtain easily (2.2) with $\alpha^0 = 1/2$, provided (2.3) holds. In general, we must come back to the original function w, and this changes the constant $1/2$ to a constant α^0, which depends only on the ratio between the highest and lowest eigenvalue of the matrix γ.

We can now prove the inequality (2.1).

Let $0 < r < R < \min\{\mathrm{dist}(x^0, \partial\Omega), R^0/2\}$, and let θ be a cut-off function: $0 \le \theta \le 1$, $\theta = 1$ on B_r, $\mathrm{supp}(\theta) \subset B_R$. If u is a minimizer of (1.1) and if we set

$$v = u - \theta(u - \mu), \qquad \mu = \text{constant},$$

we have

$$E\left(u, B\left(x^0, R\right)\right) \le E\left(v, B\left(x^0, R\right)\right)$$

and writing $w = u - v$:

$$E(u,B) = \int_{B_R} \gamma^{\alpha\beta} g_{ij} D_\alpha w^i D_\beta w^j \, dx + 2\int_{B_R} \gamma^{\alpha\beta} g_{ij} D_\alpha v^i D_\beta w^j \, dx + \int_{B_R} \gamma^{\alpha\beta} g_{ij} D_\alpha v^i D_\beta v^j \, dx$$

where the coefficients g_{ij} are calculated at $(x,u(x))$. Using the previous lemma:

$$E\left(u, B_R\right) \ge 1/2\, \alpha^0 \int_{B_R} |Dw|^2 \, dx - C \int_{B_R} |Dv|^2 \, dx \quad ,$$

and hence

$$\int_{B_R} |Dw|^2 \, dx \le C \int_{B_R} |Dv|^2 \, dx \quad .$$

On the other hand:

$$\int_{B_R} |Dw|^2 dx \ge \int_{B_r} |Du|^2 dx$$

since $w = \theta(u-\mu)$ and $\theta = 1$ on B_r. Moreover

$$\int_{B_R} |Dv|^2 dx \le C \int_{B_R} (1-\theta)^2 |Du|^2 dx + C\,(R-r)^{-2} \int_{B_R} |u-\mu|^2 dx \le$$

$$\le C \int_{B_R - B_r} |Du|^2 dx + C\,(R-r)^{-2} \int_{B_R} |u-\mu|^2 dx\,.$$

In conclusion:

$$\int_{B_R} |Du|^2 dx \le C \int_{B_R - B_r} |Du|^2 dx + C\,(R-r)^{-2} \int_{B_R} |u-\mu|^2 dx$$

and the required inequality (2.1) follows from the hole-filling Lemma 1.1 in [3].

As a consequence of (2.1) and of the Gehring-Giaquinta-Modica theorem [2], [5] on L^p integrability of functions satisfying a reverse Hölder inequality, we conclude that u belongs to $W^{1,p}$ for some $p > 2$, with the estimate

(2.4)
$$\left(\int_{B_R} |Du|^p dx \right)^{1/p} \le C \left(\int_{B_{2R}} |Du|^2 dx \right)^{1/2}$$

for any two concentric balls $B_R \subset B_{2R} \subset\subset \Omega$.

It is now possible to apply the methods of [3], proving the following theorem:

Theorem 1.1 *Let* u *be a minimizer of* $E(u, \Omega)$, *with coefficients satisfying* (1.2). *Then there exists a positive constant* e^0 *(depending only on the bound and modulus of continuity of the coefficients) such that if for some* $x^0 \in \Omega$ *and some* $R < \min\{R^0, \mathrm{dist}(x^0, \Omega)\}/6$ *we have*

$$R^2 < e^0$$

(2.5)
$$U(x^0, R) = \int_{B(x^0,R)} |u - u_{x^0,R}|^2 dx < e^0$$

then u *is Hölder-continuous, with any exponent* $\delta < 1$, *in a ball* B^0 *centered at* x^0.

Moreover, the radius r^0 *of* B^0 *can be estimated in terms of* e^0, R *and the* L^2*-norm of* u, *but does not depend otherwise on the function* u *and on the coefficients.*

The proof of the first part of the theorem follows closely that of Theorem 5.1 of [3], with Lemma 1.1 replacing the ellipticity assumption (5.2) of [3]. We shall therefore limit ourselves to the estimate of the radius r^0.

Let e^0 be the constant given by the first part of the theorem, and suppose that u satisfies (2.5) with e^0 replaced by $e^0/2$.

Considering U(x,R) as a function of x, it is easy to see that

$$|DU(x,R)| \le \int_{B(x,R)} |Du(z)|dz$$

and therefore, using (2.1):

$$|DU(x,R)| \le C R^{-1} U(x,2R) \le C' R^{-1} U(x^0,3R) \le C'' R^{-n-1} \|u\|^2_{L^2}$$

provided $|x - x^0| < R$. Since $U(x^0,R) < e^0/2$ we obtain $U(x,R) < e^0$ for every $x \in B(x^0,r^0)$, with

(2.6) $$r^0 = \min\left\{ R, \frac{1}{2} e^0 R^{n+1}/C'' \|u\|^2_{L^2} \right\} .$$

By the first part of the theorem, the function u is Hölder-continuous in a neighborhood of any such x, and therefore in $B(x^0,r^0)$.

In particular, u is Hölder-continuous near every point x^0 such that

$$\liminf_{R\to 0} U(x^0,R) = 0 .$$

Since the above relation holds for all x^0 outside a set Σ whose Hausdorff dimension is less than $n-2$, we may conclude that u is regular (i.e. Hölder-continuous) in an open set Ω^0, with

(2.7) $$\dim(\Omega - \Omega^0) < n - 2 .$$

If the coefficients γ and g are regular, it is possible to prove the regularity of u in Ω^0 by means of the standard elliptic estimates. We have:

Theorem 2.2. *If the coefficients are of class* $C^{k,\alpha}$, *then* u *is of class* $C^{k+1,\alpha}(\Omega^0)$. *Moreover, the norm of* u *in a compact subset* K *of* Ω^0 *can be estimated in terms of* k, $\|u\|_{L^2}$ *and* $\mathrm{dist}(K,\partial\Omega^0)$.

Finally, from the above results and the estimate (2.6) for r^0 we deduce the following convergence theorem:

Theorem 2.3. *Let* $\gamma^{\alpha\beta(\nu)}$ *and* $g_{ij}^{(\nu)}$ *be two sequences of coefficients satisfying* (1.2) *and suppose that* $\gamma^{(\nu)}$ *converge to* γ *uniformly on compact subsets of* Ω, *whereas* $g^{(\nu)}$ *converge to* g *on compact subsets of* $\Omega \times R^N$.

Let $u^{(\nu)}$ *be a sequence of minimizers of the corresponding functional*

$$E_\nu(w,\Omega) = \int_\Omega \gamma^{\alpha\beta(\nu)}(x)\, g_{ij}^{(\nu)}(x,w) D_\alpha w^i D_\beta w^j\, dx \quad ,$$

and suppose that $u^{(\nu)}$ *converge weakly in* L^2_{loc} *to some function* u.

Then u *minimizes the limit functional*

$$E(w,\Omega) = \int_\Omega \gamma^{\alpha\beta}(x) g_{ij}(x,w) D_\alpha w^i D_\beta w^j\, dx \quad .$$

Moreover, if for each ν, x_ν *is a singular point for* $u^{(\nu)}$ *and* $x_\nu \to x^0 \in \Omega$, *then* x^0 *is a singular point for* u. *Finally, if the coefficients are equibounded in* $C^{k,\alpha}$, *and* K *is a compact set contained in the regular set* Ω^0 *of* u, *then* $u \to u$ *in* $C^{k+1,\alpha}(K)$.

3. Functionals with splitting coefficients

The results of the preceding paragraph hold more generally for quadratic functionals of the type

$$\int_\Omega A_{ij}^{\alpha\beta}(x,u) D_\alpha u^i D_\beta u^j\, dx$$

with coefficients verifying the condition

$$A_{ij}^{\alpha\beta}(x,u)\xi_\alpha^i \xi_\beta^j \geq |\xi|^2 \quad . \tag{3.1}$$

Actually, the original proofs dealt with such functionals ([3], [4], [8]); we have shown here the simple modifications necessary to carry over the proofs in our case, when (3.1) is replaced by (1.2).

In this section we exploit the special form of the coefficients, and we obtain better estimates for the set $\Sigma = \Omega - \Omega^0$. These are based on a monotonicity formula, similar to that of minimal surfaces, which is interesting in itself.

We assume as usual that the coefficients γ and g are bounded and uniformly continuous. Since we will consider only neighborhoods of a fixed point x^0, we shall assume that it is the origin of the coordinates; moreover as in lemma 2.1 we will suppose that

$$\gamma^{\alpha\beta}(x) = \delta^{\alpha\beta} + \sigma^{\alpha\beta}(x)$$

with $|\sigma^{\alpha\beta}(x)| \leq \sigma(|x|)$, and

$$\int_0^1 t^{-1}\sigma(t)dt < +\infty \quad .$$

Lemma 3.1. [4] *Let* u *be a minimizer of the functional* E, *and suppose that the above assumptions hold. Then for every sufficiently small* r *and* R, *with* $0<r<R$, *we have*

$$\int_{\partial B} |u(Rx)-u(rx)|^2 dH_{n-1} \le C \log(R/r)\left[\Phi(R)-\Phi(r)\right] \tag{3.2}$$

where

$$\Phi(t) = t^{2-n} \exp\left[C\int_0^t s^{-1}\,\sigma(s)dx\right] E(u,B_t) \ .$$

With the help of the above lemma we can prove the first result:

Theorem 3.1. *Let* $n=3$, *and let* u *be a bounded minimizer of* $E(u,\Omega)$. *Then* u *can have at most isolated singularities in* Ω.

Proof. Suppose on the contrary that a sequence of singular points x_ν converges to a point in Ω, that we assume to be the origin. It is clear from theorem 2.3 that 0 is singular. Let $R_\nu = 2\,|\,x_\nu\,|$, and let

$$u^{(\nu)}(x) = u(R_\nu x) \ .$$

The function $u^{(\nu)}$ minimizes a suitable functional E_ν, whose coefficients converge uniformly to $\delta^{\alpha\beta} g_{ij}(0,u)$ in compact subsets of $\Omega \times R^N$. Moreover, each $u^{(\nu)}$ has a singular point y_ν, with $|\,y_\nu\,| = 1/2$. Since u is bounded, the sequence $u^{(\nu)}$ is bounded in L^2_{loc}, and therefore we may assume (passing possibly to a subsequence) that $u^{(\nu)}$ converges weakly in L^2_{loc} to a function v. In addition, we may suppose that $y_\nu \to y^0$.

The limit function v is a minimizer of the limit functional, and y^0 is a singular point for v.

On the other hand, if we write the estimate (3.2) with $R = bR_\nu$ and $r = aR_\nu$ $(0<a<b)$, we obtain

$$\int_B |u^{(\nu)}(ax)-u^{(\nu)}(bx)|^2 dH_{n-1} \le C \log(b/a)\left[\Phi(bR_\nu)-\Phi(aR_\nu)\right] . \tag{3.3}$$

The estimate (3.2) implies that $\Phi(t)$ is an increasing function, and therefore it has a finite limit as $t \to 0$. In addition, Caccioppoli's inequality (2.1) implies that $u^{(\nu)} \to v$ strongly in L^2_{loc}. Passing to the limit in (3.3) we obtain

$$\int_B |v(ax)-v(bx)|^2 dH_{n-1} = 0$$

for almost every $a<b$. That means that $v(x)$ is homogeneous of degree zero.

Since y^0 is a singular point for v, the whole segment joining y^0 with the origin should then be made of singular points of v. But this contradicts the conclusion of theorem 2.1 (in particular the estimate (2.7)), since the singular set Σ would have dimension equal to 1.

When the dimension of the set Ω is greater than 3, we have the following result:

Theorem 3.2. [4] *Let* u *be a minimizer of the functional* E. *Then the dimension of its singular set* $\Omega - \Omega^0$ *cannot exceed* $n-3$.

The proof of theorem 3.2 follows the same ideas of the previous one, but is rather technical. We remark that both proofs are adaptations of similar arguments used in the theory of minimal surfaces. For details, see [7], chapter 11.

The analogy with minimal surfaces extends to boundary regularity. For the Dirichlet problem, Jost and Meier [10] have proved that every minimizer u is regular in a suitable neighborhood of the boundary.

4. Harmonic maps into a sphere

We have already remarked that functionals of the form (1.1) describe in local coordinates the energy of a map u between Riemannian manifolds. Of particular interest is the case of a map from a disk D^r into a sphere $S^N \subset R^{N+1}$, $n \le N$.

If the map u omits a neighborhood of the south pole, and if we use the stereographic coordinates on S^N, the energy is given by:

$$F(u,D) = \int_D |Du|^2 \left(1 + |u|^2\right)^{-2} dx \quad . \tag{4.1}$$

The fact that the Gauss map from a minimal surface into the unit sphere is harmonic explains the similarity of methods and of results between the two theories (minimal hypersurfaces and harmonic mappings), a similarity particularly evident when the target manifold is the unit sphere.

The functional (4.1) admits the "equator map" $u = x/|x|$ as a stationary point (more precisely, $u^\alpha = x^\alpha/|x|$ for $\alpha = 1,2,\dots,n$, and $u^\alpha = 0$ for $\alpha > n$). We have:

Theorem 4.1. [9] *The equator map is energy minimizing if* $n \ge 7$, *whereas it is unstable (and hence it is not a minimizer) for* $n < 7$.

As a consequence of the second part of the theorem, we can improve in the case under consideration the conclusions of theorems 3.1 and 3.2.

Theorem 4.3. [6], [13] *Let* u *be a minimizer of the energy functional* (4.1). *Then:*

i) *if* $n = 7$ *it can show at most isolated singularities;*

ii) *if* $n > 7$ *the dimension of its singular set cannot exceed* $n - 7$.

The proof of these results proceeds as in section 3, and it is based on the fact that the functional (3.1), or better the quantity

$$R^{2-n}\, F(u,B_R)$$

is invariant under a change of scale of the independent variables.

The dimension 7 occurring in theorem 4.3 depends strongly on the fact that the target manifold is a sphere. If we change it into an ellipsoid:

$$S_a = \left\{ u \in R^{N+1} : u_1^2 + u_2^2 + \dots + u_{N-1}^2 + u_N^2 / a^2 = 1 \right\}$$

then the equator map is stable if and only if [1]:

(4.2) $$4(n-1) \le a^2 (n-2)^2 \quad .$$

In particular, if a is close to 1 we have regularity in dimension $n \le 6$, but when a is large we may have singularities already in dimension 3. It is not clear whether the equator map is energy minimizing if (4.2) is satisfied, except of course when $a = 1$.

5. The behaviour of minimizers at singular points

The last topic of this paper concerns the behaviour of minimizers in the vicinity of singular points. Here the results are less complete, and concern only the least dimension in which the singularities may appear: $n = 3$ for general functionals (1.1), $n = 7$ for the energy functional F defined in (4.1).

A common feature of minimizers in the lowest singular dimension is that singularities may occur only at isolated points. This fact makes possible to describe quite precisely the behaviour of the minimizing function $u(x)$ as x approaches to a singular point. We have:

Theorem 5.1. [8] *Let* $n = 3$ *and let the coefficients* γ *and* g *be of class* $C^{k,\alpha}$. *Let* u *be a bounded minimizer of the functional*

$$E(u,\Omega) = \int_\Omega \gamma^{\alpha\beta}(x) g_{ij}(x,u) D_\alpha u^i D_\beta u^j \, dx$$

and let x^0 *be a singular point for* u, *necessarily isolated. Then for every* $h \le k+1$ *we have:*

$$\limsup_{x \to x^0} |x - x^0|^h |D^h u(x)| < +\infty .$$

The same result holds in dimension 7 for minimizers of F(u,D), and for minimal hypersurfaces in R^8. In this case we have:

$$\limsup_{x \to x^0} |x - x^0|^m |\delta^m v(x)| < +\infty$$

for every integer m, where v is the unit normal to the surface, and δ denotes the tangential gradient.

It is not known whether similar results hold in higher dimensions. Here the situation is complicated from the fact that the homogeneous minimizer resulting from the blow-up procedure described in section 3 might have more singularities than the original minimizer. However, some results can be obtained for x converging to x^0 outside the cone of singularities of the limit function.

I shall conclude this paper by the discussion of an open problem, connected with the Bernstein problem for minimal graphs. Let $u : R^n \to S^N$ be an energy-minimizing harmonic map, and suppose that the image of u be contained in the upper hemisphere $u^N > 0$.

Then u is constant, provided either $n < 7$ or $u(R^n)$ stays away from the equator.

It is likely that non-constant solutions exist for $n \geq 7$, but no example is known. Of course, if we accept that the image of u be contained in the closed hemisphere, then the equatorial map will do. However, this is not a good example, since it corresponds (in the minimal surface case) to the cylinder over a minimal cone.

A result of this sort would strengthen the ties between minimal surfaces and harmonic maps.

References

[1] **A. Baldes,** Stability and Uniqueness Properties of the Equator Map from a Ball into an Ellipsoid, preprint.

[2] **F.W. Gehring,** The L^p-integrability of the Partial Derivatives of a Quasi-conformal Mapping, Acta Math. 130 (1973), 265-277.

[3] **M. Giaquinta and E. Giusti,** On the Regularity of the Minima of Variational Integrals, Acta Math. 148 (1982), 31-46.

[4] **M. Giaquinta and E. Giusti,** The Singular Set of the Minima of Certain Quadratic Functionals, Ann. Scuola Norm. Sup. Pisa (IV) 11 (1984), 45-55.

[5] **M. Giaquinta and G. Modica,** Regularity Results for Some Classes of Higher Order Nonlinear Elliptic Systems, J. Reine und Ang. Math. 311/312 (1979), 145-169.

[6] **M. Giaquinta and J. Soucek,** Harmonic Maps into a Hemisphere, Ann. Scuola Norm. Sup. Pisa (IV) 12 (1985), 81-90.

[7] **E. Giusti,** Minimal Surfaces and Functions of Bounded Variation, Birkhauser Basel 1984.

[8] **E. Giusti,** On the Behaviour of the Derivatives of Minimizers near Singular Points, to appear in Arch. Rat. Mech. & Anal.

[9] **W. Jager and H. Kaul,** Rotationally Symmetric Harmonic Maps from a Ball into a Sphere and the Regularity Problem for Weak Solutions of Elliptic Systems, J. Reine und Ang. Math. 343 (1983), 146-161.

[10] **J. Jost and M. Meier,** Boundary Regularity for Minima of Certain Quadratic Functionals, Math. Ann. 262 (1983), 549-561.

[11] **R. Schoen and K. Uhlenbeck,** A Regularity Theory for Harmonic Maps, J. Diff. Geo. 17 (1982), 308-335.

[12] **R. Schoen and K. Uhlenbeck,** Boundary Regularity and Miscellaneous Results on Harmonic Maps, preprint.

[13] **R. Schoen and K. Uhlenbeck,** Regularity of Minimizing Harmonic Maps into the Sphere, Inv. Math. 78 (1984), 89-100.

MINIMAL SURFACES OF FINITE INDEX IN MANIFOLDS OF POSITIVE SCALAR CURVATURE

Robert Gulliver
School of Mathematics
University of Minnesota
127 Vincent Hall, 206 Church Street S.E.
Minneapolis, MN 55455 (U.S.A.)

Dedicated to Hans Lewy

In the study of minimal surfaces, it is of particular interest to understand relations between *geometric* and *analytical* properties of the surface. A particularly striking example concerns the Jacobi operator $L = -\Delta + 2K$ on a parametric minimal surface M^2 in $\mathbf{R}^3$, which governs the second variation of area. It turns out that M has finite index, that is, L has only finitely many negative eigenvalues, if and only if M has finite total Gauss curvature (Theorem 1 below). In the more general context of an ambient Riemannian manifold N^3 having scalar curvature $S \geq 0$, we show in Theorem 2 that the norm squared of the second fundamental form of M must be integrable, a conclusion which reduces to finite total Gauss curvature when N is flat.

As a precise definition, the *index* of M is the maximum number of negative eigenvalues of the elliptic operator controlling the second variation of area, with Dirichlet boundary conditions, on compact subsets of M. In particular, M is *stable* if it has index zero. For an immersed minimal surface M^2 in a Riemannian manifold N^3, the first variation of the area of M vanishes for all compactly supported variations, by definition of minimality. Let a minimal surface M be varied to form nearby surfaces M_ε at distance $\varepsilon f(x)$ along the unit normal vector $\nu(x)$ to M, where $f : M \to \mathbf{R}$ has compact support. The *second variation* of area is

$$(1) \qquad d^2A(M_\varepsilon)/d\varepsilon^2\big|_0 = \int_M \Big(|\nabla f|^2 - \big(\mathrm{Ric}(\nu) + |B|^2\big)f^2 \Big)\, dA \ .$$

Here Ric denotes the Ricci curvature of N, B is the second fundamental form of M in N, and dA is the area integrand of M in the metric induced from N([D], p. 286). In the special case $N = \mathbf{R}^3$, we have $\mathrm{Ric}(\nu) = 0$ and $|B|^2 = -2K$, so that the second variation becomes $\int_M f\, Lf\, dA$, with $L = -\Delta + 2K$, where K is the Gauss curvature.

The following theorem was first proved in [G]. A quite independent proof appeared shortly thereafter in [F-C], along with significant generalizations to positively curved N^3. Although Theorem 1 is a logical consequence of Theorem 2 below, we shall first present the complete proof in this simpler case, in order to allow the ideas to exhibit themselves more clearly.

Theorem 1. *Let* M^2 *be a complete immersed minimal surface in* $\mathbf{R}^3$. *If* M *has finite index, then it has finite total curvature, and conversely.*

If a minimal surface M in $\mathbf{R}^3$ has finite total curvature, then it follows that M has quite simple geometric structure. In fact, Osserman has shown that the Gauss map $g : M^2 \to S^2$ extends to a branched conformal mapping $\hat{g} : \hat{M} \to S^2$, where $\hat{M}$ is a compact Riemann surface and M is conformally identified with $\hat{M}$ minus a finite set (cf. [O], p. 82). In particular, M has finite topological type and a well-defined asymptotic normal direction at each end. Moreover, Schoen has shown that each embedded end is uniformly approximated either by a catenoid or a plane ([S], p. 801). This implies, for example, that M is the union of a finite number of overlapping graphs.

The *converse* of Theorem 1 may be proved using Osserman's theorem. In fact, under the Gauss map g, the Jacobi operator $L = -\Delta + 2K$ corresponds to the operator $-\hat{\Delta} - 2$ on S^2, where $\hat{\Delta}$ is the spherical Laplacian. Let the spherical metric be pulled back to a singular metric on $\hat{M}$, and write $\hat{L} = -\hat{\Delta} - 2$. Using an argument of Lewy ([L], pp. 39f.), one shows that on a small neighborhood of each singular point, the degenerate operator $\hat{L}$ has a positive first eigenvalue. It follows via well-known continuation arguments that $\hat{L}$ has finite index on $\hat{M}$ (see [G], pp. 209f.). But this is exactly the index of the minimal surface M.

The case of a minimal surface in $\mathbf{R}^3$ has somewhat special character. For example the Gauss curvature K is never positive, and moreover $K = -1/2\,|B|^2$. The intrinsic condition that M have finite total Gauss curvature therefore coincides with the natural, extrinsic condition that $|B|^2$ be integrable. For a general ambient manifold N^3, this extrinsic condition appears to be the more relevant of the two.

Theorem 2. *Let* M *be a complete immersed minimal surface of finite index in a real-analytic manifold* N^3 *of scalar curvature* $S \geq 0$. *Then* M *has quadratic area growth, finite topological type and* $\int_M |B|^2 dA < \infty$.

A result related to Theorem 2 was given by Doris Fischer-Colbrie in [F-C]. She showed under these hypotheses that M is conformally equivalent to a compact Riemann surface minus a finite set. She also proved that $|B|^2$ is integrable under the stronger hypothesis that $\mathrm{Ric} \geq 0$. It might be noted that the intrinsic condition $\int_M |K| dA < \infty$ may easily fail if N has negative scalar curvature. For example, if N is the Riemannian product $M \times \mathbf{R}$, then $M \times \{0\}$ is totally geodesic and minimizes area; but M may be chosen with $K < 0$ and $\int_M K\, dA = -\infty$. It is also of interest to compare the conclusion that the area of M has *at most* quadratic growth with a consequence of the well-known monotonicity lemma: if N^n has *non-positive* sectional curvatures, then any stationary minimal surface has area *at least* πR^2 in the ball of radius R.

The theorems are based on four fundamental lemmas:

Lemma 1. *If* M *has finite index for an operator* $L = -\Delta + q(x)$, *then for some compact* $\Omega \subset M$, L *is positive semi-definite on* $M \setminus \Omega$.

Proof. Choose Ω with index equal to the index k of M. If L is not ≥ 0 on $M \setminus \Omega$, then there is a function v with $\int v\, Lv\, dA < 0$, having compact support F disjoint from Ω. Therefore, v is orthogonal to the k independent eigenfunctions with negative eigenvalues and support Ω. It follows that $\Omega \cup F$ has index $\geq k+1$, a contradiction.

Definitions. For Lemmas 2 and 3, we assume the metric of M is *real analytic.* Choose a point $x_0 \in M$, and let the intrinsic balls $B_r = B_r(x_0)$. Write $r(x)$ for the intrinsic distance from x_0 to x. Let $d\sigma$ be the element of arc length along ∂B_R, and define $L(R) := \int_{\partial B_R} d\sigma$, the *length* of ∂B_R; $\overline{K}(R) = \int_{\partial B_R} K\, d\sigma$; and $\Gamma(R) := \int_{\partial B_R} k_g\, d\sigma$, where the *inward geodesic curvature* k_g is understood to include a distributional term which contributes the sum of exterior angles to $\Gamma(R)$.

Lemma 2. $\overline{K}(R) = d[2\pi\chi(B_R) - \Gamma(R)]/dR$, *as a distributional derivative.*

Proof. The Gauss-Bonnet formula for B_R yields

$$\int_0^R \overline{K}(r)dr + \Gamma(R) = 2\pi\,\chi(B_R) \quad .$$

Lemma 3. $\Gamma(R) \geq dL(R)/dR$, *as a distributional derivative.*

Proof. We may compute $dL(R)/dR = \int_{\partial B_R} \Delta r\, d\sigma$. Now wherever r is smooth, we have

$k_g = \Delta r$. Since r is a subanalytic function, it is smooth except on a one-dimensional semi-analytic set S, the cut locus of x_0. Further ∇r has exactly two limiting values on $S \setminus F$, for some discrete set F (*cf.* [H], pp. 62f.). For $x \in S\setminus F$, define $0 < 2\theta(x) \leq \pi$ to be the angle formed by the two limiting values of ∇r at x. Then ∂B_R has exterior angle $-2\theta(x)$ at x, and $\Delta r\, dA$ has the singular part $-2 \sin\theta\, dS$, where dS is linear Hausdorff measure restricted to S. Meanwhile, $k_g\, dA$ has singular part equal to $-2\theta\cos\theta\, dS$. But $\theta\cos\theta \leq \sin\theta$ for $0 \leq \theta \leq \pi/2$, and Lemma 3 follows.

Lemma 4. *The Euler characteristic* $\chi(B_r) \leq 1 - \tilde{b}_1(B_r) \leq 1$, *where* $\tilde{b}_1(B_r)$ *is the rank of the homology homomorphism* $H_1(B_r) \to H_1(M)$ *induced by inclusion, and is monotone increasing in* r.

Proof. B_r is connected and noncompact, so the Betti numbers $b_0(B_r) = 1$, $b_2(B_r) = 0$. The relation $\tilde{b}_1(B_r) \leq b_1(B_r)$ is clear.

Proof of Theorem 1. Choose $a > 0$ large enough so that the compact set Ω of Lemma 1 lies inside B_a. Then for any function f with compact support disjoint from B_a, we have

$$\int_M \left(|\nabla f|^2 + 2Kf^2 \right) dA \geq 0 \; .$$

For greater generality, we shall replace the coefficient 2 by a constant $\beta > 1/2$: we assume that for all such f,

$$\int_M \left(|\nabla f|^2 + \beta K f^2 \right) dA \geq 0 \; . \tag{2}$$

Given $T > 0$, we choose $f(x) = \phi(r(x))$ where $\phi(r)$ vanishes on $(-\infty, a]$ and on $[a+T+2, \infty)$; $\phi(r) = 1$ on $[a+1, a+2]$; and ϕ is linear on each of the intervals $[a, a+1]$ and $[a+2, a+T+2]$. Note that $|\nabla r| = 1$ almost everywhere, so that $|\nabla f|^2 = (\phi')^2$. Observe that f is independent of T on B_{a+1}; we write

$$C_1 = \int_{B_{a+1}} \left(|\nabla f|^2 + \beta K f^2 \right) dA \; .$$

Inequality (2) becomes (integrals w.r.t. r from $a+1$ to $a+T+2$):

$$-C_1 \leq \int_{M \setminus B_{a+1}} \left(|\nabla f|^2 + \beta K f^2 \right) dA = \int \left(\left(\phi'(r)\right)^2 L(r) + \beta \bar{K}(r) \phi(r)^2 \right) dr \; . \tag{3}$$

After the use of Lemma 2 in an integration by parts, this becomes

$$-C_1 \le -\beta\Big[2\pi\chi(B_{a+1}) - \Gamma(a+1)\Big] + \int \Big((\phi')^2 L - 2\beta\phi\phi'\big[2\pi\chi - \Gamma\big]\Big)\, dr \ .$$

Note that $\phi' \le 0$ on $[a+1, a+T+2]$, so that Lemmas 3 and 4 yield

$$(4) \quad -C_2 \le \int \Big((\phi')^2 L - 2\beta\phi\phi'\big[2\pi - 2\pi\tilde{b}_1(B_r) - L'(r)\big]\Big)\, dr = 2\pi\beta\left(1 + \int 2\phi\phi'\tilde{b}_1(B_r)\, dr\right) +$$

$$+ \int \Big[(\phi')^2 L - 2\beta L\big(\phi\phi'' + (\phi')^2\big)\Big]\, dr \le 2\pi\beta + T^{-1} 2\beta L(a+2) - T^{-2}(2\beta - 1)A(B_{a+T+2} \backslash B_{a+2}) \ .$$

In particular, since we have assumed $\beta > 1/2$, for T sufficiently large we obtain quadratic area growth:

$$(5) \qquad A(B_{a+T+2}) \le C_3 T^2 \ .$$

Returning to inequality (3) and recalling $\phi(r) \ge 1/2$ for $a+1 \le r \le a+2+T/2$, we find (since $K \le 0$)

$$-\int_{a+1}^{a+2+T/2} \bar{K}(r)dr \le \beta(C_1 + C_3)/4 \ ,$$

from which finite total curvature follows. q.e.d.

We should note that the inequality of Lemma 3 has strong consequences. For example, we may prove a stronger form of the well-known theorem of Cohn-Vossen ([C-V], p. 79). Cohn-Vossen's theorem requires the hypotheses that $\int_M |K| dA$ be finite and that M have finite topological type, which are both substantially weakened in the following corollary. We would like to thank Rick Schoen for pointing this out to us.

Corollary. *Let* M *be a connected, complete, real analytic two-dimensional Riemannian manifold, such that the principal-value integral*

$$\text{p.v.} \int_M K\, dA := \lim_{R\to\infty} \int_{B_R} K\, dA$$

exists, possibly $\pm\infty$. *Then*

$$\text{p.v.} \int_M K\, dA \le \limsup_{R\to\infty} 2\pi\chi(B_R) \ .$$

Proof. From the Gauss-Bonnet formula and Lemma 3, we have

$$\int_{B_r} K\, dA = 2\pi\chi(B_r) - \Gamma(r) \le 2\pi\chi(B_r) - dL(r)/dr \quad .$$

Integration from 0 to R yields

$$0 \le L(R)/R \le R^{-1}\int_0^R \left[2\pi\chi(B_r) - \int_{B_r} K\, dA\right] dr$$

which is dominated by lim sup $2\pi\chi(B_R) - \text{p.v.} \int K\, dA$. q.e.d.

In order to relate the hypothesis of finite index to an intrinsic analytical property, we shall rewrite the zero-order coefficient in the second-variation formula (1), as was done in [SY]. Since $b_{11} + b_{22} = 0$, the Gauss equation yields

$$K = K_{12} + b_{11}b_{22} - b_{12}^2 = K_{12} - \frac{1}{2}|B|^2 \quad ,$$

where K_{12} is the ambient sectional curvature in the tangent plane to M, (e_1,e_2) is an orthonormal basis of tangent vectors to M and $b_{ij} = B(e_i,e_j)$. Hence

$$\text{Ric}(v) + |B|^2 = \left(\text{Ric}(v) + K_{12}\right) + \left(\frac{1}{2}|B|^2 - K_{12}\right) + \frac{1}{2}|B|^2 = S - K + \frac{1}{2}|B|^2 \quad .$$

In particular, if $S \ge 0$ then the second variation

$$(6) \qquad d^2A/d\varepsilon^2 \le \int_M \left(|\nabla f|^2 + Kf^2 - \frac{1}{2}|B|^2 f^2\right) dA \quad .$$

Proof of Theorem 2. Applying Lemma 1 and inequality (6) we have, for every f supported in $M \setminus B_a \subset M \setminus \Omega$.

$$0 \le \int_M \left(|\nabla f|^2 + Kf^2\right) dA \quad ,$$

which is inequality (2) for the constant $\beta = 1$. As in the proof of Theorem 1, this implies quadratic area growth (5). At the same time, the term

$$\int_{a+1}^{a+T+2} (-2\phi\phi')\tilde{b}_1(B_r)dr \quad ,$$

which was dropped from inequality (4), is seen to be bounded above by $2 + C_2/2\pi$ for large T. But $\tilde{b}_1(B_r)$ is monotone increasing in r and $\phi' \le 0$, so this integral is $\ge \tilde{b}_1(B_{a+2+T/2})/4$, which shows that $\tilde{b}_1(B_R)$ is bounded. For a.a. R, ∂B_R is smooth, and $b_1(B_R) + 1$ equals twice the genus of B_R plus the number of its boundary components. For each end of M and for large R, there is one boundary component of B_R which is not a boundary in M, and thus contributes to $\tilde{b}_1(B_R)$. Therefore, M has only a finite number of ends. Further, the genus of B_R is bounded as $R \to \infty$, which implies that M has finite topological type.

Returning to the proof of Theorem 1, we see that the steps between inequalities (3) and (4) show (for any real-analytic two-manifold):

$$\int_{B_{a+1}} K dA + \int_{a+1}^{a+T+2} \phi(r)^2 \overline{K}(r) dr \le 2\pi + 2L(a+2)/T - 2A(B_{a+T+2} \setminus B_{a+2})/T^2 .$$

Inequality (6), used now in its full strength, implies

$$\frac{1}{2}\int_M |B|^2 f^2 \, dA \le C_1 + A(B_{a+T+2} \setminus B_{a+2})/T^2 + \int_{a+1}^{a+T+2} \phi^2 \overline{K} \, dr ,$$

from which $\int_M |B|^2 dA < \infty$ follows as $T \to \infty$. q.e.d.

Acknowledgement. We would like to thank the Sonderforschungsbereich 72 at Bonn for extending its hospitality.

References

[C-V] **S. Cohn-Vossen,** Kürzeste Wege und Totalkrümmung auf Flächen, Compositio Math. 2 (1935), 69-133.

[D] **A. Duschek,** Zur geometrischen Variationsrechnung, Math. Z. 40 (1936), 279-291.

[F-C] **D. Fischer-Colbrie,** On complete minimal surfaces with finite Morse index, Inventiones Math. 82 (1985), 121-132.

[G] **R. Gulliver,** Index and total curvature of complete minimal surfaces, Proc. Symp. Pure Math. 44 (1986), 207-212.

[GL] **R. Gulliver and H.B. Lawson,** The structure of stable minimal hypersurfaces near a singularity, Proc. Symp. Pure Math. 44 (1986), 213-237.

[H] **R. Hardt,** Topological properties of subanalytic sets, Trans. Amer. Math. Soc. 211 (1975), 57-70.

[L] **H. Lewy,** Aspects of the Calculus of Variations, University of California Press, Berkeley 1939.

[O] **R. Osserman,** A survey of minimal surfaces, Van Nostrand-Reinhold, New York 1969.

[S] **R. Schoen,** Uniqueness, symmetry and embeddedness of minimal surfaces, J. Differential Geom. 18 (1983), 791-809.

[SY] **R. Schoen and S.-T. Yau,** Existence of incompressible minimal surfaces and the topology of three-dimensional manifolds with non-negative scalar curvature, Annals of Math. 110 (1979), 127-142.

REMARKS ABOUT THE MATHEMATICAL THEORY OF LIQUID CRYSTALS

Robert Hardt, David Kinderlehrer, and Mitchell Luskin
School of Mathematics - University of Minnesota
206 Church Street S.E.
Minneapolis - Minnesota 55455 (U.S.A.)

Dedicated to Hans Lewy

. Liquid crystals and liquid crystal theory

A liquid crystal is a mesomorphic phase of a material which occurs between its liquid and solid hases. Frequently the material is composed of rod like molecules which display orientational order, unlike liquid, but lacking the lattice structure of a solid. It may flow easily and so may also be thought of as an nisotropic fluid. This anisotropy is evident in the way it transmits light; for example, a nematic liquid rystal is optically uniaxial. We take this opportunity to discuss a few of the analytical and computational ssues which arise in the attempt to study static equilibrium configurations. One attractive feature of this ubject is that it has a well developed continuum description in the Ericksen-Leslie theory [9].

Some of the questions have significance in the context of harmonic mappings into the sphere and we hall attempt to clarify these connections. Indeed, at this point in time, we are able to answer questions bout harmonic mappings which we had not conceived of asking several years ago. This is especially true f the study of singularities, cf. sect. 4 and the recent work of Brezis, Coron, and Lieb [2], [3] and the uthors [18].

A general exposition of the static theory is given in Ericksen [9]. The flow theory is developed in ,eslie [22], cf. also [23]. Accessible books about this subject have been written by de Gennes [8] and :handresekhar [6]. What we describe here is primarily work with F.-H. Lin [15],[16]. The numerical spects have been studied with R. Cohen and S. Y. Lin [7]. A more complete exposition of the topics iscussed here may be found in [14]. The volume [11] is a source of contemporary issues in liquid crystal esearch. A particular experiment which has served as a focus for us and to which we have returned requently to test our ideas and our results is related by Williams, Pieranski, and Cladis [27]. This is also described in the expository article of Brinkman and Cladis [4]. A retrospective view of the issues posed I this experiment is given in §6.

The kinematic variable in the nematic and cholesteric phases, which are our concern, may be take to be the optic axis, which is a unit vector field n defined in the region Ω occupied by the material. T bulk energy at fixed temperature is given by the Oseen-Frank density

$$W(\nabla n,n) = \tfrac{1}{2}\kappa_1(\operatorname{div} n)^2 + \tfrac{1}{2}\kappa_2(n\cdot\operatorname{curl} n + q)^2 + \tfrac{1}{2}\kappa_3(n \wedge \operatorname{curl} n)^2 \qquad (1.1)$$

$$+ \tfrac{1}{2}\alpha(\operatorname{tr}(\nabla n)^2 - (\operatorname{div} n)^2),$$

where the constants $\kappa_i > 0, i = 1,2,3$, q and α are real. A static equililibrium configuration corresponds to an extremal of the functional defined by W, namely,

$$\delta \int_\Omega W(\nabla n,n)\,dx = 0\,. \qquad (1.2)$$

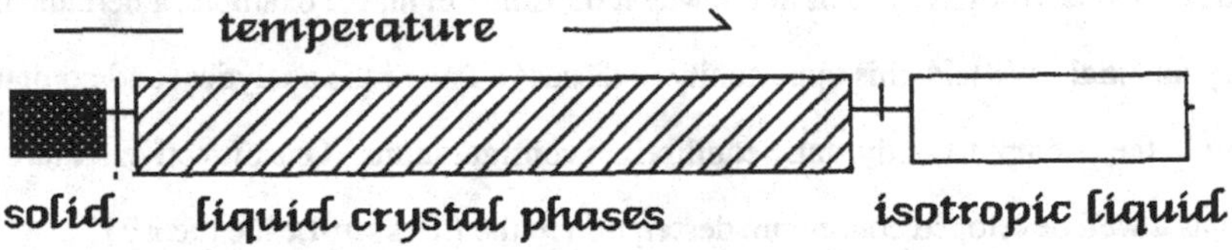

(figure)

Another means of understanding the differences between ordinary liquids and liquid crystals may I achieved by examining the stress of the material as a function of temperature, (figure). In the high temperature isotropic phase,

$$T = -p\mathbf{1}$$

where p is a constant, whereas in the liquid crystal phases T has a molecular component,

$$T = -p\mathbf{1} + T_M$$

For a nematic or cholesteric,

$$T_M = W\mathbf{1} - \left(\partial W/\partial \nabla n\right)\nabla n^T,$$

cf. [9]. For a solid, the stress is rather complicated.

With (1.2) in mind, it is reasonable to seek equilibrium configurations by a variational principle. A typical situation, called *strong anchoring* in the literature, is to prescribe the optic axis on the boundary

$\partial\Omega$ of the region $\Omega \subset \mathbb{R}^3$ occupied by the fluid. So let n_0 be defined on $\partial\Omega$ with $|n_0| = 1$ and

$$\mathcal{A}(n_0) = \{ u \in H^1(\Omega;\mathbb{R}^3) : |u| = 1 \text{ in } \Omega \text{ and } u = n_0 \text{ on } \partial\Omega \}. \quad (1.3)$$

Problem. *Find*

$$n \in \mathcal{A}(n_0) : \int_\Omega W(\nabla n, n)\, dx = \inf_{\mathcal{A}(n_0)} \int_\Omega W(\nabla u, u)\, dx . \quad (1.4)$$

We have been rather glib in our choice of admissible functions. It is not obvious that $\mathcal{A}(n_0)$ is not empty nor that W is coercive in a suitable sense. We return to this in section 2. In as much as the variational problem (1.4) is constrained, some observations may be in order.

1. In the special case

$$\kappa_1 = \kappa_2 = \kappa_3 = \alpha = 1 \text{ and } q = 0,$$

$$W = \tfrac{1}{2} |\nabla n|^2 , \quad (1.5)$$

which is the integrand for a harmonic mapping of Ω into S^2. The equilibrium equation here is the system

$$\Delta n + |\nabla n|^2 n = 0 \text{ in } \Omega, \quad (1.6)$$

which exhibits quadratic growth in $|\nabla n|$. This is a known peril in the study of elliptic systems, [5],[13].

2. The constraint on n seems to be very mild considering those imposed on the determinant of a deformation gradient in finite elasticity, but it leads to an equilibrium equation which, unlike (1.6), is not always elliptic. This is because the Lagrange multiplier generally involves second derivatives. The implications of this are not well understood. However, it seems evident that continuity methods will be of little use to find and analyze solutions.

3. When the degree of $n_0 : \partial\Omega \to S^2$ differs from zero, it does not admit any smooth extensions to Ω. Thus any element of $\mathcal{A}(n_0)$ must have singularities. Singularities are called *defects* in the liquid crystal literature. They give insight into the behavior of the material.

4. There can be many solutions for given boundary data. Circumstances which ensure uniqueness are known only in special cases.

5. Other fields may be imposed, especially magnetic or electric ones, cf. Ericksen [9], de Gennes [8],or [14],[16] for a mathematical treatment.

2. Existence

To begin our discussion we state

Proposition 2.1. *If* $\partial\Omega$ *is Lipschitz and* n_0 *is Lipschitz, then* $A(n_0) \neq \emptyset$.

The proof of this is given in [16], Lemma 1.1. A more complete, if more technical discussion, of this poir will be given later.

In order to have a reasonable theory, we would prefer that a solution $n \in H^1(\Omega;S^2)$. Are there aı conditions for which

$$W(\nabla u,u) \geq \text{const.} |\nabla u|^2 \ ? \tag{2.1}$$

If such a constant exists in a class of functions, the boundary conditions of this class must play a role. The unconstrained functional is not coercive on $H^1(\Omega;S^2)$. Brief consideration of the cases $\alpha = 0$ and α very large serve to convince one of this. To resolve this issue, observe that

$$\begin{aligned} S(\nabla u) &= \tfrac{1}{2}(\mathrm{tr}\,(\nabla u)^2 - (\mathrm{div}\, u)^2) \\ &= \tfrac{1}{2}\,\mathrm{div}(\nabla u\, u - \mathrm{div}\, u\ u) \end{aligned} \tag{2.2}$$

is a divergence. Indeed, as observed by Ericksen [10] and Oseen, it is a surface energy in the sense that

$$\int_\Omega S(\nabla u)\, dx = \int_{\partial\Omega} f(\nabla_{tan} u,u)\, dx = \int_{\partial\Omega} f(\nabla_{tan} n_0,n_0)\, dx$$

depends only on n_0 for $u \in A(n_0)$. Obviously,

$$-\,\mathrm{div}\ S_p(\nabla u) = 0 \quad \text{in } \Omega \quad \text{for any } u,$$

so this term offers no contribution to the equilibrium equations. For this reason it is often dropped and, in fact, there are no known ways to measure the constant α experimentally. For the same motive, it may be added to W. Choosing

$$\alpha = \min \{K_1, K_2, K_3\} > 0,$$

it is elementary to calculate ([16], Lemma 1.4) that

$$W(\nabla n,n) \geq \tfrac{1}{2}\alpha |\nabla n|^2 .$$

Thus by addition of a constant to W, it becomes coercive on $A(n_0)$. We assume this to be done, whic leads to the

Theorem 2.2. *There is at least one* $n_0 \in A(n_0)$ *satisfying* (1.4).

3. Examples

If Ω is diffeomorphic to a ball and the degree of the data $n_0 : \partial\Omega \to S^2$ is not zero, then an: solution must have singularities. Moreover, Hardt and Lin [20] give an example of an n_0 with degree zero for which any minimizing n must have singularities.

Solutions of the equilibrium equations (1.6), the equal constant or harmonic mapping case, may be constructed by homogeneous extension of conformal mappings of S^2 onto itself. Such mappings are essentially rational functions of z or $\bar{z}$. More precisely, let Π denote stereographic projection of S^2 from the north pole onto an equatorial plane and f a rational function of z so that

$$\Pi^{-1}f(\Pi\xi) \quad \text{and} \quad \Pi^{-1}f(\overline{\Pi\xi}), \qquad \xi \in S^2,$$

are conformal and anticonformal maps of S^2 onto itself. Then for any $\Omega \subset S^2$,

and
$$n(x) = \Pi^{-1}f(\Pi(x/|x|)) \qquad x \in \Omega \tag{3.1}$$
$$n(x) = \Pi^{-1}f(\overline{\Pi}(x/|x|))$$

are solutions of (1.6).

Write $f = p/q$ with p and q relatively prime polynomials and set

$$\deg f = \max\{\deg p, \deg q\}.$$

The solutions (3.1) have the special property that

$$\rho^{-1}\int_{B_\rho}|\nabla n|^2\,dx = 8\pi \deg f \tag{3.2}$$

for any $\rho > 0$.

Another family of solutions easy to construct is what serves as the analogue of viscometric flows in fluids or St.-Venant solutions in elasticity. These are the solutions which lie in a plane. They illustrate the instability and the non-uniqueness of solutions as well. The standard planar nematic twist

$$n(x) = (\cos qx_3, \sin qx_3, 0) \tag{3.3}$$

is a solution of (1.2) for any choice of constants. Generalizing this but confining our attention to the equal constant case, suppose that

$$n(x) = (n^1(x), n^2(x), 0), \qquad x \in \Omega, \tag{3.4}$$

is a solution of (1.2) with finite Dirichlet integral. One may infer from this, [1] for example, the existence of an angle

$$\theta \in H^1(\Omega) : -\Delta\theta = 0 \quad \text{in } \Omega \tag{3.5}$$

such that

$$n = (\cos\theta, \sin\theta, 0), \tag{3.6}$$
$$|\nabla n|^2 = |\nabla\theta|^2.$$

In particular, all planar solutions are smooth. However, they are not always stable. Suppose that $\Omega = B_{2r}$ and let

$$u(x) = \cos\eta(x)\, n(x) + \sin\eta(x)\, e, \quad x \in B_{2r}, \tag{3.7}$$

where $e = e_3$ and η is a suitable cut-off function,

$$\eta = \begin{cases} 0 & \text{on } \partial B_{2r} \\ \pi/2 & \text{in } B_r \end{cases}, \qquad |\nabla\eta| \le \pi/2r .$$

Thus $u = n$ on ∂B_r and, by (3.6),

$$|\nabla u|^2 = \cos^2\eta\, |\nabla\theta|^2 + |\nabla\eta|^2 \tag{3.8}$$

Hence for some constant $C > 0$,

$$\int_{B_{2r}} |\nabla u|^2\, dx \le \int_{B_{2r}-B_r} |\nabla\theta|^2\, dx + Cr .$$

Now n fails to minimize whenever

$$\int_{B_{2r}-B_r} |\nabla\theta|^2\, dx + Cr < \int_{B_{2r}} |\nabla\theta|^2\, dx ,$$

or whenever

$$C < r^{-1} \int_{B_r} |\nabla\theta|^2\, dx . \tag{3.9}$$

Let θ be harmonic in all of $\mathbb{R}^3$. Then (3.9) holds for r sufficiently large unless $\theta = \text{const.}$ First, it is easily shown that for any harmonic θ,

$$M(r) = r^{-1} \int_{B_r} |\nabla\theta|^2\, dx$$

is an increasing function of r. If it remains bounded, then so does

$$M(a,r) = r^{-1} \int_{B_r(a)} |\nabla\theta|^2\, dx$$

for any $a \in \mathbb{R}^3$. On the other hand, by the subharmonic character of $|\nabla\theta|^2$,

$$|\nabla\theta(a)|^2 \le |B_r|^{-1} \int_{B_r(a)} |\nabla\theta|^2\, dx$$

$$\le (4/3)\pi\, r^{-2} M(a,r) \to 0, \quad \text{as } r \to \infty,$$

whenever $M(a,r)$ remains bounded. Thus $\theta = \text{const.}$

We have shown that

a harmonic mapping of $\mathbb{R}^3$ *into* S^2 *whose image lies in a plane and which minimizes energy with respect to its boundary values on arbitrarily large balls is constant.*

Consider now a smooth harmonic function $\Theta(x)$ for which $n(x)$ defined by (3.6) does not minimize energy on a ball B. By the existence theorem, there is some minimizing $n^* \in \mathcal{A}(n)$ and if it is also planar, there is a harmonic function $\Theta^* \in H^1(B)$ for which $\cos\Theta = \cos\Theta^*$ and $\sin\Theta = \sin\Theta^*$ on ∂B. Hence

$$\Theta - \Theta^* \in H^1(B) \text{ and } \Theta - \Theta^* \text{ has discrete values on } \partial B.$$

But this can only happen if $\Theta - \Theta^*$ is constant, whence $n = n^*$. So the minimum n^* is not planar, and has a nonzero component n^{*3}. Since the energies of n^* and $(n^{*1}, n^{*2}, -n^{*3})$ are the same, the minimum is not unique and there are at least three solutions of the equilibrium equation (1.6) in B with the same boundary values on ∂B, [19].

Planar solutions have also been studied by E. MacMillan [24]. The instability of the solution (3.3) has been checked numerically.

4. Regularity theory

We begin our brief discussion of the regularity theory for liquid crystals by stating our principal result.

Theorem 4.1. *Let* n *be a minimum of* W, *that is, a solution of* (1.4). *Then*

(1) *There exist constants* σ *and* C *(depending on* κ_i *) such that if*

$$r^{-1}\int_{B_r(a)} W(\nabla n, n)\, dx < \sigma,$$

then

(i) $$\rho^{-1}\int_{B_\rho(a)} W(\nabla n, n)\, dx \le C\rho, \quad \rho \le r/2, \text{ and}$$

(ii) n *is Holder continuous in a neighborhood of* a.

(2) *There is a relatively compact subset* $Z \subset \Omega$ *such that*

$n|_{\Omega - Z}$ *is analytic and* $\mathcal{H}^1(Z) = 0$.

In the statement above, $\mathcal{H}^t(X)$ denotes the t-dimensional Hausdorff measure of X. The proofs of these statements are in Corollary 2.5 and Theorem 2.6, [16]. Part (2) may be extended to the boundary of Ω, Theorem 5.6 [16]. Analogous theorems hold when the effects of magnetic or electric fields are included.

The main idea in the proofs is to use the blow-up method, which has a long history dating at least to Almgren, DeGiorgi, and Giusti-Miranda. The method is also closely related to the Schoen-Uhlenbeck [26] argument for harmonic mappings, but it differs significantly from this even though the equal constant case concerns harmonic mappings. There are three elements to the argument which we found unusual:

1. The blow-up sequence is normalized by energy instead of radius.
2. The blow up limit is the solution of a linear system with two dependent variables.
3. An estimate due to Hardt and Lin,[21], cf also Lemma 2.3 [16].

Rather than give technical details of proofs, let us describe the Hardt-Lin estimate and illustrate some of its additional consequences. Let n be a minimizer of W in Ω and suppose that $B_{2\rho} \subset \Omega$. Then there is a constant $C_0 = C_0(\kappa_1, \kappa_2, \kappa_3)$ such that

$$\int_{B_\rho} |\nabla n|^2 \, dx \le C_0 \Big(\int_{\partial B_\rho} |\nabla n|^2 \, dS \int_{\partial B_\rho} |n - \bar{n}|^2 \, dS \Big)^{1/2} \qquad (4.1)$$

for any $\bar{n} \in \mathbb{R}^3$.

This estimate reads precisely as the Schoen-Uhlenbeck estimate of [26], but there it is required that the right hand side of (4.1) be very small before the inequality be valid. Here are some applications:

1. Upper density bound *There is a number* $M = M(\kappa_1, \kappa_2, \kappa_3)$ *such that*

$$\limsup_{\rho \to 0} \rho^{-1} \int_{B_\rho(a)} W(\nabla n, n) \, dx \le M, \qquad a \in \Omega, \qquad (4.2)$$

for any minimizer n *of* W *in* Ω.

It is to be emphasized that the number M does not depend on the boundary values of n nor on the domain Ω, but only on the material constants $\kappa_1, \kappa_2, \kappa_3$. As a special case, consider the solutions of the equilibrium equation (1.6) which are given by homogeneous extension of conformal maps of S^2 into itself, cf. (3.1), for which (3.2) holds. They are smooth mappings except at $x = 0$, where they have singularities of degree equal to deg f. If the deg f is large enough that

$$8\pi \deg f > M,$$

then the associated n is not a minimizer. *There is a maximum possible degree of a stable singularity of a harmonic mapping.*

We prove this statement, which follows in an elementary way from the estimate (4.1). Restricting our attention to the equal constant case, set $\bar{n} = 0$. Note that

$$\int_{\partial B_\rho} |n|^2 \, dS = |\partial B_\rho| = 4\pi \rho^2$$

implies that

$$\int_{B_\rho} |\nabla n|^2 \, dx \leq C\rho \left(\int_{\partial B_\rho} |\nabla n|^2 \, dS\right)^{1/2} .$$

We square this and integrate it from r to $2r$, using the most elementary estimates for the various terms.

$$\left(\int_{B_\rho} |\nabla n|^2 \, dx\right)^2 \leq C^2\rho^2 \int_{\partial B_\rho} |\nabla n|^2 \, dS ,$$

$$r\left(\int_{B_r} |\nabla n|^2 \, dx\right)^2 \leq 4C^2 r^2 \int_{B_{2r}} |\nabla n|^2 \, dx ,$$

$$\left(r^{-2}\int_{B_r} |\nabla n|^2 \, dx\right)^2 \leq 4C^2 r^{-1} \int_{B_{2r}} |\nabla n|^2 \, dx. \tag{4.3}$$

Set

$$E(\rho) = \rho^{-1} \int_{B_\rho(a)} |\nabla n|^2 \, dx ,$$

which allows us to rewrite (4.3) as

$$E(r)^2 \leq M\, E(2r) \quad \text{or} \quad E(\tfrac{1}{2}r)^2 \leq M\, E(r) \quad \text{whenever } B_{2r} \subset \Omega.$$

This may be iterated in the following way:

$$E(2^{-k}r)^{2^k} \leq M^{2^k} E(r) \quad \text{or}$$

$$E(2^{-k}r) \leq M\, E(r)^{-2^k} .$$

Letting $k \to \infty$, we obtain that $E(0) \leq M$.

Now the M established by our estimate is very general in that it holds for any liquid crystal integrand, even when contributions are given by other energies. In the harmonic mapping case, it ought to be possible to give a precise statement, as was suggested when the conference on which this note is based was presented. In the meantime, our numerical results had shown that the minimizer associated to the boundary values determined by $f(z) = z^2$ had two singularities of degree one and markedly less energy than its homogeneous extension, Cohen et al [7].

This was taken up by Brezis, Coron, and Lieb [2],[3] who proved the remarkable result that in the harmonic mapping case $M/8\pi = 1$, so that only singularities of degree one are present in minima, and moreover such singularities are, asymptotically, homogeneous extensions of rotations.

2. Higher integrability. *There is a* $q > 2$ *such that*

$$n \in H^{1,q}_{loc}(\Omega;S^2) \tag{4.4}$$

for any minimizer n *of* W *in* Ω.

The estimate (4.1) permits us to show (4.4) by a reverse Holder inequality, cf. [12],[13],[25]. As a consequence, $H^{1-\delta}(Z) = 0$ for some $\delta > 0$, slightly improving (2) of Theorem 4.1. There remains the question of whether the singular set Z consists in fact of only isolated points as in the harmonic mapping case [26].

3. Convergence of minimizers. *If* $\{n_j\}$ *is a sequence of minimizers of* W *in* Ω *which converges weakly in* $H^1(\Omega;S^2)$ *to* n, *then a subsequence* $\{n_{j'}\}$ *converges strongly in* $H^1(\Omega;S^2)$ *to* n *and* n *is also a minimizer of* W.

For the proofs of these statements, we refer to [18].

A major step in the proof of (4.1) is to show that for any $v \in H^1(B)$, with $|v| = 1$ on ∂B, or equivalently, for any $v \in H^{1/2}(\partial B)$ with $|v| = 1$, there is a $\zeta \in H^1(B)$ satisfying

$$|\zeta| = 1 \quad \text{in } B \quad \text{and}$$

$$\int_B |\nabla\zeta|^2\,dx \le 32\left(\int_{\partial B} |\nabla_{tan} v|^2\,dS \int_{\partial B} |v - c|^2\,dS\right)^{1/2}, \tag{4.5}$$

for any $c \in \mathbb{R}^3$. In this way we may place the issue of admissibility in its proper functional setting:

4. Admissibility. *If* Ω *is diffeomorphic to a ball and* $n_0 \in H^{1/2}(\partial\Omega)$, *then* $A(n_0) \neq \emptyset$.

5. Computational results

Two numerical algorithms have been employed to compute minimum energy configurations. The results of these experiments illustrate that the functions $n_f(x)$ defined by (3.1) with $f(z) = z$ minimizes the energy and $f(z) = z^2$, $f(z) = \bar{z}^2$, $f(z) = z^3$, and $f(z) = \frac{1}{2}z$ do not minimize energy, all in the equal constant case, (1.5). The unusual features of these algorithms are that they must accomodate the nonconvex constraint $|n| = 1$ and the possible singularities of the configuration. Details appear in [7].

The first method consists in applying the relaxation method to a discretized version of the functional

$$\tfrac{1}{2}\int_\Omega |\nabla n|^2 \, dx,$$

or more generally,

$$\int_\Omega W(\nabla n, n)\, dx$$

subject to the constraint that

$$|n| = 1 \text{ in } \Omega$$

and the boundary condition

$$n = n_0 \quad \text{on } \partial\Omega.$$

For the Gauss-Seidel relaxation, this method has the desirable stability property that it is energy decreasing in the usual sense. That is, if $\hat{W}(n^k)$ denotes the energy of the kth iterate n^k, then

$$\hat{W}(n^{k+1}) \le \hat{W}(n^k). \tag{5.1}$$

Also, $|n^k| = 1$ at all mesh points. Improved rates of convergence have been obtained with a version of the over-relaxation method modified to enforce the pointwise constraint. Unfortunately, the energy decreasing property (5.1) is lost with over-relaxation.

The second algorithm is a fractional step method which is motivated by penalty methods for achieving constraints. One might think of approximating minima of

$$\int_\Omega W(\nabla n, n)\, dx$$

by penalization of the constraint. For example, if $f(t)$, $t \ge 0$, is a smooth function of t satisfying

$$f(t) > 0, \quad t \ge 0, t \ne 1, \text{ and } f(1) = 0,$$

a penalization is given by minimizing the functional

$$E_\lambda(n) = \int_\Omega W(\nabla n, n)\, dx + \lambda \int_\Omega f(|n|)\, dx \tag{5.2}$$

in $H^1(\Omega;\mathbb{R}^3)$. It is not difficult to show that the minima n_λ of E_λ converge suitably to a minimum of (1.4) as $\lambda \to \infty$.

A computational technique to construct a minimum of the functional (5.2) is to solve to steady state the time dependent problem

$$\partial n/\partial t = \operatorname{div} \partial W/\partial p - \partial W/\partial n - \lambda \partial f/\partial n. \tag{5.3}$$

The algorithm that we found most efficient for computation approximated the penalty term in (5.2) and (5.3) by an "indicator" function and used a fractional step algorithm for the energy and the penalty to advance in time [7].

Figures 1 and 2 are indicative of our results in the equal constant case. The calculations were performed on the unit cube $Q = (0,1) \times (0,1) \times (0,1)$ with $(21)^3$ grid points. Figure 1 is a picture of

$$n(x) = n_f(x) = \Pi^{-1} f \Pi({}^{x-a}/_{|x-a|}) \tag{5.4}$$

where $f(z) = z^2$ and $a = (\frac{1}{2},\frac{1}{2},\frac{1}{2})$ in the plane $x_3 = \frac{1}{2}$. Clearly visible is a singularity of degree 2 at the center. Figure 2 is a picture of our 'solution' with given boundary data (5.4) in the plane $x_3 = \frac{1}{2}$, after 152 time steps of the fractional step method. At this number of iterations,

$$\hat{W}(n^{k+1} - n^k) < \delta \quad \text{for} \quad \delta = 10^{-7}.$$

The graphical output indicates that the solution has two singularities of degree 1. Moreover,

$$\hat{W}(n^K) = 24.651 < 27.978 = \hat{W}(n_f), \; K = 152.$$

The computations were executed on the Cray-2 at the University of Minnesota.

6. Furthur remarks: the Williams, Pieranski, and Cladis experiment

When nematic material is introduced into a capillary tube which has been prepared to render the director perpendicular to the tube walls, two point singularities, of degree +1 and −1, may appear along the axis of the tube. Taking account of the faces of the tube, and assuming its axis to be the x£-axis, the boundary orientation of $n(x)$ is roughly

$$n_0(x) = ({}^{x_1}/_r, {}^{x_2}/_r, 0), \; r^2 = x_1{}^2 + x_2{}^2,$$

which has degree zero. The two defects gradually move closer together, then coalesce, resulting in a configuration free of singularities. This experiment is described in [27]. This was our first exposure to singularities and was related to us by J. L. Ericksen.

One is led to surmise the existence of a family of solutions of (1.4) whose limit is a smooth configuration when the defects coalesce. Two points of view have been adopted to investigate this.

In [3], as described by Brezis in this volume and to which we refer for a complete explanation, the location and degree of each singularity in a configuration are assumed known and the minimum of energy with respect to this constraint is studied. It is not achieved.

In [17] it is shown that if a minimizing configuration is sufficiently close to one which is free of defects, then it too is free of defects. This result employs the regularity theory we have described as well as section 4.3, the Convergence of minimizers.

Both results suggest that static theory alone does not suffice to explain this phenomenon. It remains a challenging problem to analyse the dynamical equations, cf. [22],[23], pertinent to this situation.

Acknowledgements

The preparation of this manuscript owes much to our collaborator, F.-H. Lin. We wish to acknowledge our indebtedness to J. L. Ericksen and we thank H. Brezis for his continued interest.

Partially supported by NSF grants DMS 85-11357, MCS 83-01345, DMS 83-51080 and DMS 83-01575

References

[1] **F. Bethuel and X. Zheng,** Sur la densité des fonctions régulières entre deux variétés dans des espaces de Sobolev, CRAS Paris , 303, (1986), 447-449.

[2] **H. Brezis, J. M. Coron and E. Lieb,** Estimations d'énergie pour des applications de R^3 a valeurs dans S^2, CRAS Paris, 303, (1986), 207-210.

[3] **H. Brezis, J. M. Coron and E. Lieb,** Harmonic maps with defects, IMA preprint 253, to appear, Comm. Math. Physics.

[4] **W. Brinkman and P. Cladis,** Defects in liquid crystals, Physics Today, May 1982, 48-54.

[5] **S. Campanato,** Sistemi ellitici in forma divergenza. Regolarità all'interno, Quaderni della SNS Pisa (1980).

[6] **S. Chandrasekhar,** Liquid Crystals, Cambridge (1977).

[7] **R. Cohen, R. Hardt, D. Kinderlehrer, S.-Y. Lin and M. Luskin,** Minimum energy configurations for liquid crystals: computational results, Theory and applications of liquid crystals, IMA Volumes in math. and appl. 5 (J.L. Ericksen and D. Kinderlehrer, eds.).

[8] **P. G. de Gennes,** The physics of liquid crystals, Oxford (1974).

[9] **J. L. Ericksen,** Equilibrium theory of liquid crystals, Adv. in liquid crystals 2, (G. H. Brown, ed.) Academic Press (1976), 233-298.

[10] **J. L. Ericksen,** Nilpotent energies in liquid crystal theory, Arch. Rat. Mech. Anal. 10 (1962), 189-196.

[11] **J. L. Ericksen and D. Kinderlehrer,** Theory and application of liquid crystals, IMA Volumes in math. and appl. 5.

[12] **F. W. Gehring,** The L^p integrability of the partial derivatives of a conformal mapping, Acta Math. 130 (1973), 55-72.

[13] **M. Giaquinta,** Multiple integrals in the calculus of variations and nonlinear elliptic systems, Annals of Math. Studies 105, Princeton, 1983.

[14] **R. Hardt and D. Kinderlehrer,** Mathematical questions of liquid crystal theory, Theory and application of liquid crystals, IMA Volumes in math. and appl. 5 (J.L. Ericksen and D. Kinderlehrer, eds.) (1987).

[15] **R. Hardt, D. Kinderlehrer and F.-H. Lin,** Existence et régularité des configurations statiques des cristaux liquides, CRAS Paris 301, (1985), 577-579.

[16] **R. Hardt, D. Kinderlehrer and F.-H. Lin,** Existence and partial regularity of static liquid crystal configurations, IMA preprint 175, Comm. Math. Physics 105 (1986), 547-570.

[17] **R. Hardt, D. Kinderlehrer and F.-H. Lin,** A remark about the stability of smooth equilibrium configurations of static liquid crystals, IMA preprint 231, Mol. Cryst Liq. Cryst. 136 (1986).

[18] **R. Hardt, D. Kinderlehrer and F.-H. Lin,** to appear.

[19] **R. Hardt, D. Kinderlehrer and F.-H. Lin,** to appear.

[20] **R. Hardt and F.-H. Lin,** A remark on H^1 mappings, Manus. math. 56, (1986), 1-10.

[21] **R. Hardt and F.-H. Lin,** Mappings that minimize the p th power of the gradient, to appear.

[22] **F. Leslie,** Theory of flow phenomena in liquid crystals, Adv. Liq. Cryst. 4 (G. H. Brown, ed.) Academic Press (1979), 1-81.

[23] **F. Leslie,** Theory of flow phenomena in nematic liquid crystals, Theory and application of liquid crystals, IMA Volumes in math. and appl. 5, (J.L. Ericksen and D. Kinderlehrer, eds.) (1987).

[24] **E. MacMillan,** The statics of liquid crystals, Thesis, Master of Science, The Johns Hopkins University, 1982.

[25] **N. G. Meyers and A. Elcrat,** Some results on regularity for solutions of nonlinear elliptic systems and quasiregular functions, Duke Math J. 42 (1975), 121-136.

[26] **R. Schoen and K. Uhlenbeck,** A regularity theory for harmonic maps, J. Diff. Geometry 17 (1982), 307-335.

[27] **C. Williams, P. Pieranski and P. E. Cladis,** Nonsingular S = +1 screw disclination lines in nematics, Phys. Rev. Letters 29 (1972), 90-92.

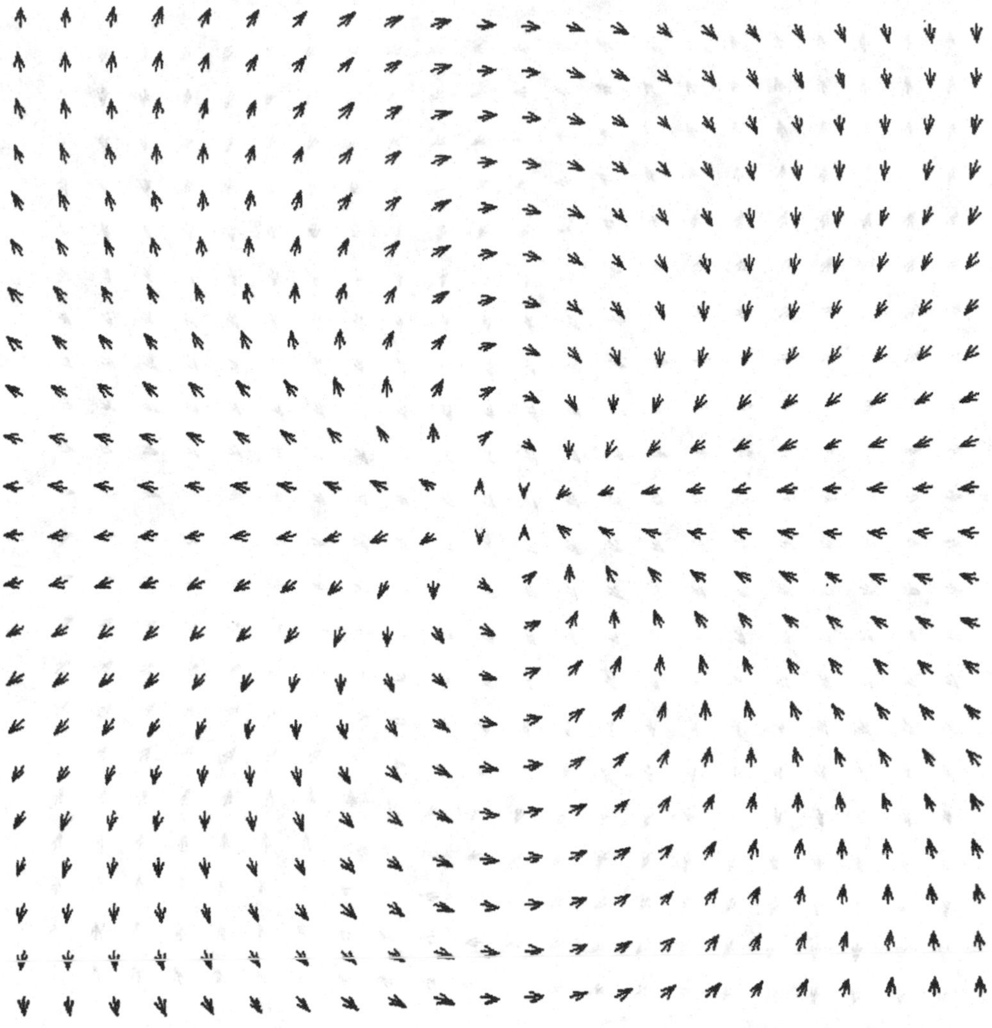

Figure 1

n_f for $f = z^2$

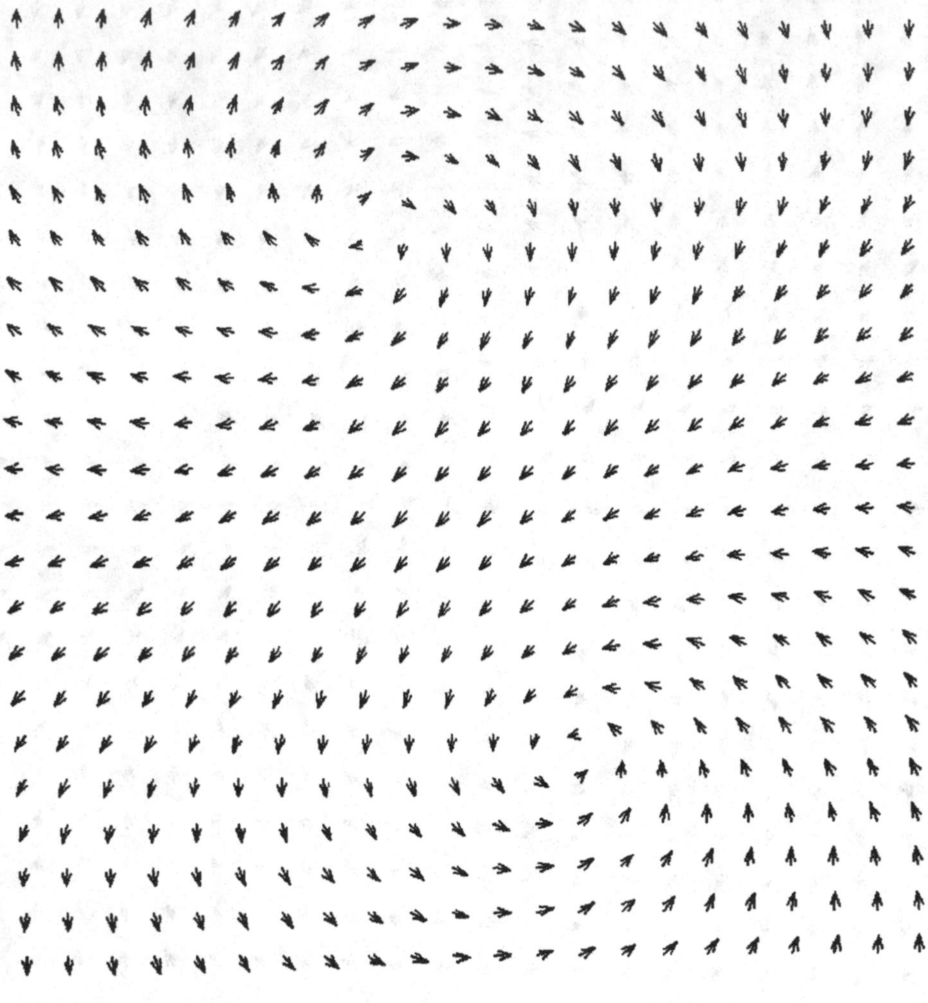

Figure 2

n^K for $K = 152$ and with boundary data n_f

ON QUASI-MINIMAL SURFACES

Erhard Heinz
Mathematisches Institut der
Georg-August-Universität
Bunsenstraße 3-5
D-3400 Göttingen (Fed. Rep. Germany)

Dedicated to Hans Lewy

Quasi-minimal surfaces were first introduced by I. Marx and M. Shiffman in order to classify unstable minimal surfaces with polygonal boundaries. Their investigations were announced in Courant's book [1, pp. 235-236] and later published in the paper [8]. However the proofs of the propositions outlined there are insufficient. In this lecture we shall give a summary of some recent results, which can be regarded as generalizations and improvements of the results stated in [8]. For more specific historical comments we refer to the bibliography at the end of this article, in particular to [6].

1. Definition of a quasi-minimal surface

The simplest way to define a quasi-minimal surface is to consider it as a solution of a certain variational problem. To formulate it we have to introduce some notations.

Let B be the unit disk $|w| < 1$ $(w = u + iv = re^{i\phi})$ and let Γ be a simple closed polygon in $\mathbf{R}^p$ $(p \geq 2)$ with $(N+3)$ vertices

$$e_1 = \left(e_1^1, \dots, e_1^p\right), \dots, e_{N+3} = \left(e_{N+3}^1, \dots, e_{N+3}^p\right)$$

arranged in cyclical order $(e_0 = e_{N+3}, e_{N+4} = e_1)$. We assume that the vectors $e_k - e_{k-1}, e_{k+1} - e_k$ $(k = 1, \dots, N+3)$ are linearly independent. Let T be the domain in $\mathbf{R}^N$ characterized by the inequalities

(1.1) $\quad 0 < \tau_1 < \dots < \tau_N < +\pi \quad .$

For abbreviation let us put

$\tau_{N+k} = \frac{\pi}{2}(1+k)$ $(1 \le k \le 3)$, $\tau_{N+4} = \tau_1 + 2\pi$, and let Q denote the set of points $w = e^{i\tau_k}$ $(K = 1,\dots,N+3)$. For fixed $\tau \in T$ consider the set $\mathcal{F}(\tau)$ of vector-valued functions

$$x = x(w) = x(u,v) = (x^1(u,v),\dots,x^p(u,v))$$

of class $C^2(B) \cap C^0(\bar{B})$, which map the circular arcs

$$\gamma_k = \left\{ w = e^{i\phi} : \tau_k < \phi < \tau_{k+1} \right\} \quad (k = 1,\dots,N+3) \tag{1.2}$$

into the straight lines $e_k + \Gamma_k$, where

$$\Gamma_k = \left\{ x \in \mathbf{R}^p : x = (e_{k+1} - e_k)t,\ t \in \mathbf{R} \right\} \ . \tag{1.3}$$

In particular, we have

$$x\left(e^{i\tau_k}\right) = e_k \quad (k = 1,\dots,N+3) \ .$$

Applying the direct methods of the calculus of variations [2, § 2] it can be shown that for any fixed $\tau \in T$ there exists a uniquely determined minimal vector $x = x(w,\tau) \in \mathcal{F}(\tau)$ such that

$$\theta(\tau) = \inf_{x \in \mathcal{F}(\tau)} D(x) = D\left(x(w,\tau)\right) < +\infty \tag{1.4}$$

holds, where $D(x)$ is the Dirichlet integral

$$D(x) = \iint_B \left(x_u^2 + x_v^2\right) du\, dv \ . \tag{1.5}$$

Furthermore it can be proved that for fixed $\tau \in T$ the function $x = x(w,\tau)$ is analytic in $\bar{B} \setminus Q$ and is a solution of the linear elliptic boundary value problem

$$\Delta x = 0 \qquad (w \in B) \tag{1.6}$$

$$x(w) \in e_k + \Gamma_k \qquad (w \in \gamma_k) \tag{1.7}$$

$$x_r(w) \perp \Gamma_k \qquad (w \in \gamma_k) \ , \tag{1.8}$$

where $k = 1,\dots,N+3$. According to [8] such a function is called a quasi-minimal surface. Thus for a fixed polygon $\Gamma \subset \mathbf{R}^p$ the class $\mathcal{H}$ of quasi-minimal surfaces $\{x(w,\tau)\}_{\tau \in T}$ is an N-parametric family. It can be shown [2, § 2] that this class $\mathcal{H}$ is equicontinuous on $T' \times B$, where T' is any compact subset of T. For the differentialgeometric applications to be made in Section 2 we have to determine the local behaviour of the surface $x = x(w, \tau)$ at a point $\zeta \in B$ for fixed $\tau \in T$. First suppose that $\zeta \notin Q$. Then in a neighborhood of ζ we have a power series expansion

(1.9) $$x_w(w, \tau) = a(w - \zeta)^m + \dots ,$$

where $a \in C^p$, $a \neq 0$, and m is a nonnegative integer. In the singular case $\zeta \in Q$ an asymptotic representation

(1.10) $$x_w(w,\tau) = a(w-\zeta)^{m+\rho} + 0\left(|w-\zeta|^{m+\rho+\varepsilon}\right) \quad (w \to \zeta, w \in \overline{B})$$

holds, where a and m satisfy the same conditions and ρ and ε are real numbers such that $-1 < \rho \leq 0$ and $\varepsilon > 0$ [4, § 2]. The expansions (1.9)-(1.10) show that $m = m(\zeta)$ $(\zeta \in \overline{B})$ is uniquely determined by the surface $x : \overline{B} \to R^p$. $\zeta \in \overline{B}$ is called a branch point of order $m(\zeta)$ iff $m(\zeta) > 0$. Clearly the quantity

(1.11) $$\kappa(x) = \sum_{|\zeta|<1} m(\zeta) + \frac{1}{2} \sum_{|\zeta|=1} m(\zeta)$$

is finite and nonnegative. Moreover the estimate

(1.12) $$\kappa(x) \leq \frac{1}{2} N$$

holds for any surface $x \in \mathcal{H}$ (see [7]).

2. Applications to Plateau's problem

There is a close connection between the class $\mathcal{H}$ of quasi-minimal surfaces and solutions $\mathcal{H}'$ of Plateau's problem for the polygon Γ. Define $\mathcal{H}^*$ to be the set of vector-valued functions $x \in C^2(B) \cap C^0(\overline{B})$ satisfying (1.6)-(1.7) for some $\tau \in T$ and the conformality relations

(2.1) $$x_u^2 = x_v^2, \quad x_u x_v = 0 \quad (w \in B) .$$

Then the inclusion

(2.2) $$\mathcal{H}' \subseteq \mathcal{H}^* \subseteq \mathcal{H}$$

is easily established [3]. The following theorem furnishes an analytical characterization of $\mathcal{H}^*$ within the class $\mathcal{H}$:

Theorem 1. *The function* $\theta(\tau)$ *is analytic in* T. *Moreover a vector function* $x = x(w, \tau) \in \mathcal{H}$ *belongs to* $\mathcal{H}^*$ *iff the equation* $\nabla\theta(\tau) = 0$ *holds.*

This theorem can be easily deduced from the following two lemmas.

Lemma 1. *Let* $x = x(w, \tau) \in \mathcal{H}$. *Then for* $(w, \tau) \in B \times T$ *we have the equation*

$$(2.3) \qquad w^2 x_w^2 = \frac{i}{8\pi} \sum_{k=1}^{N+3} R_k(\tau) \frac{e^{i\tau_k} + w}{e^{i\tau_k} - w} ,$$

where $R_k(\tau)$ *are real-valued functions in* T *satisfying the relations*

$$(2.4) \qquad \sum_{k=1}^{N+3} R_k(\tau) = 0$$

and

$$(2.5) \qquad \sum_{k=1}^{N+3} e^{i\tau_k} R_k(\tau) = 0 .$$

Moreover $\theta(\tau) = D(x(w, \tau))$ *belongs to* $C^1(T)$, *and we have*

$$(2.6) \qquad \theta_{\tau_k}(\tau) = R_k(\tau)$$

for $\tau \in T$ *and* $k = 1,\dots,N$.

Lemma 2. *For fixed* $w \in B$ *the function* $x_w(w, \tau)$ *is analytic in* T.

The proof of Lemma 1 given in [3] is based on variational arguments used by Courant [1, Chap. VI] in his classical investigations on unstable minimal surfaces with polygonal boundaries. An alternative proof depending on the results of [2] can be found in [5]. Lemma 2 is contained in a more general proposition concerning the analytical dependence of the minimal vectors $x(w, \tau)$ on the parameters $\tau_1,\dots,\tau_N$ (see [2, Satz 2]). Its proof depends on the close connection between the boundary value problem (1.6)-(1.8) and the Riemann-Hilbert problem for the system

$$(2.7) \qquad \frac{dZ}{dw} = Z \sum_{k=1}^{N+3} \frac{A_k}{w - \zeta_k}$$

of ordinary differential equations in the complex domain. Here $Z = Z(w)$ and $A_1,\dots,A_{N+3}$ are square matrices of order p, and we have

$$(2.8) \qquad \sum_{k=1}^{N+3} A_k = 0 .$$

It should be noted that the usual perturbation procedures cannot be applied here, since the quantities $\tau_1,\dots,\tau_N$ do not enter analytically into the boundary relations (1.7)-(1.8). Thus there is some need for investigating the analytical dependence of solutions of more general elliptic systems on parameters, for which (1.6)-(1.8) is a typical case.

From Theorem 1 it is evident that for the structure of the set $\mathcal{H}^*$ the Hessian

(2.9) $$M = \left(\left(\theta_{\tau_j \tau_k}(\mathring{\tau}) \right) \right)_{j,k=1}^{N}$$

is of particular importance. In what follows we assume that $p = 3$ and $x = x(w, \mathring{\tau}) \in \mathcal{H}^*$. Thus $\tau = \mathring{\tau}$ is a critical point of the function $\theta(\tau)$. Let A be the Schwarz operator

(2.10) $$A = -\Delta + 2KE$$

with domain

(2.11) $$D(A) = \{ \phi \in C^2(B) \cap C^0(\bar{B}) : \phi|\partial B = 0, -\Delta\phi + 2KE\phi \in L^2(B) \} ,$$

where $E = x_u^2$ and $K = K(w)$ is the Gauss curvature of the minimal surface $x = x(w, \mathring{\tau})$. The differential operator is regular in B, but becomes singular near the set $Q \subset \partial B$. Applying perturbation methods it can be shown [4, § 3] that A is essentially selfadjoint and semibounded from below. Moreover it has a discrete spectrum, and all eigenfunctions belong to $C^1(\bar{B})$. Let $s(A)$ and $s(M)$ be the number of negative eigenvalues of A and M, respectively, and let us denote by $r(M)$ the rank of the matrix M. Then the following theorem holds.

Theorem 2. *Let* $p = 3$ *and let* $x = x(w, \tau) \in \mathcal{H}^*$. *Then we have the relations*

(2.12) $$\dim \ker A + r(M) + 2\kappa(x) = N$$

and

(2.13) $$s(A) = s(M) \quad .$$

The relation (2.12) was established in the paper [4]. It can be viewed as a refinement of the inequality (1.12) for the case where $p = 3$ and $x \in \mathcal{H}^*$. In particular it implies that M is nondegenerate iff $\lambda = 0$ is not an eigenvalue of A and the minimal surface x has no branch point in $\bar{B}$. The second statement of the theorem is due to Sauvigny ([10], [11]), who even proved it in the case $p > 3$, where the Schwarz operator A has to be suitably generalized. Sauvigny's paper [12] contains a further application of this result to Plateau's problem. To formulate the theorem we have to introduce some definitions. A polygon $\Gamma \subset \mathbb{R}^p$ is called extremal, if all vertices e_k lie on the boundary of a bounded convex set in $\mathbb{R}^p$. Let $\eta_m \in (0, \pi)$ be the outer angle of Γ at the vertex e_m. Then the quantity

(2.14) $$k(\Gamma) = \sum_{m=1}^{N+3} \eta_m$$

is called the total curvature of Γ. With these definitions we have

Theorem 3. *Let* $\Gamma \subset \mathbf{R}^p$ *be an extremal polygon of total curvature*

$$k(\Gamma) < \frac{10}{3}\pi \quad (p = 4,5,\ldots) \quad \text{or} \quad k(\Gamma) < 4\pi \quad (p = 3) .$$

Then Γ *bounds exactly one minimal surface* x. *This minimal surface has no branch points in* $\overline{B}$.

For the case $p = 3$ the theorem is due to Nitsche [9] who used a different method. It is an open question, whether the extremality condition can be dropped. More generally, it would be desirable to determine classes of polygons for which the function $\theta(\tau)$ has only finitely many critical points in T. According to Theorem 1 this would imply that these polygons only bound finitely many minimal-surfaces. The solution of this problem would certainly require a deeper analysis of the function $\theta(\tau)$ in its dependence on the polygon Γ.

References

[1] **R. Courant,** Dirichlet's principle, conformal mapping, and minimal surfaces, Interscience publishers, New York 1950.

[2] **E. Heinz,** Über die analytische Abhängigkeit der Lösungen eines linearen elliptischen Randwertproblems von Parametern, Nachr. d. Akad. Wiss. in Göttingen, II. Math.-Phys., Kl. Jahrgang 1979, 1-20.

[3] **E. Heinz,** Über eine Verallgemeinerung des Plateauschen Problems, Manuscripta math. 28, (1979), 81-88.

[4] **E. Heinz,** Minimalflächen mit polygonalem Rand, Math. Z. 183 (1983), 547-564.

[5] **E. Heinz,** Zum Marx-Shiffmanschen Variationsproblem, J. Reine u. Angew. Math. 344 (1983), 196-200.

[6] **E. Heinz,** Zum Plateauschen Problem für Polygone. Zum Werk Leonhard Eulers, 197-204, Birkhäuser-Verlag, Basel-Boston-Stuttgart 1984.

[7] **E. Heinz,** An estimate for the total number of branch points of quasi-minimal surfaces, Analysis 5 (1985), 383-390.

[8] **I. Marx,** On the classification of unstable minimal surfaces with polygonal boundaries, Comm. P. Appl. Math. 8 (1955), 235-244.

[9] **J.C.C. Nitsche,** Uniqueness and non-uniqueness for Plateau's Problem - one of the last major questions, Obata Morio, Minimal submanifolds and geodesics, Tokyo (1978), 143-161.

[10] **F. Sauvigny,** Die zweite Variation von Minimalflächen im $\mathbf{R}^p$ mit polygonalem Rand, Math. Z. 189 (1985), 167-184.

[11] **F. Sauvigny,** Ein Eindeutigkeitssatz für Minimalflächen im $\mathbb{R}^p$ mit polygonalem Rand, J. Reine u. Angew. Math. 358 (1985), 92-96.

[12] **F. Sauvigny,** On the Morse index of minimal surfaces in $\mathbb{R}^p$ with polygonal boundaries, Manuscripta math. 53 (1985), 167-197.

A SURVEY OF RECENT REGULARITY RESULTS FOR SECOND ORDER QUEER DIFFERENTIAL EQUATIONS

Peter Laurence
Dept. of Mathematics
The Pennsylvania State University
University Park, PA 16802
(U.S.A.)

and

Edward Stredulinsky
Dept. of Mathematics
Oregon State University
Corvallis, OR 97331
(U.S.A.)

Dedicated to Hans Lewy

1. Introduction and formulation of the variational problem

Queer differential equations (QDE's) arise in the model of quasi-static equilibrium of confined plasmas due to Harold Grad. Several dissipative (ie long time scale) plasma simulation codes in use throughout the world today compute solutions to these equations. Many others use techniques closely related to the structure of these equations. The queer differential equations which are relevant to confined plasma research involve second order derivatives of the increasing rearrangement function. Thus, consider what is widely known as the plasma equation,

(1) $$\Delta\psi(x) = -\frac{d}{d\psi}p(\psi)$$

in $\Omega \subset \mathbf{R}^2$.

Ω is to be thought of as a two dimensional cross section of an axisymmetric toroidal containment device (see [T], appendix for a derivation of (1) with a slightly different elliptic operator on the left).

H. Grad proposed the following form for $p(\psi)$,

(2) $$p(\psi) = \mu(\psi)\psi^{*'2} \quad .$$

We refer to [G1,2 and G-H-S] for further physical motivation. Here $\mu(\psi)$ is a positive smooth function and $\psi^*(v)$ is the increasing rearrangement of $\psi(x)$,

(3) $$\psi^*(V) = \inf\{s : V_\psi(s) \geq V\}$$

(4) $$V_\psi(s) = |\{x : \psi(x) < s\}| \quad .$$

Introducing (2) into (1) leads to the QDE:

(5) $$\Delta\psi(x) = \mu'(\psi)\,\psi^{*'2} + 2\mu(\psi)\,\psi^{*''} \quad .$$

For physical reasons (related to the preservation of total flux) the boundary value problem associated with (3) involves boundary condition on both $\psi(x)$ are ψ^*, typically:

$$\psi = 1 \qquad \text{at } \partial\Omega$$

(6) $$\inf_{\Omega} \psi = \psi^*(0) = 0\,, \quad \sup_{\Omega} \psi = \psi^*(|\Omega|) = 1 \quad .$$

As first noted by Grad the class of equations (5) admits a variational formulation.

Minimize,

(7) $$J(\psi) = \int_{\Omega} |\nabla\psi|^2 dx + \int_0^{|\Omega|} \mu(\psi)\psi^{*'2} dV$$

subject to the condition 6).

For simplicity in the subsequent presentation, we hereafter consider the case,

(8) $$\mu(\psi) = \frac{\lambda}{2} > 0 \,.$$

When appropriate we comment on the situation for more general $\mu(\psi)$.

Using (8) the equations to be considered are

(9) $$\Delta\,\psi(x) = -\,\lambda\psi^{*''}(V)$$

(λ is not an eigenvalue, it is prescribed.)

with associated functional,

(10) $$J(\psi) = \int_{\Omega} |\nabla\psi|^2 dx + \int_0^{|\Omega|} \lambda\psi^{*'2} dv \quad .$$

It can easily be shown that (9) is invariant under the action:

$$\psi \to a\psi + d$$

a,d real constants.

This invariance allows us the (as it turns out) convenient freedom to consider instead of (6) the boundary conditions,

(11) $$\psi = 0 \qquad \text{at } \partial\Omega$$

(12) $$\psi^*(0) = 0\,, \quad \psi^*(\,|\,\Omega\,|\,) = 1 \quad .$$

In [LS1] the existence of minimizers of $J(\psi)$ given by (10) was established in the class

$$W = \left\{ \underset{\psi^*(0)\le 0,\ \psi^*(|\Omega|)\ge 1}{\psi(x)\in H_0^1(\Omega),\ \psi^*(V)\in H^1\left(0,|\Omega|\right),} \right\} .$$

Actually in [LS1] this was only established for $\lambda = 1$, but arbitrary positive λ or even $\mu(\psi)$ with $0 < c_1 < \mu(\psi) < c_2$, causes no additional difficulties.

A weak maximum principle [1] then implies that for the minimizer (s),

$$\psi^*(0) = 0 \quad , \quad \psi^*(|\Omega|) = 1 .$$

The broader class defined by W is however convenient as it allows a broader set of admissible variations of $J(\psi)$ to be made.

2. Difficulties in variational formulation

Two of the most challenging and novel features of the QDE variational principle are the following:

1) It is not possible with only the information immediately available about minimizers, i.e. that they belong to the class W, to make arbitrary (admissible) variations of the functional so as to obtain a weak equation to be satisfied by minimizers. Only special variations can be made for which it is possible to justify the directional derivative of the QDE term,

$$\lambda \int_0^{|\Omega|} \psi^{*'2} dV .$$

The difficulties encountered here surround the delicate nature of the directional derivative of the increasing rearrangement map which in turn are due to a complicated interaction with the set of critical points of $\psi(x)$, $S = \{x : \nabla\psi = 0\}$.

So assume that $\zeta \in L^\infty(\Omega)$ is given for which,

(14) $$\psi + \varepsilon\zeta \in W, \quad \forall\ \varepsilon > 0, \text{ sufficiently small.}$$

[Characterizing such ζ is in itself not a trivial matter, as $\psi \in W$ does not automatically imply ψ is continuous, and ζ must preserve the non local boundary conditions (6) as well as the condition $(\psi + \varepsilon\zeta)^* \in H^1(0, |\Omega|)$].

We may then try to apply what is (to our knowledge) the most powerful (in this general setting) result on the directional derivative of the rearrangement map due to J. Mossino and R. Temam. To state this result note that there may exist a countable family of numbers t_i, $i \in I$ such that sets

$\{\psi = t_i\}$ or $\{\psi^* = t_i\}$ have positive measures. Denote these sets respectively by P_i and P_i^*. P_i^* consists of an interval (V_i, V_{i+1}). Then we have,

$$\frac{(\psi + \varepsilon\zeta)^*(V) - \psi^*(V)}{\varepsilon} \xrightarrow{L^\infty \text{weak}*} W(V)$$

where,

(15) $$W(V) = \frac{d}{dV}\int_{\psi<\psi^*(V)} \zeta dV, \text{ if } V \notin P_i^*$$

(16) $$W(V) = \frac{d}{dV}\left[\int_{\psi<\psi^*(V)} \zeta dV + \int_0^{V-V_i} \zeta_i^* \, dv\right]$$

$$\text{if } V \in \left[V_i, V_{i+1}\right] = P_i^* \ ,$$

where here ζ_i denotes the restriction of ζ to P_i, and ζ_i^* denotes its increasing rearrangement on $(0, V_{i+1} - V_i)$.

As Mossino and Temam have shown that

$$W(V) = \int_{\psi<\psi^*(V)} \zeta, \text{ is a Lipschitz function} \ ,$$

(15) is defined a.e. in V.

However using only this result it is not possible to derive a weak equation. Consider the difference quotient for the QDE term,

$$\frac{1}{\varepsilon}\left\{\int_0^{|\Omega|} (\psi + \varepsilon\zeta)^{*'2} dV - \int_0^{|\Omega|} \psi^{*'2} dV\right\} = \int_0^{|\Omega|} \left[\frac{(\psi + \varepsilon\zeta)^{*'} - \psi^{*'}}{\varepsilon}\right]\left[(\psi + \varepsilon\zeta)^{*'} + \psi^{*'}\right] dV$$

To use the above result we could try for example to bring the derivatives of the difference quotient (integrating by parts) onto the sum so as to get

$$\int_0^{|\Omega|} \frac{(\psi + \varepsilon\zeta)^* - \psi^*}{\varepsilon}\left[(\psi + \varepsilon\zeta)^{*'} + \psi^{*'}\right]' dV \ .$$

As convergence is in L^∞ weak *, we would need,

(16) $(\psi + \varepsilon\zeta)^{*''} + \psi^{*''} \in L^1(0, |\Omega|)$,

to pass to the limit, and this is very difficult to prove. One of the *advantages* of the free boundary approximation approach described in the next section is that at least *formally* it is easier to establish a weak equation as a limit of approximating weak equations.

2) Even when additional assumptions are made which allow us to derive a weak equation, it is difficult to pass from this weak equation to the Euler equation (9). We do not enter into detail about this here, other than to say that even if minimizers are known to be very smooth this passage is not straightforward. The difficulties encountered surround the possibly complex structure of the Lebesgue-Stietjes measure induced by

$$df(t) = d\,|\{u \le t\} \cap \{\nabla u = 0\}|$$

and its relation to the coarea formula. We refer to [LS1, Lemma 1.1] for some results in a simpler context.

In the next section we describe in a general way the regularity expected for the QDE's.

3. Expected Regularity and a "formal" gradient bound

H. Grad has conjectured that solutions should be analytic if $\mu(\psi)$ is analytic and positive in a bounded smooth set Ω. The only known explicit solution of the QDE, $\Delta\psi(x) = -\lambda\psi^{*''}$, that is presently available, is in the radial case and is indeed analytic. The solution is given by,

$$\psi(|x|) = c.\log\left(\pi|x|^2 + \frac{\lambda}{4\pi}\right) + d \ , \tag{17}$$

where c and d can be adjusted to give the desired boundary conditions on $\psi(|x|)$ and ψ^*.

Grad's conjecture has not yet been thoroughly investigated for non convex Ω. For convex Ω, there is both numerical and analytic evidence that minimizers in [LS1] of a modified variational principle wherein minimizers are a priori restricted to have convex level sets are indeed analytic. This idea has first been used for certain free boundary problems by Acker and Caffarelli and Spruck [A, C-S]. We conjecture that minimizers in the restricted class are *also minimizers* to the original variational problem but this still awaits proof. We refer to the restricted variational problem in the sequel as the "modified" variational problem. Now, certain minimizers of the modified variational problem can be obtained as a limit of a sequence of free boundary problems as will be described in the next section. For such minimizers it is possible to show that the maximum is achieved at only one point where

$\psi \in C^1$ and $\nabla\psi = 0$. This fact turns out to be very important for regularity considerations. So we make the following conjecture.

Conjecture. Given Ω convex with C^2 boundary, if $\psi(x)$ is a minimizer to the modified variational problem that can be obtained as a limit of free boundary problems (of the type described in the next section), then $\psi(x)$ is analytic.

At present the free boundary approach appears to be well suited to establish the existence of Lipschitz minimizers. Other methods such as the bootstrap argument described in section can be used to establish higher regularity assuming the solution in C^2. No bridges between Lipschitz and C^2 regularity have as yet been constructed.

As it is self contained and instructive we next show how to obtain for convex Ω a gradient bound for solutions of,

(18)
$$\Delta\psi = -\psi^{*\prime\prime}, \text{ in } \Omega$$
$$\psi = 0 \text{ at } \partial\Omega$$
$$\psi^*(0) = 0, \quad \psi^*(|\Omega|) = 1 ,$$

if $\psi(x)$ is known to have convex level sets, $\psi(x)$ is a minimizer of (7) and $|\nabla\psi| > c > 0$ and if it is already known that $\psi(x)$ is a *classical* solution.

First note that, see [LS1].

$$V''(t) = \int_{\{\psi=t\}} \frac{\kappa(x)}{|\nabla\psi|^2}\, dl + \int_{\{\psi=t\}} \frac{\Delta\psi}{|\nabla\psi|^3}\, dl$$

where $\kappa(x)$ is the curvature of $\{\psi(x) = t\}$ at x. Inserting this now into (18), using the fact that $\Delta\psi$ is constant on level sets and solving for $\Delta\psi$ we get,

$$\Delta\psi(x) = \left\{ 1 + \frac{\displaystyle\int_{\{\psi=t\}} \frac{1}{|\nabla\psi|^3}\, dl}{\left(\displaystyle\int_{\{\psi=t\}} \frac{1}{|\nabla\psi|}\, dl\right)^3} \right\} \left(\int_{\{\psi=t\}} \frac{\kappa}{|\nabla\psi|^2}\, dl\right) \psi^{*\prime 3} .$$

As $\kappa(x)$ is positive this shows that ψ is superharmonic. Thus

$$\Delta\psi(x) = f_\psi(\psi) ,$$

where f is negative. Then we are in a position to apply the apriori estimate of J. Mossino (a generalization to non local equations of that of Payne) [MO, P-P], to obtain,

$$\max |\nabla\psi|^2 \leq \int_0^1 \psi^{*''} dt = \int_0^{|\Omega|} \psi^{*''}\psi^{*'}\, dv = \frac{1}{2}\left(\psi^{*'2}(|\Omega|) - \psi^{*'2}(0)\right) \quad .$$

Now for minimizers of (10) we have, (see [LS1], Thm 1.2)

$$0 \leq \psi^{*'}(v) \leq C \leq J(\psi) \quad ,$$

so we are done. We now go on to discuss in more detail the free boundary approach.

4. An approximation variational problem

The analysis of minimizers of $J(\psi)$ involves some formidable difficulties. In order to sidestep some of these we have introduced an approximation scheme in the setting where Ω is convex and all functions ψ have convex level sets. Contingent on certain conjectures about a basic free boundary problem we have used this scheme to prove the existence of a superharmonic minimizer ψ satisfying the Lipschitz bound $|\nabla\psi| \leq \inf J$. In addition we have shown that ψ achieves its maximum at a single point, where it is continuously differentiable with $\nabla\psi = 0$.

To describe our method of approximation consider partitions: $T = \{t_0, t_1, \ldots, t_n\}$ of order n such that $0 = t_0 < t_1 < \ldots < t_n = 1$. Also consider the collection of functions

(A) $C = \{\psi \in W^{1,2}(\Omega) \mid \psi^* \in W^{1,2}(0, |\Omega|), \psi = 0$ on $\partial\Omega, \psi^*(0) = 0, \psi^*(|\Omega|) = 1$, and $\{\psi = t\}$ is a convex curve for all but a possibly countable set of $t, 0 \leq t \leq 1\}$

(such ψ are continuous) and define,

$$J_T(\psi) = \int_\Omega |\nabla\psi|^2 + \sum_{i=1}^{n} \frac{(\Delta t_i)^2}{\Delta V_i}$$

where $\Delta t_i = t_i - t_{i-1}$ and $\nabla V_i = V_i - V_{i-1}$, $V_i = |\{\psi < t_i\}| = V_\psi(t_i)$. Note that $J(\psi) = J_T(\psi)$ if ψ^* is linear on the intervals (V_{i-1}, V_i). In [LS1], [LS2] we considered $t_i = i/n$, $i = 0, \ldots, n$ and general t_i are considered in [S].

When one minimizes $J_T(\psi)$ over C one gets a minimizer ψ which is continuous, and harmonic between the convex curves $\Gamma_i = \{\psi = t_i\}$, $i = 0, \ldots, n-2$, $\Gamma_{n-1} = \partial\{\psi = t_{n-1}\}$. Note that $\{\psi = t_{n-1}\}$ is a convex set of positive Lebesgue measure.

In [LS2], [S] we have proved that our approximation scheme is a reasonable one. Specifically if T_k is a sequence of partitions of order k with mesh size $\mathrm{Max}_i\, t_{i,k} \to 0$ as $k \to \infty$ then $\inf J_{T_k} \to \inf J$. In addition if ψ_k are minimizers of J_{T_k} (over C), $k = 1, 2, \ldots,$ then on a uitable subsequence $\psi_k \to \psi$ strongly in $W^{1,2}(\Omega)$ where ψ is a minimizer of J. This is ccomplished by showing on certain "good" annuli of the form $A_i = \{t_{i-1} < \psi < t_i\}$ that our inearization of the $\int \psi^{*'2} dv$ term is approximately true (ψ a minimizer of J_T) and that there are 'anishingly few "bad" annuli as the mesh size $\to 0$.

In [LS1] we consider superharmonicity and Lipschitzness of minimizers of J_T, $t_i = i/n$, $= 0,\ldots,n$. However it has not been possible in this setting to prove that the Lipschitz bound is uniform n n. To establish this we consider in [S] the quantity

$$J^n = \sup_{o(T)=n} \inf_{\psi \in C} J_T(\psi)$$

$o(T) = n$ if $T = \{t_0 < t_1 < \ldots < t_n\}$).

This is motivated by the need to do "range variations" where one varies the levels t_i. The ipparent need for such variations arises from the fact that the Lipschitz bound for minimizers ψ_T of J_T involves a discrete version of $\psi_T^{*'}$ and that one can show $\psi^{*'}$ is bounded for ψ a ninimizer of J by using range variations

$$\psi_\varepsilon = f_\varepsilon(\psi), \quad f_\varepsilon(t) = t + \varepsilon\eta(t) \ .$$

In fact we can show that there exist partitions T_n where the supremum J^n is achieved and hat for such partitions certain variations lead to the condition

$$\text{(A')} \qquad \int_{\Gamma_i} |\nabla\psi| + \frac{\Delta t_i}{\Delta V_i} = \inf J_{T_n} \le \inf J$$

$i = 1,\ldots,n$, for any minimizer ψ of J_{T_n}. Here $\Gamma_i = \{\psi = t_i\}$ for any $t, t_{i-1} < t < t_i$. Thus he difference quotients $\Delta t_i / \Delta V_i$ are uniformly bounded.

In addition minimizers ψ formally satisfy the free boundary conditions

$$\text{(B)} \qquad |\nabla\psi_{E_i}|^2 - |\nabla\psi_{I_i}|^2 = \left(\frac{\Delta t_i}{\Delta V_i}\right)^2 - \left(\frac{\Delta t_{i-1}}{\Delta V_{i-1}}\right)^2$$

across $\Gamma_i = \{\psi = t_i\}$ where $\nabla\psi_{E_i}$ and $\nabla\psi_{I_i}$ are gradients of ψ calculated from the exterior nd interior of Γ_i respectively. Given such free boundary conditions and the fact that harmonic unctions with convex level curves are convex on curves of steepest ascent one gets the estimate

(C) $$|\nabla\psi|^2 \le \left(\frac{\Delta t_n}{\Delta V_n}\right)^2 .$$

Combining this with (A') yields

(D) $$|\nabla\psi| \le \inf J .$$

One should note that our result that minimizers ψ of J_T are superharmonic as well as the uniform Lipschitz bound (D) both depend on the free boundary condition (B). In addition the condition (A') requires a certain uniqueness condition concerning our free boundary problem. At the present time these results are only conjectured although we have partial results for proving condition (B). It turns ou that one need only consider a free boundary problem on a double annulus which we describe in the next section.

A free boundary problem

In the previous section we described results depending on two conjectures dealing with the following problem.

Let

(E) $$J_t(\psi) = \int_\Omega |\nabla\psi|^2 = \frac{t^2}{|A_0|} + \frac{(1-t)^2}{|A_1|}$$

where $A_0 = \{0 \le \psi < t\}$, $A_1 = \{t \le \psi < 1\}$, and Ω is an annular region with exterior and interior convex boundary components Γ_0, Γ_1 respectively. As before we minimize our functional over functions ψ with convex level sets $\{\psi = s\}$. In addition we assume the boundary conditions $\psi = i$ on Γ_i, $i = 0,1$, and consider t satisfying $0 < t < 1$.

If ψ is a minimizer of J_t then $\Gamma = \{\psi = t\}$ is a convex curve and ψ is harmonic ir the interiors of the annuli A_0, A_1 (bounded by Γ_0, Γ and Γ, Γ_1 respectively). Analogous to (B) we have the free boundary condition

(F) $$|\nabla\psi_E|^2 - |\nabla\psi_I|^2 = \left(\frac{1-t}{|A_1|}\right)^2 - \left(\frac{t}{|A_0|}\right)^2$$

$\nabla\psi_E$, $\nabla\psi_I$ being exterior and interior gradients of ψ defined on Γ.

Conjecture 1. The free boundary condition (F) is satisfied for all minimizers of J_t.

We have proved that Γ is C^1 and that the conjecture is true on Γ away from possible line segments which are subsets of Γ.

The condition (A') from the previous section can be established contingent on the following conjecture.

Conjecture 2. There exists a unique minimizer of J_t.

The following conjecture is all that is needed to establish condition A' for some minimizer.

Conjecture 3. Given ψ a minimizer of J_t and associated free boundary Γ, there exist sequences $t_k^1 \downarrow t$, $t_k^2 \uparrow t$ and minimizers ψ_k^i, of $J_{t_k^i}$ $i = 1,2$, such that the free boundaries $\Gamma_k^i = \{\psi_k^i = t_k^i\}$, $i = 1,2$, both converge to Γ.

5. Higher regularity

Another approach, which yields higher regularity assuming solutions are C^2 has been developed recently. In any region delimited by level sets where the magnitude of the gradient is bounded below by a positive (arbitrarily small) constant, this approach allows us to conclude that solutions are C^∞. We now state this more precisely in the following

Theorem 5.1. *Given* $\Omega \subset R^2$, $\partial\Omega \in C^2$.

Let $\psi(x)$ *be a solution of*

$$\Delta \psi(x) = -\mu' \psi^{*'2} - 2\mu(\psi)\psi^{*''}, \text{ in } \Omega$$

$$\psi = 0 \quad \text{at } \partial\Omega$$

Assume

$$\mu(t) \in C^\infty(\Omega), \exists \varepsilon_1, : 0 < \varepsilon_1 \leq \mu(t) \ .$$

Then in any subregion $\Omega_{t,t'} = \{x : t < \psi(x) < t'\}$, $0 \leq t < t' \leq 1$, *where in addition* $\exists \varepsilon_2$ such that,

$$|\nabla\psi| > \varepsilon_2 > 0 \ ,$$

we have,

$$\psi(x) \in C^2(\Omega_{t,t'}) \Rightarrow \psi(x) \in C^\infty(\Omega_{t,t'}) \ .$$

Corollary 5.1. *Let* Ω *be convex, with* C^3 *boundary, and* R_{min} *denote the minimum radius of curvature of the boundary.*

Define,

$$t_{R_{min}} = \max t : \max_{\{\psi(x)=t\}} d(x,\partial\Gamma) \leq R_{min}$$

and

$$\Omega_{0,t_{R_{min}}} = \left\{ x : 0 \le \psi \le t_{R_{min}} \right\} .$$

Then, $\forall\ \varepsilon > 0$, *if in addition* $\psi(x)$ *is a minimizer of* $J(\psi)$ (see [LS1]), *then,*

$$\psi(x) \in C^{2,1}\left(\Omega_{0,t_{R_{min}}-\varepsilon}\right)$$

$$\Rightarrow \psi(x) \in C^{\infty}\left(\Omega_{0,t_{R_{min}}-\varepsilon}\right) .$$

Corollary 5.2. *If* Ω *is symmetric and convex in* 2 *orthogonal directions, and if* x_s *denotes the center symmetry, then for any* $\varepsilon > 0$, *if in addition* $\psi(x)$ *is a minimizer of* $J(\psi)$, *then*

$$\psi(x) \in C^{2,1}\left(\Omega \setminus \{x_s\}\right)$$

$$\Rightarrow \psi(x) \in C^{\infty}\left(\Omega \setminus \{x_s\}\right) .$$

Conjecture 5.1. If Ω is convex, $\partial\Omega \in C^2$ and $\psi(x)$ minimizes (7), then there exists a unique point x_0, such that for any $\varepsilon > 0$,

$$\psi(x) \in C^{2}\ \left(\Omega \setminus \{x_s\}\right)$$

$$\Rightarrow \psi(x) \in C^{\infty}\left(\Omega \setminus \{x_s\}\right) .$$

Proof of Corollary 5.1. By a recent result of C. Bandle [B] which develops the Gidas-Ni-Nirenberg result [GNN], $|\nabla\psi| \neq 0$ in $D_{R_{min}}$. By uniform continuity in $D_{R_{min}-\varepsilon}$, for any for any $\varepsilon > 0$, there therefore exists $\delta > 0$, st.

$$|\nabla\psi| > \delta(\varepsilon)$$

$$\text{in } D_{R_{min}-\varepsilon} .$$

Using that ψ is in $C^{2,1}$ and the fact that ψ is a minimizer it can be shown that $G(t) = \psi^{*''}(V_\psi(t))$ is locally Lipschitz.

Proof of Corollary 5.2. By a result of Kawohl [K], the gradient of $\psi(x)$ vanishes at only one point the center of symmetry. Then the same arguments as above apply.

Justification of Conjecture 5.1. There is now strong evidence that for convex domains Ω, minimizers of (10) have convex level sets and that their gradient vanishes only at one point which corresponds to the maximum of $\psi(x)$.

Sketch of proof of Theorem 4.1. The key to the proof of the theorem is an identity which expresses the third derivative of the distribution function of $\psi(x)$ in terms of integrals on level sets of $\psi(x)$ of second and first order derivatives weighted by reciprocal powers of the gradient. Of course since our solutions are only assumed to be $C^2(\Omega)$ a weak (difference quotient) form of this identity is needed and established in [LS3, LS4]. Here, we assume $\psi(x) \in C^3(\Omega)$; and $|\nabla\psi| > c > 0$. Also, to simplify the exposition we carry out the steps only for $\mu(\psi) = 1$.

The first step in establishing the needed identity is a special case of certain new calculus type formulas for level set analysis developed in [L1, L2], which show that, if $\psi(x) \in C^3(\Omega)$, $|\nabla\psi| > c > 0$,

$$V_\psi'''(t) = \int_{\{\psi=t\}} \left\{ -\frac{\nabla\psi \cdot \nabla(\Delta\psi)}{|\nabla\psi|^5} + \frac{|H|^2}{|\nabla\psi|^5} - \frac{\nabla\psi^t H^2 \nabla\psi}{|\nabla\psi|^7} \right\} dl \tag{19}$$

where H is the Hessian matrix,

$$|H|^2 = \sum_{i,j} \psi_{x_i x_j}^2$$

$$H^2 = HH,$$

and a superscript t denotes the transpose.

Next, we take a normal derivative of the equation (9) (for $\lambda = 1$) to obtain,

$$\frac{\nabla\psi}{|\nabla\psi|} \cdot \nabla\Delta\psi = \left(\frac{\nabla\psi}{|\nabla\psi|} \cdot \nabla \right) \Delta(-\psi^{*\prime\prime})$$

$$= \left(\frac{V'''(t)}{V'^3} - \frac{3V''^2}{V'^4} \right) |\nabla\psi|, \tag{20}$$

where for brevity we write V', V'' instead of V_ψ', $V_\psi''(t)$. We may solve for V''' to obtain,

$$V''' = \left\{ 1 + \left(\int_{\psi=t} \frac{1}{|\nabla\psi|^3} \right) V'^3 \right\} \times \left\{ \frac{3V''^2}{V'^4} \int_{\psi=t} \frac{1}{|\nabla\psi|^3} dl + \int_{\psi=t} \left[\frac{|H|^2}{|\nabla\psi|^5} - \frac{\nabla\psi^t H^2 \nabla\psi}{|\nabla\psi|^7} \right] dl \right\}.$$

To see the idea of the bootstrap argument, assume now that $\psi \in C^{2,\alpha}(\Omega)$. It is then possible to show under our assumption on $|\nabla\psi|$, that the right hand side of (21) is in $C^{0,\alpha}(0,1)$. Thus $V''' \in C^{0,\alpha}(0,1)$. This fact combined with equation (20), and elliptic regularity theory easily implies $\psi(x) \in C^{3,\alpha}$. The argument is then repeated (bootstrapped).

References

[A] **A. Acker,** On the convexity of equilibrium plasma configurations, Math. Methods Appl. Sci. 3 (1981), 435-443.

[B] **C. Bandle and B. Scarpellini,** On the location of the maxima in nonlinear Dirichlet problems, Expo. Math. 4 (1986), 75-85.

[C-S] **L. Caffarelli and J. Spruck,** Convexity properties of solutions to some classical variational problems, Communications in Partial Differential Equations 7 (11) (1982), 1337-1379.

[G-N-N] **B. Gidas, W.-M. Ni and L. Nirenberg,** Symmetry and related properties via the maximum principle, Comm. Math. Phys. 68 (1979), 209-243.

[G1] **H. Grad,** Magnetic confinement fusion energy research, Proceedings of Symposia in Applied Mathematics 21 (1979), 3-40.

[G2] **H. Grad,** Survey of $1^{1/2}$D Transport Codes, (00-307)-154, MF-93, October, 1978.

[G-H-S] **H. Grad, P.N. Hu and D.C. Stevens,** Adiabatic evolution of plasma equilibria, Proc. Nat. Acad. Sci. 72 (1975), 3789-393.

[K] **B. Kawohl,** A geometric property of solutions to semilinear elliptic Dirichlet problems, Appl. Anal. 16 (1983), 229-234.

[L1] **P. Laurence,** Une nouvelle formule pour dériver par rapport au niveau et applications a'la convexité de fonctionelles géométriques des solutions d'equations elliptiques, submitted to Comptes Rendus de l'Académie des Sciences, Oct. 86.

[L2] **P. Laurence,** On some convexity properties of geometric functionals of levels for solutions of certain elliptic equations and an isoperimetric inequality for P capacity in the plane, submitted to Journal de Mathématiques et de Physique appliquées, (ZAMP).

[LS1] **P. Laurence and E. Stredulinsky,** A new approach to queer differential equations, Communications on Pure and Applied Mathematics 38 (1985), 333-355.

[LS2] **P. Laurence and E. Stredulinsky,** Convergence of a sequence of free boundary problems related to the Grad variational problem in plasma physics, submitted to Communications on Pure and Applied Mathematics.

[LS3] **P. Laurence and E. Stredulinsky,** Une méthode de "bootstrap" pour les équation aux dérivées "queer" de la physique des plasmas, to appear in Compte Rendees de l'Académie des Sciences, Oct. 1986.

[LS4] **P. Laurence and E. Stredulinsky,** A bootstrap argument for queer differential equations, submitted to Indiana University Mathematics Journal.

[Mo] **J. Mossino,** A priori estimates in plasma confinement, Applicable Analysis 13 (1982), 185-207.

[Mo-Te] **J. Mossino and R. Temam,** Directional Derivative of the increasing rearrangement mapping and application to a queer differential equation in plasma physics, Duke Math. J. 43 (1981), 475-496.

[Pa-Ph] **L. Payne and G.A. Philippin,** Some maximum principles for nonlinear elliptic equations in divergence form with applications to capillary surfaces and to surfaces of constant mean curvature, Nonlinear analysis, Theory, Methods and Appl. 3 n. 2 (1979).

[S] **E. Stredulinsky,** Lipschitz variational solutions for Grad's generalized differential equations as the limit of solutions of a sequence of free boundary problems, in preparation.

[Te] **R. Temam,** The shape at equilibrium of a confined plasma, Arch. Ration. Mech. Analysis 60 (1975), 51-73.

ON THE DIFFUSION COEFFICIENT OF A SEMILINEAR NEUMANN PROBLEM*

Chang-Shou Lin
Department of Mathematics
University of California
La Jolla, CA 92093
(U.S.A.)

and

Wei-Ming Ni
School of Mathematics
University of Minnesota
Minneapolis, MN 55455
(U.S.A.)

Dedicated to Hans Lewy

1. Introduction

The purpose of this paper is to study the existence and nonexistence of *positive nontrivial* solutions of the following semilinear *Neumann* problem

$$\text{(1.1)} \qquad \begin{cases} d\Delta u - u + u^p = 0 & \text{in } \Omega , \\ \dfrac{\partial u}{\partial \nu} = 0 & \text{on } \partial\Omega , \end{cases}$$

where $\Delta = \sum_{i=1}^{n} \frac{\partial^2}{\partial x_i^2}$ is the usual Laplacian, $d > 0$, $p > 1$ are constants, Ω is a bounded smooth domain in $\mathbf{R}^n$, $n \geq 2$ and ν is the unit outer normal to $\partial\Omega$. This problem arises naturally in various recent studies of pattern formations in mathematical biology. For instance, a positive nontrivial solution of (1.1) would give rise to a steady state of the Keller-Segal's chemotaxis model – a two-by-two parabolic system – of cellular slime molds (amoebae) with a logarithmic sensitivity function (see e.g. [LNT] and the references therein). The problem (1.1) is also known as the "shadow system" of an activator-inhibitor type system (again, a two-by-two semilinear system) due to Gierer-Meinhardt (see e.g. [LNT] and the references therein).

Mathematically, it seems interesting to compare (1.1) to its Dirichlet counterpart

$$\text{(1.2)} \qquad \begin{cases} d\Delta u - u + u^p = 0 & \text{in } \Omega , \\ u = 0 & \text{on } \partial\Omega . \end{cases}$$

* Research supported in part by the National Science Foundation.

Here again, we are only interested in positive solutions. This Dirichlet problem has been studied quite extensively and much is known. A well-known basic result is that (1.2) *has a positive solution if and only if* $p < (n+2)/(n-2)$, *provided* Ω *is star-shaped.* In general, existence theorems of (1.2) depend on the exponent p and on the geometry of the domain Ω, but are *independent* of the "diffusion coefficient" d. Also note that any positive solution of (1.2) is nontrivial.

The story of (1.1) seems quite different. In studying (1.1), the variational approach seems fairly successful at least in case $p < (n+2)/(n-2)$. (See [LNT].) Applying the Mountain-Pass Lemma of Ambrosetti and Rabinowitz to the variational functional associated with the problem (1.1) in $H_1(\Omega)$ (i.e. the Sobolev space of all square-integrable functions with square-integrable distributional derivatives)

$$(1.3) \qquad J_d(u) = \frac{1}{2}\int_\Omega \left(d\,|Du|^2 + u^2\right) - \frac{1}{p+1}\int_\Omega (u_+)^{p+1}$$

where $u_+ = \max\{u,0\}$, we obtain easily the existence of a positive solution u_d of (1.1), for each $d > 0$. However it is not clear that this solution is different from the constant solution $u \equiv 1$. In fact, in [LNT] we showed that *in case* $p < (n+2)/(n-2)$ *all positive solutions of* (1.1) *must be the constant solution* 1 *if* d *is sufficiently large.* (Thus in particular, $u_d \equiv 1$ for d large.) On the other hand, we also proved ([LNT]) that *for* $p < (n+2)/(n-2)$, $J_d(u_d)$ *is of the order* $d^{n/2}$ *for* d *near* 0, which implies that *there exists a* $d_0 > 0$ *such that for all* $d < d_0$, *the solution* u_d *is nontrivial* since $J_d(1) = \left(\frac{1}{2} - \frac{1}{p+1}\right)|\Omega|$ is independent of d. Moreover this estimate on $J_d(u_d)$ also enables us to prove that $u_d \to 0$ *in measure but the* L^∞*-norm of* u_d *remains bounded (always greater than* 1) *as* d *tends to* 0, which gives rise to a "point-condensation" solution of (1.1).

In this paper we shall estimate the number d_0 and also investigate the "super-critical" case $p \geq (n+2)/(n-2)$. We shall prove the following two theorems.

Theorem 1.4. $d_0 \geq (p-1)/\lambda_2$ where λ_2 *is the second eigenvalue of the Laplacian* Δ *on* Ω *with zero Newmann boundary data.*

Theorem 1.5. *Let* Ω *be a ball. Then for every* $p > (n+2)/(n-2)$ *there exists a positive number* d_1 *such that* (1.1) *possesses a nontrivial positive radial solution if* $d < d_1$, *and* (1.1) *does not possess any nontrivial positive radial solution if* $d > d_1$.

The latter seems to suggest the following:

Conjecture 1.6. Let Ω *be a bounded smooth domain in* $\mathbf{R}^n$ *and* $p > 1$. *Then* (1.1) *possesses a nontrivial positive solution if* d *is small, and* (1.1) *does not possess any nontrivial positive solution if* d *is large.*

In addition to Theorem 1.5, this conjecture can also be verified for radial solutions in annuli (see Section 4 below).

We would like to point out that in contrast to the Dirichlet problem (1.2), the primary parameter for the existence of nontrivial positive solutions of (1.1) seems to be the diffusion coefficient d instead of the exponent p. However, we should remark that the exponent p does seem to play a perhaps secondary role here. For, the nontrivial positive radial solutions guaranteed by Theorem 1.5 seem to exhibit a similar but different behaviour to the "point-condensation" character that the "Mountain-Pass solutions" u_d's have in the case $p < (n+2)/(n-2)$. It seems that as d tends to 0 there is a sequence of solutions guaranteed by Theorem 1.5 tending to 1 almost everywhere while their L^∞-norms tend to ∞. We hope to report our progress in this direction in a future paper. In fact, most of the results presented here are not definitive and they seem to deserve further studies.

This paper is organized as follows. Theorem 1.4 is proved in Section 2, Theorem 1.5 is established in Section 3, and Section 4 contains miscellaneous remarks.

Finally we remark that some of the arguments in [LNT] which are needed in this paper are only sketched. The references cited here of other related mathematical or biological aspects of this problem are by no means complete, we refer the interested readers to [LNT] and the references therein for further details.

2. A lower estimates for existence

In this section we shall prove an extension of Theorem 1.4. We consider the following problem

$$(2.1)\qquad \begin{cases} d\Delta u - u + f(u) = 0 & \text{in } \Omega , \\ \dfrac{\partial u}{\partial \nu} = 0 & \text{on } \partial\Omega , \end{cases}$$

where $f(u)$ satisfies the following hypotheses:

(2.2) $f(u) = 0$ for $u \le 0$, $f(u) = o(|u|)$ near $u = 0$ and f is C^2 in $u > 0$,

(2.3) $\frac{f(u)}{u}$ is strictly increasing in $u > 0$ and $\frac{f(u)}{u} \to \infty$ as $u \to \infty$.

(2.4) $f(u) = O(u^p)$ near $u = \infty$, where $1 < p < (n+2)/(n-2)$. (In case $n = 2$, $f(u) = O(\exp \alpha(u))$ near $u = \infty$ where $\alpha(u) = o(u^2)$ near $u = \infty$).

(2.5) There exist constants $\theta \in \left(0, \frac{1}{2}\right)$ and $U > 0$ such that

$$F(u) \equiv \int_0^u f(t)dt \le \theta u f(u) \quad \text{for all} \quad u \ge U .$$

Under the above hypotheses, Theorem 2 and Proposition 2.2 in [LNT] (note that (h_5) in [LNT] is automatic here due to (2.2) and (2.3) above) guarantee that for each $d > 0$ (2.1) possesses a positive solution u_d which is a critical point of the variational functional

$$J_d(u) = \frac{1}{2}\int_\Omega \left(d\,|Du|^2 + u^2\right) - \int_\Omega F(u) \tag{2.6}$$

in $H_1(\Omega)$, and the corresponding critical value $c_d = J_d(u_d)$ is given by

$$c_d = \inf_{h\in\Gamma} \max_{0\le t\le 1} J(h(t)) \tag{2.7}$$

where

$$\Gamma = \left\{ h \in C\left([0,1], H_1(\Omega)\right) \mid h(0) = 0, h(1) = e \right\} \tag{2.8}$$

and $e \in H_1(\Omega)$, $e > 0$ in Ω with $J_d(te) \le 0$ for all $t \ge 1$ (but otherwise arbitrary). Moreover, c_d is of the order $d^{n/2}$ for d sufficiently small. Therefore there is a $d_0 > 0$ such that

$$J_d(u_d) < J_d(1) = \left(\frac{1}{2} - F(1)\right) |\Omega|$$

for all $d < d_0$. This certainly also gives an estimate on d_0. To obtain a better estimate on d_0, we consider the eigenvalue problem of the linearized equation of (2.1) at $u = u_d$

$$\begin{cases} \left(d\Delta - 1 + f'(u_d)\right)\phi + \lambda\phi = 0 & \text{in } \Omega , \\ \dfrac{\partial\phi}{\partial\nu} = 0 & \text{on } \partial\Omega . \end{cases} \tag{2.9}$$

We shall denote the eigenvalues by $\lambda_1(u_d), \lambda_2(u_d), \ldots$ and the corresponding eigenfunctions by $\phi_1, \phi_2, \ldots$. It is well-known that $\lambda_1(u_d) < \lambda_2(u_d) \le \ldots$ and $\phi_1 > 0$ in Ω .

Lemma 2.10. $\lambda_1(u_d) < 0$ *for all* $d > 0$.

Proof. By Green's identity, we have

$$0 = d\int_\Omega \left(\phi_1 \Delta u_d - u_d \Delta\phi_1\right) = \int_\Omega \left\{ \left[u_d f'(u_d) - f(u_d)\right]\phi_1 + \lambda_1(u_d)\phi_1 u_d \right\} > \lambda_1(u_d)\int_\Omega \phi_1 u_d$$

by (2.3). Thus $\lambda_1(u_d) < 0$. Q.E.D.

The following theorem is crucial in the proof of Theorem 1.4, and seems of independent interest.

Theorem 2.11. $\lambda_2(u_d) \geq 0$ *for all* $d > 0$.

A few lemmas are in order.

Lemma 2.12. *The function defined by* $g(t) \equiv J_d(te)$, *for all* $t \geq 0$, *has exactly one critical point which is the maximum point of* g *in* $t \geq 0$.

Proof. Compute

$$g(t) = \frac{t^2}{2}\int_\Omega \left(d|De|^2 + e^2\right) - \int_\Omega F(te) ,$$

$$g'(t) = t\int_\Omega \left(d|De|^2 + e^2\right) - \int_\Omega ef(te) ,$$

$$g''(t) = \int_\Omega \left(d\,|De|^2 + e^2\right) - \int_\Omega e^2 f'(te) .$$

By our assumptions on f and e, we see that $g(0) = 0$, $g(t) > 0$ for $t > 0$ and small, and $g(t) \leq 0$ for $t \geq 1$. Thus g must have at least one critical point. Now let t_0 be an arbitrary critical point of g, i.e. $g'(t_0) = 0$. Then

$$\int_\Omega \left(d|De|^2 + e^2\right) = \frac{1}{t_0}\int_\Omega f(t_0 e)e .$$

Substituting into $g''(t_0) = 0$, we obtain

$$g''(t_0) = \int_\Omega e^2\left[\frac{f(t_0 e)}{t_0 e} - f'(t_0 e)\right]$$

which is negative by (2.3). Therefore $g(t_0)$ must be a strict maximum and hence t_0 is the unique critical point. Q.E.D.

Remark 2.13. Since $u_d > 0$ on $\overline{\Omega}$ (by the maximum principle, see the proof of Theorem 2 in [LNT]) *we can now choose* $e = t_1 u_d$ in our variational formulation (2.8), where $t_1 > 0$ is the number giving $J_d(t_1 u_d) = 0$ (by Lemma 2.12, such a number is unique). It is not hard to prove that the critical value c_d given by (2.7) is left unchanged (see the proof of Proposition 2.14 in [DN]). *It then follows that* c_d *is realized by the path* $\{te \mid 0 \le t \le 1\}$ and the maximum of $J_d(te)$, $t \ge 0$, is assumed at $t = 1/t_1$, i.e. the maximum of J_d on the ray $\{tu_d \mid t \ge 0\}$ is $J_d(u_d)$.

For an arbitrary $v \in C^1(\overline{\Omega})$, there exists a $\delta > 0$ such that $u_d + \varepsilon v > 0$ on $\overline{\Omega}$ for all $\varepsilon \in (-\delta, \delta)$ since $u_d > 0$ on $\overline{\Omega}$. Fix this v for the time being and define

$$\text{(2.14)} \qquad G(t,\varepsilon) = \frac{t^2}{2}\int_\Omega \left[d\,|D(u_d + \varepsilon v)|^2 + (u_d + \varepsilon v)^2 \right] - \int_\Omega F\big(t(u_d + \varepsilon v)\big)$$

for $t \ge 0$ and $-\delta < \varepsilon < \delta$.

Lemma 2.15. *There exists a* δ' *in the interval* $(0, \delta)$ *such that for every* $\varepsilon \in (-\delta', \delta')$ *there is a unique* $t = t(\varepsilon)$ *satisfying* $G_t(t(\varepsilon), \varepsilon) = 0$. *Moreover,* $t(\varepsilon) \in C^2(-\delta', \delta')$.

Proof. It is obvious that $G_t(1,0) = 0$ since u_d is a solution of (2.1). Exactly the same proof of Lemma 2.12 shows that $G_{tt}(1,0) \ne 0$ (see also (2.19) below). Thus the Implicit Function Theorem implies that there exists a unique function $t = t(\varepsilon)$ with $t(0) = 1$ such that $G_t(t(\varepsilon), \varepsilon) = 0$ for all $\varepsilon \in (-\delta', \delta')$ where δ' is some number between 0 and δ. The conclusion that $t(\varepsilon)$ is C^2 follows from the fact that $G_t(t, \varepsilon)$ is C^2 near the point $t = 1$, $\varepsilon = 0$. Q.E.D.

We now come to the

Proof of Theorem 2.11. First observe that for each fixed ε in $(-\delta', \delta')$, $G(t, \varepsilon)$, as a function of $t \ge 0$, assumes its unique maximum at $t = t(\varepsilon)$ by Lemma 2.12. Set $k(\varepsilon) = G(t(\varepsilon), \varepsilon)$ for $\varepsilon \in (-\delta', \delta')$. From (2.7), Remark 2.13 and the arguments in the proof of Proposition 2.14 in [DN], we see that $k(\varepsilon) \ge k(0) = G(1,0)$ for $\varepsilon \in (-\delta', \delta')$. Thus $k'(0) = 0$ and $k''(0) \ge 0$. Using Lemma 2.15, we compute

$$0 = \frac{d}{d\varepsilon} G(t(\varepsilon),\varepsilon)\Big|_{\varepsilon=0} = G_\varepsilon(1,0) \ ;$$

and, since $G_t(1,0) = 0$,

(2.16) $$0 \le \frac{d^2}{d\varepsilon^2} G\big(t(\varepsilon),\varepsilon\big)\big|_{\varepsilon=0} = [G_{tt}t'^2 + G_t t'' + 2G_{t\varepsilon} t' + G_{\varepsilon\varepsilon}]\big|_{\varepsilon=0} =$$

$$= G_{tt}(1,0)[t'(0)]^2 + 2G_{t\varepsilon}(1,0)t'(0) + G_{\varepsilon\varepsilon}(1,0) \ .$$

On the other hand, differentiating $G_t(t(\varepsilon), \varepsilon) \equiv 0$ with respect to ε and evaluating at $\varepsilon = 0$, we derive

(2.17) $$G_{tt}(1,0)t'(0) + G_{t\varepsilon}(1,0) = 0 \ .$$

Combining with (2.16), we have

(2.18) $$-G_{tt}(1,0)[t'(0)]^2 + G_{\varepsilon\varepsilon}(1,0) \ge 0 \ .$$

Straightforward computation gives

(2.19) $$G_{tt}(1,0) = \int_\Omega \left[u_d f(u_d) - u_d^2 f'(u_d)\right] \ ,$$

(2.20) $$G_{\varepsilon\varepsilon}(1,0) = \int_\Omega \left[\left(d\,|Dv|^2 + v^2\right) - v^2 f'(u_d)\right] \ ,$$

(2.21) $$G_{t\varepsilon}(1,0) = \int_\Omega v\left[f(u_d) - u_d f'(u_d)\right] \ .$$

Recall that the condition (2.3) implies that $f(u_d) - u_d f'(u_d) < 0$ in Ω and thus $G_{tt}(1,0) < 0$.

Setting

$$V = \left\{ v \in H_1(\Omega) \mid \int_\Omega v[f(u_d) - u_d f'(u_d)] = 0 \right\} \ ,$$

we conclude from (2.17) and (2.21) that $t'(0) = 0$ if $v \in C^1(\bar{\Omega}) \cap V$. (2.18) then implies that $G_{\varepsilon\varepsilon}(1,0) \ge 0$, i.e.,

(2.22) $$\int_\Omega \left[\left(d\,|Dv|^2 + v^2\right) - v^2 f'(u_d)\right] \ge 0 \ ,$$

for all $v \in V \cap C^1(\bar{\Omega})$. Since $C^1(\bar{\Omega})$ is dense in $H_1(\Omega)$ and u_d is bounded in Ω, (2.22) holds for all $v \in V$.

By the variational characterization of $\lambda_2(u_d)$ (see [CH])

$$\lambda_2(u_d) = \sup_{w \in H_1(\Omega)} \inf \left\{ \frac{\int_\Omega \left[d\,|Dv|^2 + v^2 - f'(u_d)v^2 \right]}{\int_\Omega v^2} \;\middle|\; 0 \neq v \in H_1(\Omega) \text{ and } \int_\Omega vw = 0 \right\}$$

our conclusion follows. (Just choose $w = f(u_d) - u_d f'(u_d)$. The fact that $w \in H_1(\Omega)$ follows from (2.2) and that $u_d > 0$ in $\bar{\Omega}$). Q.E.D.

From (2.3) it is easy to see that for every $d > 0$, the Neumann problem (2.1) has a *unique positive constant solution,* which we donote by u_c (note that u_c is independent of d); i.e. $f(u_c) = u_c$ and $u_c > 0$. Similarly, for each $d > 0$, let $\lambda_2(u_c;d)$ be the second eigenvalue of the linearized problem of (2.1) at u_c, i.e. for $j = 1,2,\ldots$,

$$\text{(2.23)} \qquad \begin{cases} \left[d\Delta - 1 + f'(u_c) \right] \psi_j + \lambda_j(u_c;d)\psi_j = 0 & \text{in } \Omega, \\[2ex] \dfrac{\partial \psi_j}{\partial \nu} = 0 & \text{on } \partial\Omega. \end{cases}$$

Since $f'(u_c)$ is a constant, the first eigenfunction of (2.23) is a constant function, and the second eigenfunction ψ_2 of (2.23) is also a second eigenfunction of Δ in Ω with zero Neumann boundary data. Rewrite (2.23) as follows:

$$\begin{cases} \left(\Delta + \dfrac{f'(u_c) + \lambda_2(u_c;d) - 1}{d} \right) \psi_2 = 0 & \text{in } \Omega, \\[2ex] \dfrac{\partial \psi_2}{\partial \nu} = 0 & \text{on } \partial\Omega, \end{cases}$$

it must then follow that

$$\lambda_2 = \left[f'(u_c) + \lambda_2(u_c;d) - 1 \right] / d$$

where λ_2 is the second eigenvalue of Δ in Ω with zero Neumann boundary data. Thus

$$\lambda_2(u_c;d) = d\lambda_2 + 1 - f'(u_c)$$

which is negative if $d < (f'(u_c) - 1)/\lambda_2$. Therefore we have proved the following extension of Theorem 1.4.

Theorem 2.24. *If* $d < (f'(u_c) - 1)/\lambda_2$ *where* u_c *is the unique positive zero of* $f(u) - u = 0$, *then* $u_d \neq u_c$ *where* u_d *is any critical point of* J_d *(defined by* (2.6)) *with* $J_d(u_d) = c_d$ *(defined by* (2.7)).

Remark. If $f(u) = u^p$, then $u_c = 1$ and $f'(u_c) = p$.

3. The super-critical case

This section is devoted to the study of radial solutions of (1.1) in the case Ω is a ball and the exponent $p > (n+2)/(n-2)$. For simplicity, we shall be dealing with the nonlinearity u^p instead of the more general one as in Section 2 although our analysis does generalize.

Let B_r be the ball centered at the origin with radius r. Consider (1.1) in the unit ball

$$(3.1)\qquad \begin{cases} d\Delta u - u + u^p = 0 \quad \text{in } B_1 , \\ u > 0 \quad \text{in } B_1 , \\ \dfrac{\partial u}{\partial \nu} = 0 \quad \text{on } \partial B_1 . \end{cases}$$

To establish Theorem 1.5, it suffices to prove the following

Theorem 3.2. *If* $p > (n+2)/(n-2)$, *then* $u \equiv 1$ *is the only radial solution of* (3.1) *provided* d *is sufficiently large.*

Indeed, a change of variables $y = x/\sqrt{d}$, $\tilde{u}(y) = u(x)$ brings (3.1) to

$$(3.3)_d\qquad \begin{cases} \Delta\tilde{u} - \tilde{u} + \tilde{u}^p = 0 \quad \text{in } B_{1/\sqrt{d}} , \\ \tilde{u} > 0 \quad \text{in } B_{1/\sqrt{d}} , \\ \dfrac{\partial \tilde{u}}{\partial \nu} = 0 \quad \text{on } \partial B_{1/\sqrt{d}} . \end{cases}$$

Then Theorem 1.5 follows from Theorem 1.6 in [N], Theorem 3.2 above and the continuous dependence on initial values. Furthermore, Theorem 3.2 follows from

Theorem 3.4. *If* $p > (n+2)/(n-2)$, *then* $\tilde{u} \equiv 1$ *is the only radial solution of* $(3.3)_d$ *provided* d *is sufficiently large.*

The proof of Theorem 3.4 consists of several steps.

Lemma 3.5. *There exists* $\tilde{d} > 0$ *such that if* $(3.3)_d$ *possesses a nontrivial radial solution* $\tilde{u}$ *for some* $d > \tilde{d}$, *then* $\tilde{u}(0) > 1$.

Proof. We assert that if $\tilde{u}$ is a radial solution of $(3.3)_d$ for some d with $\tilde{u}(0) < 1$, then $d^{-1/2} \geq C$ where the positive constant C is independent of d and $\tilde{u}(0)$.

Let r_0 be the point at which $\tilde{u}(r_0) = 1$ and $\tilde{u}(r) < 1$ for $r < r_0$. Observe that $\tilde{u}$ is strictly increasing in $(0,r_0)$. If $\tilde{u}(0)$ is close to 0, then the continuous dependence on initial values implies that $d^{-1/2}$ must be large. Thus we only have to prove our assertion for $\tilde{u}(0) < 1$ but close to 1. The following identity holds for all $R \in (0,d^{-1/2}]$

$$\frac{1}{2}(\tilde{u}'(R))^2 + \int_0^R \frac{n-1}{r}[\tilde{u}'(r)]^2\,dr + h(\tilde{u}(R)) - h(\tilde{u}(0)) = 0 \quad ,$$

where $h(u) = \int_0^u \left(-t + t^p\right)dt$. Choosing $\tilde{u}(R)$ to be the maximum or the minimum of $\tilde{u}$ in $(0,d^{-1/2}]$, we obtain $h(\tilde{u}(R)) < h(\tilde{u}(0))$, which implies that there exists a constant C such that

(3.6) $$|\tilde{u}(R) - 1| \leq C|\tilde{u}(0) - 1|$$

provided $\tilde{u}(0)$ is sufficiently close to 1. Now, rewrite the equation in $(3.3)_d$ as

(3.7) $$\Delta\tilde{\rho} + [(p-1) + \xi]\tilde{\rho} = 0$$

where $\tilde{\rho}(r) = [\tilde{u}(r) - 1]/[\tilde{u}(0) - 1]$ and

$$\xi(r) = \begin{cases} \dfrac{\tilde{u}^p(r) - \tilde{u}(r)}{\tilde{u}(r) - 1} - (p-1) & \text{if } \tilde{u}(r) \neq 1 \ ; \\ 0 \quad \text{if } \tilde{u}(r) = 1 \ . \end{cases}$$

It then follows from (3.6) that $\xi(r) = 0(|\tilde{u}(0) - 1|)$ *uniformly in* $r \leq d^{-1/2}$, i.e. when $|\tilde{u}(0) - 1|$ is small, $\tilde{\rho}$ in (3.7) is small perturbation of the radial solution ρ of the following linear equation in $B_{1/\sqrt{d}}$

$$\Delta\rho + (p-1)\rho = 0$$

with $\rho(0) = 1 = \tilde{\rho}(0)$. Let z_0 be the first positive zero of ρ (i.e. $p-1$ is the first eigenvalue of Δ on B_{z_0} with zero Dirichlet boundary data), and $\tilde{z}_0$ be the first positive zero of $\tilde{\rho}$.

Now the continuous dependence on parameters implies that as $\tilde{u}(0)$ tends to 1, $\tilde{z}_0$ tends to z_0. Since $\tilde{z}_0 < d^{-1/2}$, our assertion is established. Q.E.D.

Remark. We have actually proved that if $\tilde{u}$ is a radial solution of $\Delta u - u + u^p = 0$ in $\mathbf{R}^n$, then as $\tilde{u}(0)$ tends to 1 the first positive zero of $\tilde{u} - 1$ tends to a positive limit z_0 (the case $\tilde{u}(0) > 1$ is handled by exactly the same arguments).

Next, suppose that $\tilde{u}$ is a nontrivial radial solution of $(3.3)_d$ for some $d > \tilde{d}$, then there exists an $r_0 \in (0, d^{-1/2})$ such that $\tilde{u}(r) > 1$ for all $r < r_0$ and $\tilde{u}(r_0) = 1$. This implies that the Dirichlet problem

$$(3.8)_{d'} \qquad \begin{cases} \Delta w - w + w^p = 0 & \text{in } B_{1/\sqrt{d'}}, \\ w > 1 & \text{in } B_{1/\sqrt{d'}}, \\ w = 1 & \text{on } \partial B_{1/\sqrt{d'}}, \end{cases}$$

has a solution for some $d' > d$. Making the change of variables $x = y\sqrt{d'}$, $v(x) = w(y) - 1$, we derive from $(3.8)_{d'}$ that

$$(3.9) \qquad \begin{cases} d'\Delta v + f(v) = 0 & \text{in } B_1, \\ v > 0 & \text{in } B_1, \\ v = 0 & \text{on } \partial B_1, \end{cases}$$

where

$$f(v) = -(v+1) + (v+1)^p = (v+1)\left[(v+1)^{p-1} - 1\right]$$

for $v \in \mathbf{R}$. Observe that $f(0) = 0$, $f(v) \geq 0$ for $v \geq 0$ and $f'(0) = p - 1 > 0$. Thus $f(v) = 0(|v|)$ near $v = 0$ and $F(v) = 0(v^2)$ near $v = 0$ where

$$F(v) = \int_0^v f(t)dt .$$

Lemma 3.10. *If* $p > (n+2)/(n-2)$, *then* (3.9) *does not possess any solution if* d' *is sufficiently large.*

Proof. We shall make use of an idea due to Rabinowitz in [R]. Suppose that v is a solution of (3.9) for some $d' > 0$. Using the Rellich-Pohozaev identity, we have

$$(3.11) \qquad 0 < \int_{\partial B_1} d'(x \cdot \nu) |Dv|^2 = \int_{B_1} \left[2nF(v) - (n-2)vf(v)\right] .$$

Recall that $2nF(v) - (n-2)vf(v) = 0(v^2)$ near $v = 0$. Since the leading term of

$2nF(v) - (n-2)vf(v)$ near $v = \infty$ is

$$\left[\frac{2n}{p+1} - (n-2)\right] v^{p+1} ,$$

for any $\varepsilon > 0$, there exists v_0 such that

$$2nF(v) - (n-2)vf(v) \le \left[\frac{2n}{p+1} - (n-2) + \varepsilon\right] v^{p+1}$$

for all $v \ge v_0$. By our assumption on p, we can choose an $\varepsilon > 0$ so small that $2n/(p+1) - (n-2) + \varepsilon < 0$. Then choose a constant C_ε so large that

$$2nF(v) - (n-2)vf(v) \le \left[\frac{2n}{p+1} - (n-2) + \varepsilon\right] v^{p+1} + C_\varepsilon v^2$$

for all $v \ge 0$. Substituting this into (3.11), we derive

$$\text{(3.12)} \qquad \int_{B_1} v^{p+1} \le C \int_{B_1} v^2 ,$$

where the constant C is independent of d' and v. On the other hand, from equation (3.9) we have

$$\text{(3.13)} \qquad d' \int_{B_1} |Dv|^2 = \int_{B_1} vf(v) \le C \int_{B_1} \left(v^2 + v^{p+1}\right)$$

for some constant C independent of d' and v since $vf(v) = 0(v^2)$ near $v = 0$ and $vf(v) = 0(v^{p+1})$ near $v = \infty$. It then follows that

$$\text{(3.14)} \qquad \mu_1 d' \|v\|^2_{L^2(B_1)} \le d' \|Dv\|^2_{L^2(B_1)} \le C \|v\|^2_{L^2(B_1)}$$

where $\mu_1 > 0$ is the first eigenvalue of Δ on B_1 with zero Dirichlet boundary data. (3.14) implies that $d' \le C/\mu_1$ and the proof is completed. Q.E.D.

Now Theorem 3.4 follows from Lemmas 3.5, 3.10 and the arguments before Lemma 3.10.

4. Miscellaneous remarks

(I) In case Ω is an annulus, Conjecture 1.6 holds true for positive radial solutions of (1.1). The existence of nontrivial positive radial solution of (1.1) for small d follows from applying the same arguments used in [LNT] and in Section 2 above to the one dimensional problem (*for any* $p > 1$)

$$(4.1)\qquad \begin{cases} d(r^{n-1}u')' - r^{n-1}u + r^{n-1}u^p = 0 \quad \text{in} \quad (a,b) \ , \\ u > 0 \quad \text{in} \quad [a,b] \ , \\ u'(a) = u'(b) = 0 \ , \end{cases}$$

where $0 < a < b < \infty$ ($\Omega = \{x \in \mathbf{R}^n \,|\, a < |x| < b\}$). We omit the proof here. The nonexistence of nontrivial positive solutions of (4.1) for large d also follows from similar arguments used in [NT] in establishing nonexistence theorems for general domain Ω but smaller exponents p. We sketch a simple proof here.

Let u be a nontrivial solution of (4.1). Integrating the equation in (4.1) gives

$$\int_a^b u^p r^{n-1} dr = \int_a^b u r^{n-1} dr \le C \left(\int_a^b u^p r^{n-1} dr \right)^{1/p} ,$$

where the constant C depends on n, a and b. Thus

$$\int_a^b u^p r^{n-1} dr \le C \ .$$

Integrating the equation in (4.1) from a to r, we obtain

$$d r^{n-1} u'(r) = \int_a^r \left(u - u^p\right) s^{n-1} ds \ .$$

Therefore

$$d\,|u'(r)| \le C \ .$$

Since there is an $r_0 \in (a,b)$ with $u(r_0) = 1$, it follows that

$$|u(r) - 1| \le \int_{r_0}^{r} |u'(s)|\,ds \le \frac{C}{d} \ ,$$

i.e.

$$\|u\|_{L^\infty} \le 1 + \frac{C}{d} \ ,$$

where C depends only on n, a, b, and is independent of d. Now standard arguments (see e.g. the proof of Theorem 3 in [NT]) show that $d \le C$.

Remark. The above proof works for very general nonlinear term $f(u)$ (instead of the pure power u^p). The only assumption needed is that *there exists a constant* $C > 1$ *such that* $f(u) \geq Cu$ *for* u *sufficiently large.*

(II) In case Ω is a ball and $p = (n+2)/(n-2)$, Conjecture 1.6 is yet to be verified. The method used in Section 3, namely, deducing the nonexistence of solutions to Neumann problems from the nonexistence of solutions to Dirichlet problems, however, breaks down. (See [BN]).

(III) Although results in Section 3 above indicate that there are striking differences between semilinear Dirichlet and Neumann problems, some of the methods and results in this paper and in [LNT] nonetheless work equally well when applied to the semilinear Dirichlet problem (1.2). For instance, the estimate for the critical value c_d given by (2.7)

$$c_d \sim d^{n/2}, \quad \text{for } d \text{ small}, \tag{4.2}$$

also holds for the variational function J_d defined by (2.6) when restricted to a subspace $\mathring{H}_1(\Omega)$ (i.e. all functions in $H_1(\Omega)$ which are zero on $\partial\Omega$). (Recall that a nontrivial critical point of J_d restricted to $\mathring{H}_1(\Omega)$ is a positive solution of the Dirichlet problem

$$\begin{cases} d\Delta u - u + f(u) = 0 & \text{in } \Omega, \\ u = 0 & \text{on } \partial\Omega, \end{cases} \tag{4.3}$$

where f satisfies the hypotheses in Section 2). Another such example is that Theorem 2.11 also holds true if we replace (all) the zero Neumann boundary condition by the corresponding zero Dirichlet boundary condition. Both facts above may be obtained by exactly the same proofs as before except for some obvious modifications.

(IV) The estimate (4.2) above enables us to prove that in case Ω is an annulus both problems (2.1) and (4.3) possess *positive non-radial* solutions. For, if we apply the arguments in obtaining (4.2) to the variational functional J_d restricted to *radial* functions in $H_1(\Omega)$ or $\mathring{H}_1(\Omega)$, we also obtain the existence of a critical point u_d^* with the corresponding critical value c_d^* being of the order $d^{1/2}$ for d sufficiently small. Therefore u_d^* must be different from u_d, i.e. u_d must then be non-radial.

Similar results hold for the domain Ω being (solid) torus or any domain of the form

$\Omega' \times S^m$ where $1 \leq m \leq n-1$ and Ω' is a bounded smooth domain in $\mathbf{R}^{n-m}$ which does not contain the origin. In such cases, for sufficiently small d we can always find *positive* solutions of (2.1) and of (4.3) which do not possess the symmetry that Ω has.

Acknowledgement

This paper and the paper [LNT] essentially constitute the mathematical part of an expository-survey lecture that the second author gave in the conference "Calculus of Variations and Partial Differential Equations" in honor of Hans Lewy held at Trento, Italy in June, 1986. The second author whishes to thank the organizing committee of this conference for the invitation.

The final draft of this paper was written while both authors were visiting the Centre for Mathematical Analysis at Australian National University.

References

[BN] **H. Brezis and L. Nirenberg,** Positive solutions of nonlinear elliptic equations involving critical Sobolev exponents, Comm. Pure Appl. Math. 36 (1983), 437-478.

[CH] **R. Courant and D. Hilbert,** Methods of Mathematical Physics, Vol. 1, Interscience 1953, New York.

[DN] **W.-Y. Ding and W.-M. Ni,** On the existence of positive entire solutions of a semilinear elliptic equation, Arch. Rational Mech. Anal. 91 (1986), 283-308.

[LNT] **C.-S. Lin, W.-M. Ni and I. Takagi,** Large amplitude stationary solutions to a chemotaxis system, preprint.

[N] **W.-M. Ni,** On the positive radial solutions of some semilinear elliptic equations on $\mathbf{R}^n$, Appl. Math. Optim. 9 (1983), 373-380.

[NT] **W.-M. Ni and I. Takagi,** On the Neumann problem for some semilinear elliptic equations and systems of activator-inhibitor type, Trans. Amer. Math. Soc. 297 (1986), 351-368.

[P] **S.I. Pohozaev,** Eigenfunctions of the equation $\Delta u + \lambda f(u) = 0$, Soviet Math. Dokl. 5 (1965), 1408-1411.

[R] **P.H. Rabinowitz,** Variational methods for nonlinear eigenvalue problems, Indiana Univ. Math. J. 23 (1974), 729-754.

SOME ISOPERIMETRIC INEQUALITIES FOR THE LEVEL CURVES OF CAPACITY AND GREEN'S FUNCTIONS ON CONVEX PLANE DOMAINS

Marco Longinetti
Istituto di Analisi Globale e Applicazioni
Via S. Marta, 13/A
50139 Firenze (Italy)

Dedicated to Hans Lewy

1. Introduction

The purpose of this paper is to show how convexity properties lead to sharp estimates for geometric quantities related to level curves of solutions to some classical partial differential equations.

We denote by $L(t)$ and $a(t)$ the perimeter and the area, respectively, of the domain bounded by a closed level curve $\{u=t\} \equiv \{x : u(x)=t\}$, where u is a real function on a plane domain.

We start by considering the solution u of the following capacity problem in a convex plane ring $D = D_0 - D_1$, where $D_1 \subset D_0$ and D_0 and D_1 are plane convex domains:

(1.1) $\Delta u = 0$ in D,

(1.2) $u = t_0$ on ∂D_0, $u = t_1$ on ∂D_1,

with t_0 and t_1 real constants.

We derive isoperimetric inequalities involving the geometric quantities $L(t)$ and $a(t)$ related to the level curve $\{u=t\}$.

In Theorem 3.1 we prove that

(i) $\log L(t)$ *is a convex function of* t, and

(ii) $\log |a'(t)|$ *is a convex function of* t.

In the sequel ϕ is called logarithmic convex (concave) if $\log|\phi|$ is convex (concave).

In Theorems 4.1 and 4.2 we derive from (i) and (ii) further inequalities for the level sets of the Green's function g for the Laplacian in a plane convex domain D. More specifically, if $\mu(t)$ is the distribution function of g, i.e., the area of the level set $\{g \geq t\}$, we prove in Theorem 4.1 that:

(iii) $\log \mu(t)$ *is a convex function of* t.

In Theorem 4.2 a sharp upper bound for the length $L(t)$ of the level curves of g is obtained, namely, if L_0 is the perimeter of D:

(iv) $L(t) \leq L_0 \exp(-2\pi\, t)$.

In Theorem 3.2 we establish properties similar to (i)-(ii) for the following problem:

$$\text{(1.3)} \qquad \operatorname{div}(|\nabla u|^{p-2} \nabla u) = 0 \ \text{ in } D_0 - D_1,$$
$$u = t_0 \ \text{ on } \partial D_0, \quad u = t_1 \ \text{ on } \partial D_1, \quad p > 1 .$$

In Theorem 3.3 we give an explicit sharp differential inequality for the length $L(t)$ of the level curves of the solution u to the following problem:

$$\text{(1.4)} \qquad \Delta u = f(u) \ \text{ in } D_0 - D_1 ,$$
$$u = t_0 \ \text{ on } \partial D_0, \quad u = t_1 \ \text{ on } \partial D_1, \quad t_0 < t_1 ,$$

with $f \geq 0$, and f monotone non decreasing.

In Theorem 3.4 we derive a sharp upper bound for the function $L(t)$ related to the solutions of problems (1.3) and (1.4). Furthermore in Theorem 3.5 we derive a sharp upper bound for the area $a(t)$ related to the solutions of problems (1.1), (1.3) in the case where D_1 is a circle.

Finally in 5. we derive isoperimetric inequalities for the capacity problems (1.1)..(1.4) in the case where u satisfies a Bernoulli condition on the outside boundary ∂D_0, i.e.,:

$$\text{(1.5)} \qquad |\nabla u| = \text{const} > 0 \ \text{ on } \partial D_0 .$$

Usually classical isoperimetric inequalities for the distribution function of Green's function or of solutions to elliptic equations are established by analysis and symmetrization arguments on the level sets, see [2]. Here we use arguments which stress the convexity properties of the function u. In fact in [5] and [7] it is proved, for an arbitrary dimension, that the solution u to (1.3) or (1.4), respectively, has convex level surfaces. An improvement for the problem (1.4) in dimension two is given in [4]. Strict convexity properties of the Green's function g in a plane convex domain are shown in [4].

An interesting approach to a plasma physics problem, [6], shows convexity properties of the level lines of harmonic functions which imply differential inequalities which supplement (i) and (ii).

The principal idea in the present paper is the introduction of a special coordinate system related to the convex level curves of u. More precisely, we consider the coordinates (θ, t) where t is the "level" of the curves $\{u = t\}$ and $(\cos\theta, \sin\theta)$ is the direction of the exterior normal vector to the level curve $\{u = t\}$. Furthermore we consider also the support function h of any level curve $\{u = t\}$ and rewrite any geometric quantity such as $L(t)$, $a(t)$, $|\nabla u|$ and the curvature K of $\{u = t\}$, in terms of h and its derivatives with respect to the curve parameters (θ, t). When u satisfies an elliptic differential equation, calculus arguments show that h is a solution to a nonlinear elliptic equation in (θ, t) coordinates. By analyzing this equation we obtain the sharp differential inequalities for $L(t)$ and $a(t)$, corresponding to (i)-(iv).

2. Support function

In this paragraph we start by recalling the geometric definition and the principal properties of the support function h of a plane convex domain D. For the necessary proofs and other details we refer the reader to [3]. Next we consider a family of support functions associated to a family of convex level curves of a given function u and show how certain geometric properties are defined in terms of derivatives of h.

Let D be a plane convex domain, and let us choose the origin of the coordinates inside D. Let us consider the exterior normal vector to ∂D at (x_1, x_2) given by $n = (\cos\theta, \sin\theta)$, for $\theta \in S \equiv [0, 2\pi)$. The distance from the origin to the support line r supporting ∂D at (x_1, x_2) orthogonal to n is given by the *support function:*

(2.1) $$h(\theta) = x_1 \cos\theta + x_2 \sin\theta \quad .$$

If D is strictly convex r supports ∂D at only one point (x_1, x_2), h is of class C^1 and the derivative of h with respect to θ is given by

(2.2) $$h'(\theta) = -x_1 \sin\theta + x_2 \cos\theta \quad .$$

If ∂D is C^2 and has strictly positive curvature, then h is of class C^2 also and

(2.3) $$h(\theta) + h''(\theta) = R(\theta) > 0 \quad ,$$

where $R(\theta)$ is the radius of curvature of ∂D.

For the proofs of the previous statements we refer the reader to [3] where the formulas

(2.4) $$L = \int_S R(\theta)d\theta = \int_S h(\theta)d\theta \quad ,$$

$$(2.5) \qquad A = \frac{1}{2} \int_s h(\theta) R(\theta) d\theta \ ,$$

for the perimeter L and the area A of D, respectively, also appear.

Let us now consider a real function u with strictly convex level curves in a domain D. Let us suppose also that u is of class c^2 and that the derivative u_n of u along the outward normal to the level curve $\{u = t\}$ does not vanish at any point in D. For any value t in the range of u we consider the corresponding level curve of u: $\gamma_t \equiv \{u = t\}$, and for fixed t let $h(\theta, t)$ be the support function of the convex domain D_t bounded by γ_t. Furthermore, let $L(t)$, $a(t)$ and $R(\cdot, t)$ be the perimeter, the area and the radius of curvature of D_t, respectively. Partial derivatives are denoted by subscripts.

Of course by (2.1)-(2.3) the following equalities hold:

$$(2.6) \qquad h(\theta, t) = x_1 \cos \theta + x_2 \sin \theta, (x_1, x_2) \in \gamma_t \ ,$$

$$(2.7) \qquad h_\theta(\theta, t) = -x_1 \sin \theta + x_2 \cos \theta, (x_1, x_2) \in \gamma_t \ ,$$

$$(2.8) \qquad h(\theta, t) + h_{\theta\theta}(\theta, t) = R(\theta, t) > 0 \ ;$$

where (x_1, x_2) is the unique point on γ_t with normal exterior vector $(\cos \theta, \sin \theta)$. Conversely, for any point (x_1, x_2) in D we can find θ and t by using,

$$(2.9) \qquad (\cos \theta, \sin \theta) = \pm \nabla u(x_1, x_2) / |\nabla u(x_1, x_2)| \ ,$$

$$(2.10) \qquad t = u(x_1, x_2) \ .$$

The $+$ or $-$ sign in (2.9) is given by the sign of u_n. Moreover by (2.4) and (2.5) we have

$$(2.11) \qquad L(t) = \int_s R(\theta, t) d\theta = \int_s h(\theta, t) d\theta \ ,$$

$$(2.12) \qquad a(t) = \frac{1}{2} \int_s h(\theta, t) R(\theta, t) d\theta \ .$$

We now show that certain classical expressions involving the partial derivatives of u with respect to x_1 and x_2 can be rewritten as derivatives of h with respect to θ and t. More precisely we have the following:

Proposition 2.1. *If* u *has strictly convex level curves and its normal derivative,* u_n, *does not vanish at any point on* D, *then*

$$(2.13) \qquad u_n = (h_t)^{-1} \ ,$$

(2.14) $$\Delta u = \left[-h_{tt} + \left(h_{\theta t}^2 + h_t^2 \right) R^{-1} \right] h_t^{-3} \quad ,$$

Proof. We differentiate (2.6) and (2.10) with respect to t to get:

$$h_t = \frac{\partial x_1}{\partial t} \cos\theta + \frac{\partial x_2}{\partial t} \sin\theta \quad ,$$

$$1 = u_{x_1} \frac{\partial x_1}{\partial t} + u_{x_2} \frac{\partial x_2}{\partial t} \quad ;$$

expression (2.13) then follows with the use of (2.9). On the other hand by differentiating (2.7) with respect to the exterior normal direction $n = (\cos\theta, \sin\theta)$, we get

(2.15) $$h_{\theta t} \bullet t_n + h_{\theta\theta} \bullet \theta_n = -\frac{\partial x_1}{\partial n} \sin\theta - x_1 \cos\theta \bullet \theta_n + \frac{\partial x_2}{\partial n} \cos\theta - x_2 \sin\theta \bullet \theta_n \quad .$$

By using (2.6), (2.13) and the fact that

(2.16) $$t_n = u_n, \quad \frac{\partial x_1}{\partial n} = \cos\theta, \quad \frac{\partial x_2}{\partial n} = \sin\theta \quad ,$$

we can rewrite (2.15) in the form

$$h_{\theta t} \bullet h_t^{-1} + h_{\theta\theta} \bullet \theta_n = -h \bullet \theta_n \quad .$$

So by solving for θ_n and using (2.8) we derive that the curvature of the orthogonal trajectories is given by

(2.17) $$\theta_n = -h_{\theta t} h_t^{-1} R^{-1} \quad .$$

Differentiating (2.13) with respect to n yields

(2.18) $$u_{nn} = -\left(h_{tt} t_n + h_{\theta t} \theta_n \right) h_t^{-2} \quad ,$$

and so from (2.13) and (2.17) it follows that

(2.19) $$u_{nn} = -\left(h_{tt} + h_{\theta t}^2 R^{-1} \right) h_t^{-3} \quad .$$

Now (2.14) follows from (2.13) and the following expression for the Laplacian in terms of normal derivatives of u and of the curvature $K = R^{-1}$ of the level curves of u:

$$\Delta u = u_{nn} + K u_n \quad . \qquad \square$$

3. Capacity functions in convex rings

We start by considering a harmonic function u with closed convex level curves

$\gamma_t = \{x : u(x) = t\}$ in a convex ring D.

Theorem 3.1. *If* u *is a solution to* (1.1)-(1.2) *then* L(t) *is a logarithmic convex function in* t,

i.e.

(3.1) $L''L - (L')^2 \geq 0$,

and $\mid a'(t) \mid$ *is a logarithmic convex function in* t, *i.e.,*

(3.2) $a''' a' - (a'')^2 \geq 0$.

Moreover, equality holds in (3.1). *or* (3.2) *for some* t *if and only if all the level curves of* u *in* D

are concentric circles.

Proof. By assumption D_0 and D_1 are two convex domains bounding the convex level curves

$\{u = t_0\}$ and $\{u = t_1\}$. We can suppose that $t_0 < t_1$. J. Lewis (c.f. [7]) under this assumption has

proved that the function u has the following properties:

(a) $\{u = t\}$ is a strictly convex curve for $t \in (t_0, t_1)$,

(b) $\mid \nabla u \mid \neq 0$ in D.

          Therefore we can consider as in the previous section the support function $h(\theta, t)$ of the level

curves of u.

          By (2.14) the following equality holds:

(3.3) $h_{tt} = \left(h_t^2 + h_{t\theta}^2 \right) R^{-1} - h_t^{+3} \Delta u$.

Since u is harmonic we derive that

(3.4) $h_{tt} \geq h_t^2 R^{-1}$.

Moreover, by differentiating (2.11) with respect to t we get

(3.5) $L'(t) = \int_s h_t(\theta, t) d\theta$,

(3.6) $L''(t) = \int_s h_{tt}(\theta, t) d\theta$.

So from (3.4) it follows that

(3.7) $L''(t) \geq \int_s h_t^2(\theta, t) R^{-1}(\theta, t) d\theta$.

By Schwarz inequality we have

$$(3.8) \qquad \left(\int_S h_t d\theta\right)^2 \le \left(\int_S h_t^2 R^{-1} d\theta\right) \cdot \left(\int_S R d\theta\right) .$$

So from (3.5), (3.6) and (2.11) we obtain (3.1). Equality in (3.1) holds for some $\tau \in (t_0,t_1)$ if and only if equality holds in (3.4) and (3.8) for $t = \tau$. This implies:

(3.9) $\qquad h_{\theta t}(\cdot,\tau) \equiv 0$ *on* S , $h_t(\cdot,\tau)$ *is proportional to* $R(\cdot,\tau)$ *on* S .

Equivalently, we can say that

(3.10) $\qquad h_t(\cdot,\tau)$ *and* $R(\cdot,\tau)$ *are constant on* S.

From (3.10) it then follows that $\{u = \tau\}$ is a circle and $|\nabla u| = |h_t|^{-1}$ is constant on $\{u = \tau\}$. From unique analytic continuation arguments it follows that any level curve of u is a circle concentric to D_τ.

Now we establish (3.2). By differentiating (2.8) and (2.12) with respect to t, we have
$R_t = h_t + h_{\theta\theta t}$,

and

$$a'(t) = \frac{1}{2}\int_S (h_t R + hR_t) d\theta ,$$

respectively. Replacing R_t we get

$$a'(t) = \frac{1}{2}\int_S (h_t R + hh_t + h\, h_{\theta\theta t}) d\theta .$$

Integrating the last term in the previous integral two times by parts with respect to θ and using (2.8), yields

$$(3.11) \qquad a'(t) = \int_S h_t R \, d\theta .$$

By differentiating (3.11) it follows that

$$(3.12) \qquad a''(t) = \int_S (h_t R_t + R h_{tt}) d\theta ,$$

and using (3.3) it turns out that

$$a''(t) = \int_S \left(h_t h_{t\theta\theta} + 2h_t^2 + h_{t\theta}^2\right) d\theta - \int_S \left(R h_t^3 \Delta u\right) d\theta .$$

Integrating the first term in the integral above by parts we have:

(3.13) $$a''(t) = 2\int_S h_t^2\, d\theta - \int_S \left(R\, h_t^3\, \Delta u\right) d\theta$$

and so having used that $\Delta u = 0$ we derive that

(3.14) $$a'''(t) = 4\int_S h_t h_{tt}\, d\theta \ .$$

Let us now observe that from (2.13) we have $h_t < 0$; so from (3.4) and (3.14) it follows that

(3.15) $$a'''(t) \le 4\int_S h_t^3\, R^{-1}\, d\theta \ .$$

But Schwarz's inequality implies:

(3.16) $$\left(\int_S h_t^2\, d\theta\right)^2 \le \left(\int_S |h_t^3| R^{-1}\, d\theta\right) \cdot \left(\int_S |h_t| R\, d\theta\right) .$$

So putting together (3.11), (3.13), (3.15), (3.16) gives (3.2). Moreover equality holds in (3.2) if and only if equality holds in (3.16) and (3.4), i.e., if (3.10) holds for some τ. The same previous analytic continuation argument completes the proof. ❑

The same arguments of the previous theorem can also be applied to the solution u of the capacity problems (1.3) to obtain:

Theorem 3.2. *If* u *is a solution of* (1.3) *then the length* $L(t)$ *satisfies:*

(3.17) $$L \cdot L'' - \frac{1}{p-1}(L')^2 \ge 0 \ ,$$

i.e., $1/\alpha$ *is a convex function for* $\alpha = (p-2)/(p-1)$, $p \ne 2$. *Moreover the function* $a(t)$ *satisfies:*

(3.18) $$a'''\, a' - \frac{2}{p}(a'')^2 \ge 0 \ ,$$

i.e., $\frac{1}{\beta}|a'|^\beta$ *is a convex function for* $\beta = \frac{p-2}{p}$, $p \ne 2$. *Equality holds in* (3.17) *or* (3.18) *for some* t *if and only if all the level curves of* u *are concentric circles.*

Proof. Using normal coordinates, (1.3) becomes

(3.19) $$\Delta u + (p-2)u_{nn} = 0 \ .$$

So by (2.14) and (2.19) we have:

(3.20) $$h_{tt} = \left(\frac{1}{p-1}h_t^2 + h_{\theta t}^2\right) R^{-1} \ ,$$

From which it follows that

(3.21) $$h_{tt} \geq \frac{1}{p-1} h_t^2 R^{-1} .$$

If we now replace (3.4) by (3.21), then arguments along the lines of the previous theorem establish (3.17) and (3.18). □

Remark. By (3.3) the inequality (3.4) holds for any subharmonic function u with convex level curves and negative exterior normal derivative. So under this assumption the inequality (3.1) holds; moreover, by (3.12) and (3.3) one can show that $a'' > 0$.

Let us now consider the solution u to (1.4).

Theorem 3.3. *If* u *is a solution to* (1.4) *then*

(3.22) $$\left[\log L(t)\right]'' \geq \frac{1}{4\pi^2} \cdot f(t) \cdot \frac{|L'(t)|^3}{L(t)} .$$

Equality in (3.22) *holds for some* τ *if and only if* $\{u = \tau\}$ *is a circle and* $|\nabla u|$ *is constant on* $\{u = \tau\}$.

Proof. Caffarelli and Friedman (c.f. [4]) proved that u has strictly convex level curves. Moreover, $u_n = h_t^{-1} < 0$ in $D_0 - D_1$. So by (3.3) we get

$$h_{tt} \geq h_t^2 R^{-1} + f(t) |h_t|^3 .$$

By integrating the above inequality on S and by using Holder's inequality for the last term, the same arguments in the proof of (3.1) prove (3.22). □

We now show that (3.17) (respectively (3.22)) leads to a sharp upper bound for the length L(t) of the level curves of the solution u to (1.3) (respectively (1.4)).

Let D_0 and D_1 be two concentric circles with the same perimeters as D_0 and D_1. We consider the radial solutions v to (1.3) or to (1.4) with boundary conditions:

(3.23) $$v = t_0 \text{ on } \partial D_0 , \quad v = t_1 \text{ on } \partial D_1 .$$

We call v the *L-symmetrization* of u in $D_0 - D_1$.

Theorem 3.4. *If* u *is a solution of* (1.3) *or* (1.4) *satisfying* (1.2) *and* v *is the L-symmetrization of* u , *then*

(3.24) $$L(t) \leq e(t)$$

where e(t) *is the perimeter of the level circles* $\{v = t\}$.

Proof. We first consider the problem (1.3). By (3.23) we have

$$e(t_0) = L(t_0), \quad e(t_1) = L(t_1) \ .$$

Moreover by Theorem 3.2 we get that $e(t)$ satisfies

$$(3.25) \qquad e(t) \cdot e''(t) - \frac{1}{p-1} e'(t)^2 = 0 \, , \quad t \in (t_0, t_1) \ .$$

Similarly for the problem (1.4) we get

$$(3.26) \qquad \left[\log e(t)\right]'' = \frac{1}{4\pi} f(t) \cdot \frac{|e'(t)|^{\beta}}{e(t)} \, , \quad t \in (t_0, t_1) \ .$$

By comparing (3.25) and (3.26) with (3.17) and (3.22), respectively, and by standard comparison theorems for differential inequalities (c.f. [8]), we prove (3.24). ❑

We now wish to establish an upper bound for the area $a(t)$ in the problem (1.31). So let D_0^* and D_1^* be two concentric circles with the same area as D_0 and D_1. Let w be the radial solution to (1.3), with boundary conditions

$$w = t_0 \ \text{on} \ \partial D_0^* \ , \quad w = t_1 \ \text{on} \ \partial D_1^* \ .$$

We call w the *a-symmetrization of* u *in* $D_0 - D_1$.

Theorem 3.5. *Let* u *be a solution of* (1.3) *and let* w *be the a-symmetrization of* u. *If*

(3.27) $\qquad D_1$ *is a circle,*

then for any $p > 1$:

$$(3.28) \qquad a'' a - \frac{p}{2(p-1)} (a')^2 \geq 0 \ ;$$

and

$$(3.29) \qquad a(t) \leq A(t) \ ,$$

where $A(t)$ *is the area of the level circle* $\{w = t\}$.

Proof. First we prove (3.28) and (3.29) for $p = 2$. Let us set

$$(3.30) \qquad M(t) = \frac{a''(t) \cdot a(t) - (a'(t))^2}{a(t)} \ .$$

By differentiating, (3.30) becomes

$$M' \equiv a''' - 2 \frac{a'a''}{a} + \frac{(a')^3}{a^2} \ .$$

Since $a' < 0$ we get from (3.2) that

$$a''' \le \frac{(a'')^2}{a'}$$

and so

$$M' \le \frac{(a'')^2}{a'} - 2\frac{a'a''}{a} + \frac{(a')^3}{a^2} = \frac{a''}{a}\left(\frac{a''a - (a')^2}{a'}\right) + \frac{a'}{a}\left(\frac{(a')^2 - a''a}{a}\right) = \frac{M^2}{a'} \quad ,$$

from which it follows that

(3.31) $\qquad M'(t) \le 0 \quad .$

Let us compute now $M(t_1)$. By (2.12), (3.11) and (3.13) we get

$$a \cdot M = \left(\int_s h_t^2\, d\theta\right) \cdot \left(\int_s h\, R\, d\theta\right) - \left(\int_s h_t\, R\, d\theta\right)^2 .$$

By assumption (3.27) we have $R(\bullet,t_1) \equiv$ constant, say R_1, and so

$$\frac{aM(t_1)}{R_1^2} = 2\pi\left(\int_s h_t^2(\theta,t_1)\, d\theta\right) - \left(\int_s h_t(\theta,t_1) d\theta\right)^2 .$$

Schwarz inequality implies that

$$M(t_1) \ge 0 \quad .$$

So, by (3.31) we get $M(t) \ge 0$, and (3.28) follows from (3.30) for $p = 2$; by standard comparison theorems (c.f. [8]), (3.28) implies (3.29).

For $p > 1$, $p \ne 2$, by replacing (3.30) with

$$M \equiv \left[a''a - \frac{p}{2(p-1)}(a')^2\right] a^{-\alpha} \quad ,$$

where $\alpha = \dfrac{1}{p-1}$,

we can establish (3.28) and (3.29) by using arguments similar to those for the case $p = 2$. ❑

Remark. In Theorem 5.2 we prove that for $p = 2$, in the case where (3.27) is replaced by Bernoulli condition (1.5), the opposite inequality to (3.28) holds. So in general assumption (3.27) is essential for inequalities (3.28) and (3.29) to hold.

4. Isoperimetric inequalities for Green's function

Usually the point of departure in level set-analysis of a function u is the coarea formula (c.f. [2], page 52):

(4.1) $$|\mu'(t)| = \int_{u=t} \frac{ds}{|\nabla u|} ,$$

where $\mu(t)$ is the distribution function of u, i.e., the Lebesgue measure of the level set $\{u \geq t\}$. Moreover, applying Schwarz inequality to (3.25), the following inequality is usually considered:

(4.2) $$|\mu'(t)| \geq \left(\int_{u=t} ds \right)^2 \Big/ \int_{u=t} |\nabla u| ds .$$

Let us consider now the Green's function of the Laplace operator in a plane domain D. It is of the form

(4.3) $$g(x,y) = \frac{1}{2\pi} \log \frac{R(y)}{|x-y|} + H(x,y) ,$$

where H is determined such that for fixed $y \in D$

(i) $g(x,y) = 0$, for $x \in \partial D$,

(ii) $H(\bullet,y)$ is harmonic in D and continuous in D,

(iii) $H(y,y) = 0$.

$R(y)$ is called the *conformal radius* of D with respect to y. Let us define D^* the circle with the same area as D, with center at the origin of the coordinates and radius R^*. Let D_y be a concentric circle to D^*, with radius $R_y = R(y)$. We consider the functions g^* and g_y given by

(4.4) $$\begin{cases} g^*(x) = \dfrac{1}{2\pi} \log \dfrac{R^*}{|x|} , \\ g_y(x) = \dfrac{1}{2\pi} \log \dfrac{R_y}{|x|} , \end{cases}$$

respectively. For fixed y, we denote the distribution functions of $g(\bullet,y)$, g^*, g_y, by $\mu(t)$, $\mu_*(t)$ and $\mu_y(t)$, respectively. By (4.3), (4.4) it follows that

(4.5)
$$\begin{cases} \mu(t) = \pi(R_y)^2 \exp(-4\pi t) + 0\big(\exp(-4\pi t)\big) \ , \\ \mu_*(t) = \pi(R^*)^2 \exp(-4\pi t) \ , \\ \mu_y(t) = \pi(R_y)^2 \exp(-4\pi t) \ . \end{cases}$$

Since $\int |\nabla g| ds = 1$ on any level curve of g, it follows from (4.2) that

(4.6) $$|\mu'(t)| \geq L^2(t) \ .$$

From the classical isoperimetric inequality we derive

(4.7) $$|\mu'(t)| \geq 4\pi \ \mu(t) \ ,$$

and by integration, see also [2],

(4.8) $$\mu_y(t) \leq \mu(t) \leq \mu_*(t) \ .$$

If D is a convex plane domain the following theorem is an improvement of the inequality (4.7).

Theorem 4.1. *Let* D *be a convex plane domain and let* y *be a fixed point in* D, *then the following differential inequalities hold:*

(4.9) $$\mu''' \mu' - (\mu'')^2 \geq 0 \ ,$$

i.e., $\log|\mu'|$ *is a convex function;*

(4.10) $$\mu'' \mu - (\mu')^2 \geq 0 \ ,$$

i.e., $\log \mu$ *is a convex function; and*

(4.11) $$\mu'' \geq 4\pi|\mu'| .$$

Moreover D *is a circle with center at* y *if and only if equality holds in* (4.9), (4.10) *or* (4.11) *for some* t.

Proof. In [5] the authors prove that g has strictly convex level curves, so (4.9) follows directly from Theorem 3.1.

To prove (4.10) we consider a sequence $D^{(m)}$ of circles with center at y and radius going to zero as m goes to ∞. Let

$$t^{(m)} = \max_{\partial D^{(m)}} g(\bullet,y) \ , \ \tau^{(m)} = \min_{\partial D^{(m)}} g(\bullet,y) \ .$$

From (4.3) it follows that

(4.12) $$\lim_{m\to\infty} |t^{(m)} - \tau^{(m)}| = 0 \ .$$

Let $u^{(m)}$ and $v^{(m)}$ be the harmonics functions in $D - D^{(m)}$ satisfying:

$$u^{(m)} = v^{(m)} = 0 \quad \text{on} \quad \partial D \qquad u^{(m)} = t^{(m)},\ v^{(m)} = \tau^{(m)} \quad \text{on} \quad \partial D^{(m)} \ .$$

From the maximum principle we derive that

$$v^{(m)} \le g \le u^{(m)} \quad \text{in} \quad D - D^{(m)} \ ,$$

and from (4.12) it follows that $u^{(m)}$ and $v^{(m)}$ approaches g when m goes to ∞.

From (3.28) in Theorem (3.5) it follows that the area $a^{(m)}$ of the level sets of $u^{(m)}$ is logarithmic convex. Since

$$r(t) = \lim_{m \to +\infty} a^{(m)}(t)$$

we derive that r is logarithmic convex too. This proves 4.10. Finally 4.11 follows from (4.7) and (4.10). This concludes the proof. ❑

The following corollary follows from (4.10).

Corollary 4.1. *Let* $0 \le t_0 < t_1$ *be fixed constants, and* y *a fixed point in* D. *Let* D_0 *and* D_1 *be the level set* $\{g \ge t_0\}$ *and* $\{g \ge t_1\}$, *respectively. If* w *is the a-symmetrization of* g *in* $D_0 - D_1$, *then*

$$\mu(t) \le A(t) \quad \text{for} \quad t \in (t_0, t_1) \ ,$$

where $A(t)$ *is the area of the circle* $\{w = t\}$.

We now wish to establish a lower and an upper bound for the length $L(t)$ of the level curves $\{g = t\}$

Of course from the left inequality in (4.8) and the classic isoperimetric inequality we get

(4.13) $\quad 2\pi R_y \exp(-2\pi t) \le L(t)$.

Unfortunately a similar argument does not apply to the right inequality in (4.9) which will give an upper bound for $L(t)$. However, in the following theorem we give an upper bound for $L(t)$ which only depends on the perimeter L_0 of D.

Theorem 4.2. *Let* D *be a convex plane domain with perimeter* L_0 *and let* y *be a fixed point in* D. *Then* L *is logarithmic convex and*

(4.14) $\quad L(t) \le L_0 \exp(-2\pi t)$.

Equality holds in (4.14) *if and only if* D *is a circle centered at* y.

Proof. As in the proof of the previous theorem we have that g has strictly convex level curves. So Theorem (3.1) applies and $\log L$ is convex. Moreover, by (4.8) it follows that

$$\log \mu_y \le \log \mu \le \log \mu \ .$$

Since $\log \mu$ is convex and $\log \mu_y$, $\log \mu$ are linear functions with slope -4μ, it follows that

$$\lim_{t\to+\infty} \frac{\mu'(t)}{\mu(t)} = -4\pi$$

From (4.5) and the previous equality we have

$$\lim_{t\to+\infty} \frac{|\mu'(t)|}{\exp(-4\pi t)} = 4\pi^2 R_y^2 ,$$

and so from (4.6),

$$\lim_{t\to+\infty} \frac{L(t)}{\exp(-2\pi t)} \le 2\pi R_y .$$

By logarithmic convexity properties of L and the previous equality we derive (4.14) □

5. Optimal conductors

In [9] the following result is proved: given a convex domain D_1 and a constant $\gamma > 0$ there exists a unique convex domain $D_0 \supset D_1$, such that the solution u to (1.1), (1.2) satisfies the Bernoulli condition:

(5.1) $$|\nabla u| = \gamma \quad \text{on } \partial D_0 .$$

This problem arises in optimal conductors and in some classic free boundary problems, see [1].

The results of paragraph 3 can be applied to obtain isoperimetric inequalities for the *optimal conductor* $D_0 - D_1$. For simplicity, let $t_0 = 0$, $t_1 = 1$ in (1.2). So the logarithmic capacity of $D_0 - D_1$ is given by

(5.2) $$C = \int_{\partial D_0} |\nabla u| ds ,$$

and the constant γ in (5.1) is given by

(5.3) $$\gamma = C/L_0 ,$$

where L_0 is the perimeter of ∂D_0.

Theorem 5.1. *Let* L_0, L_1 *be the perimeter, and* A_0, A_1 *be the area of* D_0 *and* D_1, *respectively.*

The logarithmic capacity C *of the optimal conductor* $D_0 - D_1$ *satisfies the following inequalities:*

(5.4) $$\frac{1}{C} \ge \frac{1}{2\pi} \log \frac{L_0}{L_1} ,$$

(5.5) $$\frac{1}{C} \le \frac{1}{4\pi} \log\left[\frac{L_0^2}{L_0^2 - 4\pi(A_0 - A_1)}\right] .$$

Equality holds in (5.4) *or* (5.5) *if and only if* $D_0 - D_1$ *is a circular anulus.*

Proof. By logarithmic convexity of the function $L(t)$, Theorem 3.1, we have

(5.6) $$\log L(1) \ge \log L(0) + \frac{L'(0)}{L(0)}$$

Moreover by (3.5), (5.1)-(5.3) we have

(5.7) $$\frac{L'(0)}{L(0)} = -\frac{2\pi}{C} .$$

Now (5.4) follows by (5.6) and (5.7).

Similarly by Theorem (3.2) we derive

(5.8) $$\log |a'(t)| \ge \log |a'(0)| + \frac{a''(0)}{a'(0)} \cdot t$$

and from (3.11), (3.12) we have

(5.9) $$|a'(0)| = \frac{L_0^2}{C} , \quad a''(0) = \frac{4\pi L_0^2}{C^2} .$$

So by integrating (5.8) and by the equalities above, we get

$$A_0 - A_1 \ge \frac{L_0^2}{4\pi}\left[1 - \exp\left(-\frac{4\pi}{C}\right)\right] .$$

Inequality (5.5) is proved now by solving for $1/C$ in the inquality above. □

Remark 5.1. The length L_0 in (5.4) and (5.5) is not explicitly given, since L_0 is the length of the free boundary ∂D_0.

Moreover, by computation one can show that the upper bound given in (5.5) for $1/C$ is less than in the Carleman's inequality:

(5.10) $$\frac{1}{C} \le \frac{1}{4\pi} \log\frac{A_0}{A_1} .$$

In the following theorem we give an explicit isoperimetric inequality for the free boundary ∂D_0.

Theorem 5.2. *If* u *is a solution to* (1.1), (1.2), *satisfying* (5.1), *then*

(5.11) $$L_0^2 - 4\pi A_0 \leq L_1^2 - 4\pi A_1 \;;$$

moreover the following differential inequality hold:

(5.12) $$(L^2 - 4\pi a)' \geq 0 \;,$$

(5.13) $$a''a - (a')^2 \leq 0 \;,$$

i.e., a *is a logarithmic concave function. Equality holds in* (5.11), (5.12) *or* (5.13) *if and only if* D_1 *is a circle.*

Proof. We obtain (5.11) by comparing of the two terms on the right of (5.4) and (5.5). More simply, (5.11) follows by (5.12) which we now prove. In fact let set

(5.14) $$G(t) \equiv L^2(t) - 4\pi a(t) \;.$$

Since $\log L$ is convex, Theorem 3.1, we have that L'/L is increasing and by (5.7)

(5.15) $$L'(t) \geq \frac{-2\pi}{C} L(t) \;.$$

So by (5.14)

$$G'(t) \geq -4\pi \left[\frac{L^2(t)}{C} + a'(t) \right] \;.$$

Inequality (5.12) follows now by (4.2) and the inequality above.

Finally to prove (5.13) we consider the function M defined by (3.30). By (5.9) we have:

$$M(0) = \left(4\pi \frac{L_0^2}{C^2} A_0 - \frac{L_0^4}{C^2} \right) \frac{1}{A_0} \;.$$

So by classical isoperimetric inequality $M(0) \leq 0$ and by (3.31) we derive $M(t) \leq 0$ which implies (5.13). □

Finally, let us observe that: for the solution u to (1.4) and satisfying (5.1) the inequality (5.4) and (5.12) even hold. (5.4) and (5.13) can be also extended in a suitable form to the problems (1.3) satisfying Bernoulli condition (5.1).

Acknoledgement

This research was carried out while the author was visiting Cornell University. He wishes to thank Cornell University for the hospitality and Professor L.E. Payne for his advice and the several fruitful discussions about the problems presented here.

References

[1] **A. Acker,** A free boundary optimisation problem, SIAM J. Math. Anal. 9 (1978), 1179-1191.

[2] **C. Bandle,** Isoperimetric Inequalities and their applications, Pitman, London 1980.

[3] **T. Bonnesen and W. Fenchel,** Theorie der Konvexen Korper, Springer, Berlin 1934.

[4] **L.A. Caffarelli and A. Friedman,** Convexity of solutions of semilinear elliptic equations, Duke M. Jour. 52 (1985), 431-456.

[5] **L.A. Caffarelli and J. Spruck,** Convexity properties of solutions of some classic variational problems, Comm. in P.D.E. 7 (1982), 1337-1379.

[6] **P. Laurence and E. Stredulinsky,** A new approach to Queer Differential Equations, Communications on Pure and Applied Mathematics 38 (1985), 333-355.

[7] **J. Lewis,** Capacitary functions in convex rings, Arch. Rational Mech. Anal. 66, 3 (1977), 201-224.

[8] **M.H. Protter and H.F. Weinberger,** Maximum Principles in Differential Equations, Prentice-Hall, Inc., Englewood Cliff, N.J. 1967.

[9] **D.E. Tepper,** Free boundary optimisation problem, SIAM J. Math. Anal. 5 (1974), 841-846.

NONHOMOGENEOUS QUASILINEAR HYPERBOLIC SYSTEMS: INITIAL AND BOUNDARY VALUE PROBLEM

Pierangelo Marcati
Dipartimento di Matematica Pura e Applicata
Università - Via Roma, 33
67100 L'Aquila (Italy)

Dedicated to Hans Lewy

1. Introduction

The purpose of this paper is to present some existence results concerning the existence of weak solutions in the large for the IBVP

$$U_t + F(U)_x = G(U) \ , \quad t \geq 0, \quad x \geq 0$$

$$\text{(E)} \qquad U(x,0) = U_0(x)$$

$$B(U(0,t)) = 0 \ , \ \text{for a.e.} \quad t \geq 0$$

where $U \in \mathbf{R}^N$; $F;G : \mathbf{R}^N \to \mathbf{R}^N$ and $B : \mathbf{R}^N \to \mathbf{R}$.

Similar results can be obtained even in the case of x varying in a closed interval $[a,b]$ with a boundary condition on each one of the two end points. We shall limit, for brevity, the exposition to (E).

A weak solution to (E) is a bounded measurable function $U(x,t)$ satisfying

$$\iint\limits_{\substack{t \geq 0 \\ x \geq 0}} \left\{ U\phi_t + F(U)\phi_x - G(U)\phi \right\} dx\,dt + \int_{-\infty}^{+\infty} U_0(x)\phi(x,0)dx = 0$$

for any smooth function $\phi(x,t)$, with support in $t \geq 0$ and $x \geq 0$, such that $\phi(0,t) = 0$. Moreover we want the existence of a trace of U on the t-axis which is L^1_{loc}. This trace must satisfy the boundary condition (this is obtained via the general trace theorem in [Mi] and [Mi2])

$$B(U(0,t)) = 0 \quad \text{a.e. in} \quad t \ .$$

The theory presented here, takes its general ideas and methods from the theory, in the BV framework, of Hyperbolic Conservation Laws (see [La], [S] and [G]). The Glimm difference scheme has been modified by [Li1] and [DH] in order to handle nonhomogeneous systems. The presence of boundary conditions has been considered by some authors, see for instance [NS] and [Li3]. We shall present here some results concerning the use of Dafermos and Hsiao scheme in presence of general nonlinear boundary conditions. Some of these ideas have been used by Hsiao and the author in [HM] to study the weak solution for a problem arising in chemical engineering. As usual with the Glimm scheme, the general results will be obtained assuming small initial data. For restricted classes of 2×2 system we will be able to exhibit solution with "big" initial data. In general this fact depends, strictly, from the interactions of the geometry if the wave curves with the boundary conditions.

However the general theory is complicated by technical conditions involving the relations between the manifold $\{G = 0\}$ with the integral curves of the right eigenvectors and the hypersurface $\{B = 0\}$, so it goes beyond the goal of the present exposition. Then we confine ourselves to systems of two equations with Relaxation. Namely

$$U = \begin{pmatrix} u \\ v \end{pmatrix}, \quad F(U) = \begin{pmatrix} f(u,v) \\ g(u,v) \end{pmatrix}, \quad G(U) = \begin{pmatrix} 0 \\ h(u,v) \end{pmatrix} .$$

On h, we assume there exists a smooth curve $v = v^*(u)$ such that $h(u,v^*(u)) = 0$ and $h(u,v)$ represents a source (sink) when v is less (greater) than its equilibrium value $v^*(u)$. In this context the definition is due to T.P. Liu [Li2]. The complete results will be the object of a forthcoming author's paper.

2. The difference scheme of [DH]

We choose a mesh-length $h = \Delta x > 0$ and determine the corresponding time mesh-length $s = \Delta t = \Lambda^{-1}h$, where Λ is a fixed upper bound of the supremum over $\mathcal{B}$ of the moduli of characteristic speeds λ_i, where $\mathcal{B}$ is a prescribed domain containing all the approximating solutions, that will be choosen later. Thus $\Delta x/\Delta t$ fulfills the well-known Courant-Friedrichs and Lewy [CFL] condition, so that waves emanating from points at a distance $2h$ apart will not interact on a time interval of length s. Choose an equidistributed Random sequence $\{\alpha_n\}$ in $\Pi(-1,1)$ and define $y_{m,n} = (m + \alpha_n)h$. We will denote Random mesh points by $(y_{m,n}, n\, s)$, with $m + n$ odd. We begin the algorithm at $n = 0$, by setting

$$U_h(x,0-) = U_0(x), \quad x > 0 .$$

Set

$$S_n = \{(x,t) : x \geq 0, ns \leq t < (n+1)s\} .$$

We devide the recursive procedure into two steps:

1st Step. Assume known the approximate solution on the set

$$\bigcup_{j=0}^{n-1} S_j ,$$

then extend it to the line $t = n s$.

Denote by

$$U_{m,n} = U_h(y_{m,n}, n s^-)$$

and

$$Z_{m,n} = U_{m,n} + G_{m-1,n}\, s$$
$$W_{m,n} = U_{m,n} + G_{m+1,n}\, s$$
$$G_{m,n} = G\left(\left\{ U_h\left(m h^+, n s^-\right) + U_h\left(m h^-, n s^+\right) \right\} / 2 \right) .$$

We shall define, for all $x \in ((m-1)h, (m+1)h)$

$$U_h(x, n s) = \begin{cases} Z_{m-1,n} & \text{if } x < m h \\ W_{m+1,n} & \text{if } x > m h \end{cases} .$$

2nd Step. Define U_h in S_n, provided known its value on the half line $\{(x, n s)\}_{x \geq 0}$. This step has to be analyzed into two different subcases:

i) Assume n is *odd* then m is *odd* too, for all the mesh points in the half-line $t = n s$. In this case we define U_h in $S_n \cap \{(m-1)h < x < (m+1)h\}$ as the Riemann solution for the *homogeneous* problem related to (E), centered in the mesh point $(m h, n s)$. Moreover we define $U_h(0,t) = U_h(0, n s-)$, for all $t \in (n s, (n+1)s)$.

ii) If n is even then m is even too, for all the mesh points on the half line. We define U_h by using the above method, for $x > h$. In the rectangle $\{0 < x < h, n s \leq t < (n+1)s\}$ we assume U_h coincides with the solution to the BRP

$$U_t + F(U)_x = 0$$

$$\text{(BRP)} \qquad B(U(0,t)) = 0, \quad U(x, n s) = W_{1,n} .$$

That, for the moment, is supposed (this will be clarified in the next section) to be solvable by centered elementary waves.

Theorem (Consistency). *Assume that each* $\{\alpha_i\}$ *yealds a family* $\{U_h\}$ *of approximating solution which are defined and have (uniformly in* h *) locally bounded variation on* $[0,\infty) \times [0,T)$. *For almost every sequence* $\{\alpha_i\}$ *there exists a sequence* $\{h_l\}$ *such that*

$$U_{h_l} \to U$$

almost everywhere in x, t *and* U *is a weak solution to* (E) *of locally-bounded variation on* $[0,\infty) \times [0,T)$.

Moreover a.e.

$$B(U(0,t)) = 0 \quad .$$

The proof will be omitted since it can be reconstructed by reverting at [DH]. The boundary conditions are fulfilled owing to the boundary behaviour of the approximating solutions. Indeed one has a.e. in t

$$B(U(0,t)) = 0 \quad .$$

Remark. Since the approximating solutions are defined by exact solutions, the entropy admissibility criterion holds.

3. BRP and Interaction Estimates

We shall assume for simplicity to deal with a system having a Riemann invariants coordinate system, so that we are able to define the Hugoniot curves in large. This can be done for instance with respect to p system (see [S]) or the so-called K-class (see [DiP]). As a model of this latter class, we mention the problem

$$\text{(P)} \qquad \begin{cases} \rho_t + (\rho u)_x = 0 \\ v_t + \left(\frac{1}{2} v^2 + \rho\right)_x = \emptyset(v,\rho) \\ B\big(\rho(0,t), v(0,t)\big) = 0 \end{cases}$$

to which is related [HM]. The Riemann problem for the homogeneous system related with (P) has been investigated in [DiP] and we refer to that paper for the details.

Denote by $S_i(U_0)$ (respectively $R_i(U_0)$), $i = 1,2$, the i-th shock wave (resp. rarefaction wave) curve and by S_i^I and R_i^I the inverse wave curve. The solution of the (BRP) for (P) can be done in the following way. Let U_+ be the datum on the positive x-half line, and consider the inverse wave curve

$$W_i^I(U_+) \text{ emanating from } U_+, \quad W_i^I = S_i^I \cup R_i^I \ .$$

Assume, in order to simplify the computation, $\lambda_1 < 0 < \lambda_2$. Then check the intersection between W^I and $\{B = 0\}$. If the intersection is non-empty and reduces to a single point the BRP is well posed and we have a state U_-, such that $B(U_-) = 0$, which is connected to U_+ by an elementary wave of the first kind. Therefore the study of BRP is actually, the study of the intersections between W_i and $\{B = 0\}$. For systems of K-class due to the special geometry of the wave curves W_i^I we can, for instance, say that BRP is well posed in the following situation. Denote by s, r the Riemann invariants

i) If $\partial r/\partial s \leq -1$ along $B = 0$.

ii) If $\partial r/\partial s > 0$ along $B = 0$.

iii) If U_+ is sufficiently close to $\{B = 0\}$ and this curve admits a cartesian representation $r = \psi(s)$ in the (s,r) plane.

Similar conditions can be identified for the case of p-systems.

We continue to show the methods to study BRP by helping ourselves with the example (P). One of the main step, if we want the BV estimates required for the convergence of the Random choice method, is the study of the interactions between nonlinear waves. Since in the example (P), only the wave of first kind can interact with the boundary, indeed $\lambda_2 > 0$, the resulting analysis is greatly simplified. Consider the case (i) in the above list. Consider the interaction of a 1-shock with the boundary $x = 0$. Denoting by $\beta = (U_1, U_2, s)$ the 1-shock wave we may have two possible reflected waves. The former is a 2-shock $\beta' = (\hat{U}, U_2, s')$ where $\hat{U} \in \{B = 0\}$, the latter is a 2-rarefaction α wave joining a constant state $\hat{U} \in \{B = 0\}$ with U_2.

If we measure the strength of the elementary waves as the variation in absolute value in the directions $r + s = \text{const}$, thus in the former case $|\beta'| \leq |\beta|$ while in the latter we may have $|\alpha'| \geq |\beta|$.

The reflected wave of a 1-rarefaction wave α is a 2-rarefaction wave α' and

$|\alpha'| \le |\alpha|$. The remarkable fact of example (P), but this is true in general if $v = \phi(r+s)$, is that under Dirichlet boundary conditions $v = v_1$ const, on $x = 0$, namely $B(\rho, v) = v - v_1$, the bad case $\beta \to \alpha'$ is dropped. Summarizing it follows

Proposition. *Assume there exists* U^* *such that* $B(U^*) =$ *and* $G(U^*) = 0$ *then if* $|U_0(x) - U^*|$ *is small and* $TV(U_0)$ *is small the resulting approximating solution have a locally bounded variation (uniformly in* h*). Moreover in the case of Dirichlet boundary conditions, namely conditions on* $r+s$, *then the smallness assumptions can be removed.*

The construction of the domain $\mathcal{B}$ can be done in the same way of [HM]. For the convergence of the algorithm we can repeat, essentially, the proof given by [DH] and [HM].

Finally, we conclude with an application of these results on p-system.

Consider an open simply connected bounded set $\Omega \subset \mathbf{R}^N$ for simplicity containing zero. The dynamic equation for an elastic membrane is given by

$$
\text{(D)} \quad \begin{cases} W_{tt} - \operatorname{div}\left(\dfrac{Dw}{\sqrt{1+|Dw|^2}}\right) = f(w_t, Dw) \\ w\big|_{\partial\Omega} = w^* \\ w(x,0) = w_0(x), \quad w_t(x,0) = w_1(x) \quad . \end{cases}
$$

We want to study the existence of plane-wave solutions to (D). Assume there exist $\omega \in S^{N-1}$, such that $w_0(x) = \phi(\omega \cdot x)$, $w_1(x) = \psi(\omega \cdot x)$, thus we are able to look for solutions to (D) of the form $w(x,t) = z(\omega \cdot x, t)$, being $z = z(\xi, t)$, with $\xi \in (\min_x \omega \cdot x, \max_x \omega \cdot x) = (a,b)$. Then (D) reduces to

$$
\text{(D')} \quad \begin{cases} Z_{tt} - \left(\dfrac{Z_\xi}{\sqrt{1+Z_\xi^2}}\right)_\xi = g(Z_t, Z_\xi) \\ Z(a,t) = Z(b,t) = W^* \\ Z(\xi,0) = \phi(\xi), \quad Z_t(\xi,0) = \psi(\xi) \end{cases}
$$

which is equivalent to a p-system, in $u = Z_\xi$, $v = Z_t$

(D'')
$$\begin{cases} u_t - v_\xi = 0 \\ v_t - \left(u/\sqrt{1+u^2}\right)_\xi = g(u,v) \\ v(a,t) = v(b,t) = 0 \\ u(\xi,0) = \phi'(\xi), \; v(\xi,0) = \psi(\xi) \;\; . \end{cases}$$

This system, despite of the loss of the genuine nonlinearity in $u = 0$ and the presence of two boundary conditions, can be studied in a similar way to (P). In a recent paper [FT] investigated the initial-boundary value problem (D) using the methods of Geometric Measure theory, namely the Theory of Varifold (see [A]). However their solutions do not seem to contain the solutions obtained here, where the energy conservation principle is violated along the jump discontinuities induced by the planar shock waves.

References

[A] **W. Allard,** The first variation of a varifold, Ann. of Math. 95 (1972), 417-491.

[CFL] **R. Courant, K.O. Friedrichs and H. Lewy,** Über di Partiellen Differenzengleichungen der Mathematischen Physik, Math. Ann. 100 (1928), 32-74.

[DH] **C.M. Dafermos and L. Hsiao,** Hyperbolic systems of balance laws with inhomogeneity and dissipation, Indiana Univ. Mathematics J. 31 (1982), 471-491.

[DiP] **R.J. Di Perna,** Global solutions to a class of nonlinear hyperbolic systems of equations, Comm. Pure and Appl. Math. 26 (1973), 1-28.

[FT] **D. Fujiwara and S. Takakuwa,** A varifold solution to the nonlinear equation of motion of a vibrating membrane, preprint Dept. of Mathematics, Tokyo Institute of Technology.

[G] **J. Glimm,** Solutions in the large for nonlinear hyperbolic systems of equations, Comm. Pure and Appl. Math. 18 (1965), 697-715.

[Gi] **E. Giusti,** Minimal surfaces and functions of bounded variation, Notes on Pure Mathematics 10 (1977), Australian National University, Canberra.

[HM] **L. Hsiao and P. Marcati,** A nonhomogeneous quasilinear hyperbolic system arising in chemical engineering, submitted.

[La] **P.D. Lax,** Hyperbolic systems of conservation laws, Comm. Pure and Appl. Math. 10 (1957), 537-566.

[Li] **T.-P. Liu,** Quasilinear hyperbolic systems, Comm. Math. Phys. 68 (1979), 141-172.

[Li2] **T.-P. Liu,** Hyperbolic conservation laws with relaxation and damping, Proceedings Int. Conf. L'Aquila 1986.

[Li3] **T.-P. Liu,** The free piston problem for gas dynamics, J. Diff. Eqns 30 (1978), 175-191.

[Mi] **M. Miranda,** Distribuzioni aventi derivate misure ed insiemi di perimetro localmente finito, Ann. Scuola Norm. Sup. Pisa 18 (1964), 27-56.

[Mi2] **M. Miranda,** Comportamento delle successioni convergenti di frontiere minimali, Rend. Sem. Mat. Univ. Padova 38 (1967), 238-257.

[NS] **T. Nishida and J.A. Smoller,** Mixed problems for nonlinear conservation laws, J. Diff. Eqns 23 (1977), 244-269.

[S] **J.A. Smoller,** Schock waves and reaction-diffusion equations, Grundlehren der mathematischen Wissenschaften 258, Springer-Verlag, New York-Heidelberg-Berlin 1983.

EXISTENCE RESULTS FOR NON CONVEX PROBLEMS OF THE CALCULUS OF VARIATIONS

Elvira Mascolo
Istituto di Matematica - Università
Facoltà di Scienze
84100 Salerno (Italy)

Dedicated to Hans Lewy

Consider a functional of the calculus of variations

(I) $$F(v) = \int_G f(x,v,Dv)dx \;,$$

where G is an open subset of $\mathbf{R}^N$ with a sufficiently smooth boundary ∂G.

Let $f : (x,s,p) \in G \times \mathbf{R} \times \mathbf{R}^N \to \mathbf{R}$ be a Caratheodory function such that

(2) $$\lambda_1 \,|\, p \,|^2 - \lambda_2 \leq f(x,s,p) \leq \Lambda_1 + \Lambda_2 + \Lambda_3 \,|\, p \,|^2$$

with λ_1, λ_2, Λ_1, Λ_2, Λ_3 positive constants, and consider

(3) $$\operatorname{Inf}\,\{F(v), v \in X\}$$

where $X = H_0^{1,2}(G) + u_0$, with $u_0 \in H^{1,2} \cap L^\infty$ or

$$X = \left\{ v \in H^{1,2}(G) : v_G = \int_G v\,dx = 0 \right\},$$

which corresponds respectively to either Dirichlet boundary conditions or Neumann boundary conditions. It is well known that the sequential lower semicontinuity of functional F and the compactness of at least one minimizing sequence in the same topology ensure the existence of solutions. On the other hand the convexity of the integrand function with respect to p is a necessary and sufficient condition for the semicontinuity of F (see [1], [2], [3]). Nevertheless, the convexity of the integrand function is not a necessary condition for the existence of a solution of (3).

Moreover, problems of type (3) are related with the theory of nonlinear elasticity and in this physical context the convexity of the integrand function is not an acceptable hypothesis, (see for example, [4], [5], [6]). To overcome the difficulty that, if f is not convex in p, it is not possible to apply the direct methods, new arguments, called "relaxation methods", have been introduced. Define a new functional $\mathbf{F}$ and a new problem

$$\text{(4)} \qquad \operatorname{Inf}\left\{\mathbf{F}(v), v \in X\right\} ,$$

it is possible to show that (see, for example, [2], [3], [6])

$$\text{(5)} \qquad \operatorname{Inf}\left\{\mathbf{F}(v), v \in X\right\} = \operatorname{Min}\left\{\mathbf{F}(v), v \in X\right\} ,$$

and that the optimal solutions of (4) are the limit points of (3). In general, $\mathbf{F}$ is the greatest functional, less than or equal to F, which is lower semicontinuous in the weak topology in which we consider the problem. In some cases, we have

$$\mathbf{F}(v) = F^{**}(v) = \int_G f^{**}(x,v,Dv)dx ,$$

where f^{**} is the lower convex envelope of f with respect to p.

We are now interested in finding sufficient conditions on F to obtain solutions of non convex problems directly.

We recall the results of Ericksen [4], obtained by studying the equilibrium configurations of bars, in soft and hard device (which correspond to the Neumann and Dirichlet boundary problems), and the results by Marcellini [7] for the Dirichlet problems.

Both these studies are worked out in dimension one. The tools used in the paper of Marcellini are very interesting. In fact, starting from them, it was possible to develop the study of nonconvex problems in dimension higher than one.

Marcellini considers the problem

$$\text{(6)} \qquad \operatorname{Inf}\left\{\int_0^1 f(v')dx,\ v \in C^{0,1}(0,1), v(0) = A, v(1) = B\right\}$$

where $f : \mathbf{R} \to \mathbf{R}$ is such that

$$\lim_{t \to \pm\infty} f(t)t^{-1} = +\infty ,$$

and the relaxed problem, that is

$$\text{(7)} \qquad \operatorname{Inf}\left\{\int_0^1 f^{**}(v')dx, v \in C^{0,1}(0,1), v(0) = A, v(1) = B\right\} ,$$

where f^{**} is the lower convex envelope of f, (see the figure below).

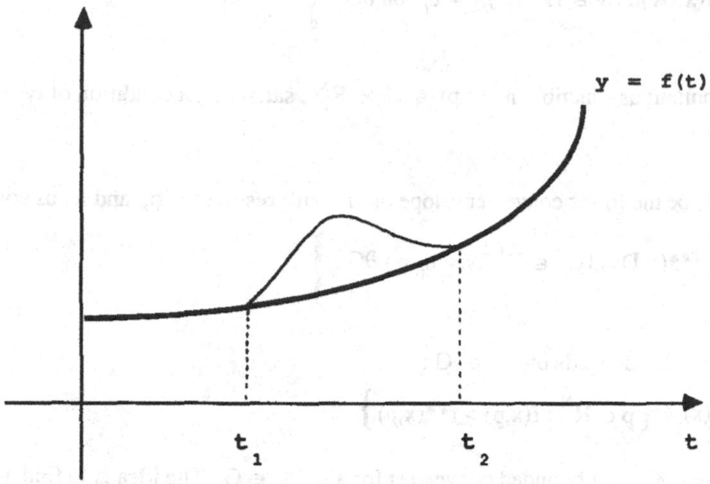

The thick line is the graph of f^{**}.

Let $K = \{t \in \mathbf{R} : f(t) > f^{**}(t)\}$, (in the case of the figure $K =]t_1,t_2[$). In [7], Marcellini shows that if $B - A \notin K$, the solution u of (7) is such that $u'(x) \in \mathbf{R}^N - K$. If $B - A \in K$, it is possible to construct a solution u of (7) such that $u'(x) \in \mathbf{R} - K$ (in particular, u' is equal to t_1 in a suitable subset of $(0,1)$ and equal to t_2 in the complementary set).

Then, as a consequence

$$\int_0^1 f^{**}(u')dx = \int_0^1 f(u')dx \ ,$$

and so u is a solution of (6) too.

In using these arguments the following two points are crucial:

(i) f is convex near infinity, in other words K is a bounded subset of $\mathbf{R}$;

(ii) f^{**} restricted to K is an affine function.

In [4] Ericksen proves the existence of solution for an analogous problem with Neumann boundary conditions, by using quite different methods. However, it is interesting to point out that existence can be obtained by proceeding as above (see [8]).

The idea presented in [7] has been developed, for dimensions higher than one, in some papers [9], [10], [11], [12] by Mascolo–Schianchi. We now want to recall the results of [12].

Let now G be a bounded subset of $\mathbf{R}^N$, $N \geq 2$, and let us consider the problem

$$(8) \quad \text{Inf}\left\{ \int_G f(x,Dv)dx, v \in H^{1,2}(G), v = u_0 \text{ on } \partial G \right\}$$

where f is a continuous function in $(x,p) \in G \times \mathbf{R}^N$, satisfying a condition of type (2) and $u_0 \in H^{1,2} \cap L^\infty$.

Let f^{**} be the lower convex envelope of f with respect to p, and let us consider

$$(9) \quad \text{Inf}\left\{ \int_G f^{**}(x,Dv)dx, v \in H^{1,2}, v = u_0 \text{ on } \partial G \right\} .$$

In this case the set K depends on $x \in G$:

$$(10) \quad K = K(x) = \left\{ p \in \mathbf{R}^N : f(x,p) > f^{**}(x,p) \right\} .$$

As above, suppose $K(x)$ a bounded convex set for a.e. $x \in G$. The idea is to find a solution **u** of (9) such that $\mathbf{Du}(x) \in \mathbf{R}^N - K(x)$ a.e. in G. However, we need the solutions of (10) to be sufficiently regular, at least a.e. differentiable in G.

We recall that f^{**} is not a strictly convex function in p, and in general it is not a C^2–function in (x,p), even if f is a C^2–function in (x,p).

So, there are no general regularity results for the solutions of (9). In [12], some regularity results for problems of type (10) are given. In particular, we consider a minimum problem with a convex integrand function which, for $|p|$ sufficiently large, is of C^2–class in (x,p) and strictly convex in p. We obtain that every solution is in $C_{loc}^{0,1}(G)$. Moreover, with additional assumptions on G and on the boundary data, every solution turns out to be in $C^{0,1}(G)$, where G is the closure of G.

Then, if f in (8) is, for $|p|$ sufficiently large, a C^2–function in (x,p) and strictly convex in p, we have that the relaxed problem (9) has a.e. differentiable solutions, because f^{**} has the same properties of f.

However, we remark that in order to find solutions for the non convex problem, we need also that f^{**} verifies an affinity condition in $K(x)$, which is always satisfied in dimension one (see (ii)). In fact, we suppose that there exist $(N+1)$ functions m_i, $1 \leq i \leq N$ and q such that

$$(11) \quad f^{**}(x,p) = \sum_{i=1}^{N} m_i(x)p_i + q(x) ,$$

for all $p \in K(x)$ and $x \in G$. In [12] we get existence for problem (8), under some additional conditions on the functions m_i.

In particular, we assume that $m_i \in C^1(G)$ and that the set

$$(12) \qquad \left\{ x \in G : \sum_{i=1}^{N} D_{x_i} m_i(x) = 0 \right\},$$

either is equal to G or has measure zero.

These results are generalized in [13], by approximating each function m_i, with a sequence of polynominals, for which the set (12) has measure zero.

From the above results it follows that all the functionals, for which the lower convex envelope of the integrand function verifies an affinity condition, are very "near" to the convex functionals, since they have very similar properties. In fact, not only existence results can be proved but also that all local minima are absolute minima. Here, we define a local minimum of problem (1) to be a function u for which there exists $\varepsilon > 0$ such that:

$$F(u) \leq F(v) \ , \ \text{for } v \in X, \ \ \| v - u \|_{H^{1,2}} < \varepsilon \ .$$

For Neumann problem and in dimension one, the result is proved in [14]. For $N \geq 2$, some partial results are contained in [15].

However, we want to know if the affinity condition (II) of f^{**} is necessary for the existence.

In this framework, some counterexamples, due to Marcellini [7], [16], are very enlightening. In fact he gives some examples in which f^{**} is not affine in K and the non convex problem has no solutions. Our idea is that an important role is played by boundary data.

Consider, in fact, the problem:

$$(12) \qquad \operatorname{Inf} \left\{ \int_G g(Dv)dx, v \in C^{0,1}(G), v = u_0 \text{ on } \partial G \right\},$$

where $g(p) = f(\,|\,p\,|\,)$, and $f : \mathbf{R}_+ \to \mathbf{R}_+$ is an increasing continuous function, convex for t near the infinity.

It is easy to prove that $g^{**}(p) = f^{**}(\,|\,p\,|\,), p \in \mathbf{R}^N$ (see [7]).

Assume that f is convex in $t = 0$, so g^{**} doesn't verify the affinity condition. For example, suppose that f is as in the above figure; in this case

$$K = \left\{ p \in \mathbf{R}^N : t_1 < |p| < t_2 \right\} .$$

Consider the relaxed problem of (12), i.e.

$$(13) \quad \operatorname{Inf} \left\{ \int_G g^{**}(Dv)dx, v \in C^{0,1}(G), v = u_0 \text{ on } \partial G \right\} .$$

Let u_0 verify the bounded slope condition with a constant L_0 (see [17]). Then, from a well known theorem of Hartman-Stampacchia (theorem 13.2 of [7]), problem (13) has at least one solution $\mathbf{u}$ such that $| \mathbf{Du}(x) | \leq L_0$ a.e. in G. In particular, if $L_0 < t_1$, we obtain that $| \mathbf{Du}(x) | < t_1$ so $\mathbf{Du}(x) \in \mathbf{R}^N - K$ a.e. in G. Then $\mathbf{u}$ is also a solution of (12).

Now we want to give a partial (eventually improvable) non–existence result for non convex problem.

To do that, we need the following uniqueness theorem for convex but not strictly convex problems, due to Marcellini [16]:

Theorem. *Let* $h : \mathbf{R}_+ \to \mathbf{R}_+$ *be a convex function with* $h(0) < h(t)$ *for every* $t > 0$. *Let* G *be a convex bounded open set of* $\mathbf{R}^N$ *with boundary* ∂G *of class* C^1. *If there is a function* $u \in C^1(G)$ *with* $Du \neq 0$ *in* G, *that minimizes the following problem:*

$$(14) \quad \operatorname{Inf} \left\{ \int_G h(|Dv|)dx, v \in C^{0,1}(G), v = u \text{ on } \partial G \right\} ,$$

then u *is the unique solution of* (14).

Let now G be as in the above theorem and consider problem (12) and its relaxed (13). Suppose that (13) has a solution $\mathbf{u} \in C^1(G)$ with $\mathbf{Du} \neq 0$ in G, so the previous uniqueness theorem can be applied.

Moreover, if u_0 verifies the bounded slope condition with a constant L_0, $t_1 < L_0 < t_2$, via Hartman-Stampacchia theorem, (13) has only one solution $\mathbf{u}$ with $| \mathbf{Du} | \leq L_0 < t_2$ a.e. in G.

Finally, we assume that u_0 does not verify the following condition:

$$(15) \qquad | u_0(x) - u_0(y) | \leq t_1 \, | x - y | , \quad \forall x,y \in \partial G .$$

We want to prove that under all these hypotheses the non convex problem (14) has no solutions. In fact, (15) is a compatibility condition on ∂G, which is necessary and sufficient for the existence of function $v \in C^{0,1}(G)$, with $| Dv(x) | \leq t_1$ a.e. in G and $v = u_0$ on ∂G. (See theorem 5.1 of [18], part (IV) and also remark 5.3).

Since (16) is not verified, necessarily there exists $G_0 \subset G$ with meas $G_0 > 0$, such that, $|Du(x)| > t_1$ in G_0. So, **u** is not a solution of (12), which therefore has no solution at all.

References

[1] **L. Tonelli,** Fondamenti di Calcolo delle Variazioni, Vol. I, Zanichelli (1921).

[2] **I. Ekeland, R. Temam,** Analyse convexe et problemes variationels, Dunod & Gauthier Villars 1974.

[3] **P. Marcellini, C. Sbordone,** Semicontinuity problems in the Calculus of Variations Nonlinear Analysis 4, n° 2 (1980).

[4] **J. Ericksen,** Equilibrium of bars, Journal of elasticity 5 (1975).

[5] **J.M. Ball,** Convexity conditions and Existence Theorems in Nonlinear Elasticity, Arch. Rat. Mech. Analysis 63 (1986).

[6] **M. Gurtin, R. Temam,** On the Antiplane Shear Problem in Finite Elasticity, Journal of elasticity 11, n° 2 (1981).

[7] **P. Marcellini,** Alcune osservazioni sull'esistenza di minimi di integrali del Calcolo delle Variazioni senza ipotesi di convessità, Rend. di matematica 13 (1980).

[8] **E. Mascolo,** A Survey of existence results for non convex problems, Conference Inst. for Math. and Appl., Univ. of Minnesota (1984).

[9] **E. Mascolo, R. Schianchi,** Existence Theorems for non convex Problems, J. de Math. Pure et Appl. 62 (1983).

[10] **E. Mascolo, R. Schianchi,** Un theoreme d'existence pour des problemes du Calcul des Variations non convexes, CR Acad. Sc. Paris 297 (1983).

[11] **E. Mascolo, R. Schianchi,** Non convex problems of the Calculus of Variations, Nonlinear Analysis 9 (1985).

[12] **E. Mascolo, R. Schianchi,** Existence theorems in the Calculus of Variations, J. of Diff. Equations 66 (1987).

[13] **E. Mascolo,** Some remarks on non convex problems, to appear on Proceedings of Symposium Year on "Material Instabilities in Continuum Mechanics", Ed. J.M. Ball.

[14] **D. Kinderlehrer,** A Mathematical Afterthought, Lecture Notes in Math. 1063 1984.

[15] **D. Kinderlehrer, E. Mascolo,** Local minima for non convex problems, to appear.

[16] **P. Marcellini,** A relation between existence of minima for non convex integrals and uniqueness for non strictly convex integrals of the Calculus of Variations, Proceedings of Congress on Math. Theory of optim., Lecture Notes in Math. (1982).

[17] **P. Hartman and G. Stampacchia,** On some nonlinear Elliptic Differential-Functional Equations, Acta Math. 115 (1966).

WIENER CRITERIA AND VARIATIONAL CONVERGENCES

Umberto Mosco
Dipartimento di Matematica - Università di Roma "La Sapienza"
P.zzale A. Moro, 2
00185 - Roma - (Italy)

Dedicated to Hans Lewy

1. Introduction

In 1969 Hans Lewy and Guido Stampacchia considered in [38] the following *thin-obstacle* problem:

Let D be a bounded open domain of $\mathbf{R}^N$, $N \geq 2$, E a closed subset of D on which a real function ϕ is given. A real function $u(x)$ in D is to be found which minimizes the Dirichlet integral of u among all functions v vanishing on the boundary ∂D and satisfying the inequality $v(x) \geq \phi(x)$ for $x \in E$.

This problem, which had been previously studied by H. Lewy in the special case $N = 2$ and E a closed straight segment, [35], was attacked in [38] the general framework of variational inequalities considered by G. Stampacchia, [52], and J.L. Lions-G. Stampacchia, [40].

A unique solution was shown to exist as an element of the Sobolev space $H^1_0(\Omega)$, the inequality restricting the solution being interpreted in a manner appropriate to this space, namely, almost everywhere in the capacity sense. The solution u is the smallest superharmonic function among the admissible elements of $H^1(\Omega)$.

We are tacitly assuming that the set of such admissible functions is not empty, which is certainly the case if ϕ is bounded from above on E (for unbounded ϕ see D.R. Adams [1]).

Lewy and Stampacchia also studied the continuity of u and of its first order partial derivatives, before proceeding to investigate the properties of the coincidence set where $u(x) = \phi(x)$. They used methods from both nonlinear PDE and potential theory.

The regularity of the derivatives of u was studied only in the special case of ϕ a smooth function on all of $\overline{D}$ and negative on ∂D, that is, ϕ a smooth *full-obstacle* in D. The basic tools were Calderon-Zygmund type inequalities for certain nonlinear partial differential equations.

Potential theory was instead the context in which the study of the continuity of the solution itself was carried out, for the general thin-obstacle problem. This was seen by Lewy and Stampacchia as an extension of the problem of the regularity of a conductor potential for the Laplace equation. In fact, if $\phi \equiv 1$ on E, then the solution u of the problem stated at the beginning of this section is nothing but the electrostatic equilibrium potential (or, capacitary potential) of the conductor E in D.

They proved that u has a continuous representative on all of D, provided the following regularity conditions are satisfied by the given E and ϕ:

(i) Every point $x \in \partial E$ is a *regular boundary point* for the Dirichlet problem in the (open) region $D - E$ of $\mathbf{R}^N$.

(ii) The function ϕ is *continuous on* E.

It is instructive to see the main lines of their proof. The solution u is represented as the l.s.c. potential of the measure $\mu = -\Delta u$, which is non-negative and supported by some closed subset of E. By the maximum principle, $u \geq z$ in D, where z denotes the harmonic extension of ϕ on $D - E$ vanishing on ∂D. By the assumptions (i) and (ii), $z(x)$ is continuous in D, hence the set $U := \{x \in D : u(x) > z(x)\}$ is open. Since $\Delta u = 0$ in U, the support of the measure μ is contained in the closed subset of E where $u(x) = z(x)$, in particular, u is continuous when restricted to the support of μ. Therefore, by Evans' theorem, u is continuous on D.

Let us mention that for the full-obstacle problem the regularity of u and of its first order derivatives was later also studied by H. Brezis and G. Stampacchia [8], by Sobolev spaces techniques and in a more general context of variational inequalities. Estimates of the modulus of continuity of u and its derivatives were also given by H. Lewy and G. Stampacchia [39] in the framework of the classical theory of superharmonic functions, U.Mosco - G. Troianiello [50] by dual estimates for weak solutions, M. Biroli [6] and J. Frehse [20] by PDE energy methods, L.A. Caffarelli and D. Kinderlehrer [10] again in a classical context of boundary behaviour of harmonic functions, M. Giaquinta [29] by comparison arguments. The thin-obstacle case was further investigated by H. Lewy [36], [37], H. Beirão da Veiga [3], H. Beirão da Veiga - F. Conti [5], D. Kinderlehrer [32], [33], [34] and J. Frehse [18], [19]. We refer in particular to [33], [19] for further references.

Let us also mention that boundary unilateral conditions of Signorini's type had been previously studied in the theory of elasticity by G. Fichera [16], [17], and that thin-obstacle problems for minimal surfaces were considered by J.C.C. Nitsche in [51], giving rise to a parallel line of investigation whose description goes beyond the scope of this exposition.

In our talk we will discuss various refinements and generalizations of the initial Lewy-Stampacchia thin-obstacle problem and of their continuity result in [38].

We will see that a theory of *pointwise regularity* can be developed for *local* variational solutions of general *two-obstacle* problems, in a unifying framework for both Dirichlet and (one-sided) obstacle problems. Our context will be that of the calculus of variations, combining methods from PDE and potential theory.

Arbitrary *Borel sets* and *Borel functions* are allowed as constraints, in particular, Dirichlet conditions on arbitrary Borel sets fit into the scheme.

The variational approach turns out also to be the natural one for studying the effect of wild *perturbations* of the constraints. This aspect will be discussed only in the special case of Dirichlet problems in highly perturbed, possibly irregular, domains. In this case a zero-order term appears in the limit equation, which involves a *Borel measure*. This term has the nature of a *penalization* on the size of the solution and must be seen as a *relaxed* (homogeneous) Dirichlet condition.

The theory is based on *structural estimates* of the type occurring in Saint-Venant's principle of linear elasticity. In their general form, these estimates involve instrinsic notions of energy and capacity and are *stable* under variational perturbations.

In particular, the classical *Wiener's criterion* and *Maz'ja's estimate* for the Dirichlet problem, suitably recasted in the variational context, keep to hold for general two-obstacle problems and for the relaxed Dirichlet problems.

2. The role of the capacity: Wiener's criterion and Maz'ja's estimate

Lewy-Stampacchia conditions (i) (ii) of Section 1 bring us back to the classical work by H. Lebesgue and others on the boundary regularity for the Dirichlet problem.

We recall that a boundary point x of an open domain D of $\mathbf{R}^N$, $N \geq 2$, is termed *regular* if whenever $h(x)$ is a given continuous function on ∂D, then the generalized solution v of the Dirichlet problem: $\Delta v = 0$ in D, $v = h$ on ∂D, is such that $v(y)$ has limit $h(x)$ as y approaches x on D.

The *im Kleinen* character of this regularity property was soon recognized. The problem was to describe the "good boundaries". A fundamental contribution in this direction came in 1924 from N. Wiener [54] and [55].

As pointed out by Wiener himself, the characterizations of regularity that had been previously given by Lebesgue and others, all suffered from "the defect of involving the geometrical character of the boundary only in a very indirect and devious manner". The novelty of Wiener's approach was the use of the electrostatic notion of *capacity* to obtain a "quasi-geometrical" characterization of the regular points.

In the form given to it by O.D. Kellog and F. Vasilesco [31], and in terms of relative capacities in order to avoid technical peculiarities of the case $N = 2$, Wiener's criterion can be written at a given point $x_0 \in \partial D$ as follows:

$$\int_0^1 \frac{\operatorname{cap}\left(E \cap B_\rho(x_0), B_{2\rho}(x_0)\right)}{\operatorname{cap}\left(B_\rho(x_0), B_{2\rho}(x_0)\right)} \frac{d\rho}{\rho} = +\infty \tag{2.1}$$

where $E = D^C$ is the complement of D in $\mathbf{R}^N$ and $B_\rho(x_0) = \{x \in \mathbf{R}^N : |x - x_0| < \rho\}$ for every $\rho > 0$.

The capacities involved are the usual harmonic capacities (the logarithmic one for $N = 2$), whose definition Wiener himself extended to arbitrary bounded closed subsets of $\mathbf{R}^N$, as required in its application.

The far reaching influence of Wiener's capacity approach on the development of potential theory and PDE certainly needs not to be stressed here. We refer for instance to the paper by M. Brelot [7] and other related papers, like the one by M. Kac, in the special issue of the Bulletin of the AMS dedicated to N. Wiener. For some topics connected with our discussion we also refer to the survey paper by J. Frehse [21], where a proof of Wiener's criterion can be found, based on energy estimates for weak solutions. A general enlightening treatment of capacity properties in Sobolev spaces and related matters can be found in the book of V.G. Maz'ja [45], see also the survey paper by V.P. Kahvin and V.G. Maz'ja [30].

It was indeed V.G. Maz'ja who in 1963 gave another important contribution to the theory of the boundary regularity for the Dirichlet problem.

He brought into light what was only implicit in Wiener's proof, namely the relationship between the rate of divergence of the integral (2.1) and the modulus of continuity of the solution at x_0, [42], [43] and [44].

He proved in particular that the weak solution $u \in H^1(D)$ of the Dirichlet problem

$$(2.2) \qquad \Delta u = 0 \ \text{ in } \ D, \quad u - h \in H_0^1(D) \ ,$$

for given $h \in C(\bar{D}) \cap H^1(D)$ satisfies the following estimate at an arbitrary boundary point x_0 of D:

$$(2.3) \qquad \underset{B_r(x_0)\cap D}{\operatorname{osc}} u \le \underset{B_R(x_0)\cap\partial D}{\operatorname{osc}} h + c\left(\underset{\partial D}{\operatorname{osc}} h\right) \exp\left(-\beta\int_r^R \frac{\operatorname{cap}\left(E\cap B_\rho(x_0), B_{2\rho}(x_0)\right)}{\operatorname{cap}\left(B_\rho(x_0), B_{2\rho}(x_0)\right)} \frac{d\rho}{\rho}\right)$$

for every $0 < r \le \alpha R$, where $E = D^C$ and $\alpha > 0$, $\beta > 0$ and c are constants depending only on N and $\operatorname{osc} = \sup - \inf$.

The proof relies on the Harnack inequality and the maximum principle and uses special barriers constructed with capacitary potentials.

Let us mention that the "structural" nature of Wiener's criterion had been previously shown by W. Littman, G. Stampacchia and H. Weinberger [41] and G. Stampacchia [53], who proved that a boundary point of a given domain D is regular with respect to the Laplace operator if and only if it is regular with respect to an arbitrary second order divergence form operator

$$(2.4) \qquad L = -\sum_{i,j=1}^{N} \frac{\partial}{\partial x_i}\left(a_{ij}(x)\frac{\partial}{\partial x_j}\right) ,$$

with measurable coefficients $a_{ij}(x)$ satisfying

$$\sum_{i,j=1}^{N} a_{ij}(x)\,\xi_i\xi_j \ge \lambda|\xi|^2 \quad \forall\, \xi \in \mathbf{R}^N \ \forall\, x \text{ a.e.} \in D \quad |a_{ij}(x)| \le \Lambda \ \text{ for all } \ x \text{ a.e.} \in D \ .$$

Maz'ja proves indeed (2.3) for any operator of this type, the constants depending then also on λ and Λ, and more generally for a class of quasi-linear operators of the type of the p-Laplacian. Further nonlinear extensions, however only qualitative ones, were given by H. Beirão da Veiga [4], R. Gariepy and W.P. Ziemer [27], [28].

Wiener's criterion and Maz'ja's estimate cannot be directly applied to obstacle problems, because the free boundary, that is the boundary of the coincidence set, which is peculiar to these problems, is not known explicitly.

However, with the classical theory in mind, it is natural to ask whether a pointwise version of Lewy-Stampacchia result holds that assures the continuity of u at a given point $x_0 \in \partial E$, under the assumption only that Wiener's condition (2.1) is satisfied at x_0 and ϕ is continuous at x_0 on E.

If x_0 is an interior point of E, then u is a local solution of the full-obstacle problem with obstacle ϕ in a neighborhood of x_0 and the problem becomes whether u is continuous at x_0 provided ϕ is such.

These questions were answered positively by J. Frehse-U. Mosco [25], [26] in the more general setting of the irregular-obstacle problems that will be described in the next section. The estimates of the modulus of continuity of the solution given in [26], however, were not as sharp as the local estimates mentioned in Section 1 and they were improved in [47] and [48].

An immediate consequence of the results in [47], [48] is the following *Maz'ja-type estimate::*

$$\text{(2.5)}\quad \operatorname*{osc}_{B_r(x_0)} u \leq \operatorname*{osc}_{B_R(x_0)\cap E} \phi + c \max\left(\sup_{B_{2R}(x_0)\cap E} \phi, 0\right) \exp\left(-\beta \int_r^R \frac{\operatorname{cap}\left(E\cap B_\rho(x_0), B_{2\rho}(x_0)\right)}{\operatorname{cap}\left(B_\rho(x_0), B_{2\rho}(x_0)\right)} \frac{d\rho}{\rho}\right)$$

for every $0 < r \leq R \leq 1/2 \operatorname{dist}(x_0, \partial\Omega)$, c and $\beta > 0$ being suitable constants that depend only on N (and on the ellipticity constants λ, Λ in case a more general operator (2.4) replaces $-\Delta$).

Note the similarity with (2.3). It appears from (2.5) that the set E supporting the unilateral constraint and the function ϕ bounding the solution on E contribute separately to the estimate, as if, like in Maz'ja's case, ϕ were a prescribed value on ∂E. The free boundary, on which the value ϕ is actually taken, plays no explicit role in (2.5).

If u is an arbitrary *local* solution of the thin-obstacle problem in a neighborhood Ω of x_0, according to the definition of the following Section 3, then (2.5) takes the form

$$(2.6)\qquad \begin{cases} \operatorname*{osc}_{B_r(x_0)} u \le \operatorname*{osc}_{B_R(x_0)\cap E} \phi + K \exp\left(-\beta \displaystyle\int_r^R \frac{\operatorname{cap}\left(E\cap B_\rho(x_0), B_{2\rho}(x)\right)}{\operatorname{cap}\left(B_\rho(x_0), B_{2\rho}(x_0)\right)} \frac{d\rho}{\rho} \right) \\ K = c \left\{ R^{-\frac{N}{2}} \|u\|_{L^2(B_{2R}(x_0))} + \max\left(\sup_{B_{2R}(x_0)\cap E} \phi, 0\right) \right\}, \end{cases}$$

for every $0 < r \le R \le 1/2 \operatorname{dist}(x_0, \partial\Omega)$.

At an *interior* point x_0 of E, (2.6) reduces to

$$(2.7)\qquad \operatorname*{osc}_{B_r(x_0)} u \le c \operatorname*{osc}_{B_R(x_0)} \phi + K\left(\frac{r}{R}\right)^\beta$$

for every $0 < r \le R \le R_0$, $B_{2R_0}(x_0) \subset \overset{\circ}{E}$.

If ϕ is *Hölder continuous* at x_0, with exponent α then, by standard iterative arguments, it follows from (2.7) that u is also Hölder continuous at x_0, with exponent $\alpha' = \min(\beta - \varepsilon, \alpha)$ for every $\varepsilon > 0$. This is the pointwise version of Frehse's result [20] mentioned before (in [20] it is also proved that if the coefficients of L are smooth then $\alpha' = \alpha$).

For arbitrary *continuous* ϕ, not necessarily with a Hölder exponent of continuity at x_0, (2.7) again differs from the similar local estimate given by Caffarelli-Kinderlehrer [10] in the presence of the exponent β. This is explained by the fact that (2.6), (2.7) hold for more general operators L of the type (2.4).

In particular (2.7) can be applied to an arbitrary local solution u of the *equation* $Lu = 0$ in Ω, by taking ϕ to be a negative constant of sufficiently large absolute value. Then we get

$$(2.8)\qquad \operatorname*{osc}_{B_r(x_0)} u \le c R^{-\frac{N}{2}} \|u\|_{L^2(B_{2R}(x_0))} \left(\frac{r}{R}\right)^\beta$$

for every $0 < r \le R \le 1/2 \operatorname{dist}(x_0, \partial\Omega)$, according to the classical result of De Giorgi and Nash, cfr. G. Stampacchia [53]. We should mention however, that the De Giorgi-Nash-Moser results are indirectly used in the proof of (2.6), *via* the Green function associated with the operator L and the estimate of its size, see [53].

3. Irregular obstacles

One-sided obstacle problems present an interesting feature: Discontinuous obstacles may still lead to continuous solutions. It is the free boundary of the problem that self adjusts to keep the solution continuous on the whole domain.

This regularizing effect suggests a more general formulation of the problem of Section 1, in which a discontinuous function, say a *Borel function* ϕ, is allowed as an obstacle.

Another extension, which is a natural one from the point of view of potential theory, consists in allowing the set E supporting the obstacle to be an arbitrary *Borel subset* of $\mathbf{R}^N$. With respect to the previous case of a compact E, now the set $D - E$ may not be open and we can no more rely on the regularity theory for the Dirichlet problem in an open domain.

A general formulation for *irregular-obstacle* problems was considered by J. Frehse-U. Mosco [22], [23], [24]. It was motivated by the *implicit* obstacles (that is, obstacles that depend on the solution itself) of certain quasi-variational inequalities connected with problems of optimal impulse control of stochastic diffusions. These obstacles have the property of being monotone functions $\psi(x)$ of $x \in \mathbf{R}^N$, with respect to the usual partial ordering of $\mathbf{R}^N$. In [22] the following rather surprising result was proved:

Arbitrary local weak solutions u of an obstacle problem with a *monotone obstacle* $\psi(x)$ in $\mathbf{R}^N$ are Hölder continuous on their domain and a common Hölder exponent exists for all such u 's.

This local continuity result was later extended in [23], [24] to more general classes of irregular obstalces. It was shown, in particular, that what affects the regularity of the solution is the whole family of *level sets* of the obstacle and that the "good obstacles" can be described in capacity terms.

The methods in [22], [24] are similar to those of Frehse's proof of Wiener's criterion in [21] and are based on integral energy estimates of Morrey's type for weak solutions.

By combining these methods with suitable integration arguments from Maz'ja [43], [44], J. Frehse-U. Mosco were then able to prove in [25], [26] that local weak solutions are continuous at every point at which the *Wiener condition* described below is satisfied. This condition was then shown in [48] to be also necessary, so that a Wiener-type criterion for obstacle problems was established. Qualitative sufficient conditions for nonlinear obstacle problems have been given by J.H. Michael-W.P. Ziemer [46] and W.P. Ziemer [56].

Let us now describe the Wiener criterion, and further estimates from [47], [48], more precisely.

By an *obstacle* ψ in $\mathbf{R}^N$, $N \geq 2$, we mean an arbitrary given function $\psi(x)$ defined for $x \in \mathbf{R}^N$, with values in the extended reals $[-\infty,+\infty]$. Since the inequality imposed to the solution is required to hold a.e. in the capacity sense, the "true" obstacle is indeed the equivalence class of ψ in the a.e. capacity sense. In the following we shall sometimes write *q.e.* to mean "a.e. in the capacity sense".

Let an obstacle ψ be given in $\mathbf{R}^N$ and let x_0 be an arbitrary point of $\mathbf{R}^N$. By

$$\{u\}_{\psi,x_0}$$

we shall denote the set of all *local weak solutions* u of the obstacle problem with obstacle ψ, that is, of all functions $u(x)$ which are solutions of the problem

$$(3.1)\qquad \begin{cases} u \in H^1(\Omega),\ u(x) \geq \psi(x) \text{ q.e. in } \Omega \\ \displaystyle\int_\Omega Du\, D(u-v)dx \leq 0 \\ \text{for every } v \in H^1(\Omega),\ v(x) \geq \psi(x) \text{ q.e. in } \Omega,\ v-u \in H_0^1(\Omega) \ , \end{cases}$$

in some arbitrary open neighborhood Ω of x_0 in $\mathbf{R}^N$.

Definition 3.1. We say that x_0 is a regular point of ψ if the set $\{u\}_{\psi,x_0}$ is not empty and every $u \in \{u\}_{\psi,x_0}$ is finite and continuous at x_0 in $\mathbf{R}^N$.

In order to characterize these points, we consider the following family of level sets of ψ:

$$(3.2)\qquad L(\varepsilon) = \left\{ x \in \mathbf{R}^N : \psi(x) \geq \overline{\psi}(x_0) - \varepsilon \right\},\ \ \varepsilon > 0 \ ,$$

where

$$\overline{\psi}(x_0) := \inf_{\rho>0} \sup_{B_\rho(x_0)} \psi \ ,$$

the supremum being in the q.e. sense.

The regularity of ψ at x_0 turns out to be related to the regularity of all level sets (3.2) at x_0, in the classical Wiener's sense.

Definition 3.2. We say that x_0 is a *Wiener point* of ψ if for every $\varepsilon > 0$ condition (2.1) of Section 2 holds with $E = L(\varepsilon)$ given by (3.2).

The *Wiener criterion* for the obstacle problem can then be stated as follows (see [48], [15]):

Theorem 3.1. *Given an obstacle* ψ *in* $\mathbf{R}^N$, $N \geq 2$, *a point* $x_0 \in \mathbf{R}^N$ *is a regular point of* ψ *according to Definition* 3.1 *if and only if* $\psi(x_0) < +\infty$ *and* x_0 *is a Wiener point of* ψ *according to Definition* 3.2.

A similar sufficient, but not necessary, local condition for the regularity of x_0 was also given by P. Charrier [11].

Let us now discuss how the modulus of continuity of $u \in \{u\}_{\psi, x_0}$ can be estimated at a regular point of ψ.

Let us first remark that, as shown in [48], Maz'ja-type estimate (2.6) extends to the case that E is an arbitrary subset of the domain Ω, with $E \subset \Omega$, and ϕ is an arbitrary real-valued function defined on E. The sup, inf and osc = sup – inf should be taken now in the q.e. sense and the capacities involving the set E should be external capacities. However, such an estimate only covers the case when ϕ is continuous at x_0 on E in the q.e. sense (that is, the q.e. oscillation of ϕ on $B_r(x_0) \cap E$ vanishes as $r \to 0$).

For arbitrary ψ we expect the oscillation of u on $B_r(x_0)$ to decrease to zero as $r \to 0$ according to the rate of divergence of the integrals (2.1), where $E = L(\varepsilon)$. Moreover, we also expect the *energy* of u on $B_r(x_0)$ to decay to zero as $r \to 0$ according to the rate of divergence of these integrals.

Indeed more general estimates hold, which can be seen as an analogue for the obstacle problem of the *structural estimates* for the energy's decay in Saint-Venant's principle of linear elasticity, *i.e.*: If u is an arbitrary local weak solution of the equation $Lu = 0$ and x_0 is an interior point of the domain, then the *ratio* of the energy

$$\mathcal{E}(r) = \int_{B_r(x_0)} |Du|^2 \, dx$$

on two concentric balls $B_r(x_0) \subset B_R(x_0)$ behaves according to

$$\mathcal{E}(r) \leq c\, \mathcal{E}(R) \left(\frac{r}{R}\right)^{N-2+2\beta} \qquad (N \geq 3) \tag{3.3}$$

$0 < r \le 1/2\,R$, where c and $\beta > 0$ are constants depending only on N and on the ellipticity constants of the operator L in (2.4).

The quantity that for an obstacle problem plays a role similar to that of the energy in (3.3) turns out to be the following *potential seminorm*

$$(3.4)\qquad \begin{cases} V(r) = \operatorname*{osc}_{B_r(x_0)} u + \left(\displaystyle\int_{B_r(x_0)} |Du|^2 |x - x_0|^{2-N} dx \right)^{\frac{1}{2}}, & \text{if } N \ge 3 \\[2ex] V(r) = \operatorname*{osc}_{B_r(x_0)} u + \left(\displaystyle\int_{B_r(x_0)} |Du|^2 \log \frac{r}{|x - x_0|} dx \right)^{\frac{1}{2}}, & \text{if } N=2 \end{cases}$$

defined for $u \in \{u\}_{\psi,x_0}$ and $B_r(x_0)$ in the domain of u.

The decay of $V(r)$ as $r \to 0$ can be conveniently described by introducing the *Wiener modulus* of ψ at x_0. This is the function

$$\omega_\sigma(r,R) \equiv \omega_{\psi,\sigma}(x_0;r,R)$$

defined for every $\sigma > 0$ (a scaling factor) and every $0 < r \le R$, by setting

$$(3.5)\qquad \omega_\sigma(r,R) = \inf \left\{ \omega > 0 : \omega \exp \int_r^R \delta(\sigma\omega,\rho) \frac{d\rho}{\rho} \ge 1 \right\},$$

where for every $\varepsilon > 0$ and $\rho > 0$

$$(3.6)\qquad \delta(\varepsilon,\rho) = \frac{\operatorname{cap}\left(E(\varepsilon,\rho), B_{2\rho}(x_0)\right)}{\operatorname{cap}\left(B_\rho(x_0), B_{2\rho}(x_0)\right)}$$

and

$$(3.7)\qquad E(\varepsilon,\rho) = \left\{ x \in B_\rho(x_0) : \psi(x) \ge \sup_{B_\rho(x_0)} \psi - \varepsilon \right\},$$

the $\sup$ being in the q.e. sense.

It is easy to see that x_0 is a Wiener point of ψ according to Definition 3.2 if and only if for every $\varepsilon > 0$ there exists $R > 0$ such that

$$\omega_\sigma(r,R) \to 0 \text{ as } r \to 0$$

for suitable values σ, possibly depending on ε, r and R, such that $\sigma\,\omega_\sigma(r,R) \le \varepsilon$ for all $0 < r \le R$.

Theorem 3.2. *There exist constants* k, c *and* $\beta > 0$ *such that for every* $u \in \{u\}_{\psi,x_0}$ *with domain* Ω *in* $\mathbb{R}^N$,

$$V(r) \le k\, V(R)\, \omega_\sigma(r,R)^\beta + c\,\sigma\,\omega_\sigma(r,R) \tag{3.8}$$

for every $0 < r \le R \le \operatorname{dist}(x_0,\partial\Omega)$ *and every* $\sigma > 0$. *Moreover, there exists a constant* c_0 *such that*

$$V(R) \le c_0 R^{-\frac{N}{2}} \|u-d\|_{L^2(B_{2R}(x_0))} \quad \textit{for every constant } d \ge \sup_{B_{2R}(x_0)} \psi \tag{3.9}$$

and for every $0 < R \le 1/2 \operatorname{dist}(x_0,\partial\Omega)$.

The constant k depends only on N, the constants c, c_0 and β depend only on N and on the ellipticity constants λ, Λ of L as in (2.4), if such a more general operator takes the place of the Laplacian.

Maz'ja-type estimate (2.6), E being an arbitrary subset of Ω, $E \subset \Omega$, and ϕ an arbitrary real-valued function on E, is obtained from (3.8), (3.9) when $\psi \equiv \phi$ on E, $\psi \equiv -\infty$ in $\mathbb{R}^N - E$ and, if $\operatorname{osc}\phi > 0$, $\sigma = \sigma(r,R)$ chosen to be

$$\sigma = \operatorname*{osc}_{B_R(x_0)\cap E} \phi \cdot \exp\left(\int_r^R \frac{\operatorname{cap}\left(E\cap B_\rho(x_0), B_{2\rho}(x_0)\right)}{\operatorname{cap}\left(B_\rho(x_0), B_{2\rho}(x_0)\right)} \frac{d\rho}{\rho} \right),$$

what leads to

$$\omega_\sigma(r,R) \le \exp\left(-\int_r^R \frac{\operatorname{cap}\left(E\cap B_\rho(x_0), B_{2\rho}(x_0)\right)}{\operatorname{cap}\left(B_\rho(x_0), B_{2\rho}(x_0)\right)} \frac{d\rho}{\rho} \right) \quad \text{and}$$

$$\sigma\,\omega_\sigma(r,R) \le \operatorname*{osc}_{B_R(x_0)\cap E} \phi\,,$$

if $\operatorname{osc}\phi = 0$, we choose $\sigma > 0$ arbitrary.

The energy's decay in (3.3) is obtained when $\psi \equiv -\infty$, by estimating the L^2-norm in (3.9) by Poincaré's inequality.

While (3.8), (3.9) are the most general estimates available for (variational) unilateral obstacle problems, yet they do not include Maz'ja's estimate (2.3) as a special case. For that, we need a unified formulation of the one-sided obstacle problems, which are special free-boundary problems, and the usual fixed-boundary Dirichlet problems. Such a general framework will be described in the next section, in the context of *two-obstacle* problems.

4. The bite of the shark: the two-obstacle problem

In a two-obstacle problem the solution u minimizing the Dirichlet integral is restricted from both below and above by two given obstacles, ψ_1 and ψ_2 respectively. In addition to the lower and upper free boundaries, where $u = \psi_1$ and $u = \psi_2$ respectively, a fixed boundary is now part of the problem, that is the boundary of the region where $\psi_1 \equiv \psi_2$, hence also $u \equiv \psi_1 \equiv \psi_2$. The admissible test functions in a neighborhood of an arbitrary point of the domain are reduced drastically and in that lies one of the main technical difficulties of the problem.

In [15], G. Dal Maso, M.A. Vivaldi and the author carry out a theory of pointwise regularity for two-obstacle problems which is comprehensive of all cases discussed up to now. In particular, a Wiener criterion and structural estimates are given, that generalize those stated in the previous sections.

Let two obstacles $\psi_1(x)$, $\psi_2(x)$ be given in $\mathbf{R}^N$, $N \geq 2$.

We say that a function $u(x)$ is a *local variational solution* of the two-obstacle problem $\{\psi_1, \psi_2\}$ in an open region Ω of $\mathbf{R}^N$, if u satisfies the following conditions

$$(4.1)\qquad \begin{cases} u \in H^1(\Omega),\ \psi_1(x) \leq u(x) \leq \psi_2(x) \ \text{q.e. in } \Omega \\[2ex] \displaystyle\int_\Omega Du\, D(u-v)dx \leq 0 \quad \text{for every } v \in H^1(\Omega) \\[2ex] \text{such that } \psi_1(x) \leq v(x) \leq \psi_2(x) \ \text{q.e. in } \Omega \text{ and } v-u \text{ in } H^1_0(\Omega) \end{cases}.$$

One-sided obstacle problems of Section 3 are obvious special cases, with $\psi_1 \equiv \psi$ and $\psi_2 \equiv +\infty$.

The Dirichlet problem (2.2) of Section 2, where for simplicity we assume now $h(x)$ to be the trace on D of a function

(4.2) $$h \in C(\mathbf{R}^N) \cap H^1(\mathbf{R}^N) ,$$

can be also formulated as a problem (4.1), by taking

(4.3) $$\psi_1 \equiv \psi_2 \equiv h \text{ on } D^C, \ \psi_1 \equiv -\infty \text{ and } \psi_2 \equiv +\infty \text{ in } D$$

and extending $u = h$ on D^C.

More generally, the *formal* Dirichlet problem

(4.4) $$\Delta u = 0 \text{ in } \Omega - E, \ u = h \text{ on } E$$

where E is a (Borel) subset of Ω and h is a (Borel) real-valued function on E, can be also rigorously formulated as a problem of the type (4.1), by taking $\psi_1 \equiv \psi_2 \equiv h$ on E, $\psi_1 \equiv -\infty$ and $\psi_2 \equiv +\infty$ on $E^C = \Omega - E$. Then (4.1) becomes equivalent to

(4.5) $$\begin{cases} u \in H^1(\Omega), \ u = h \text{ q.e. on } E \\ \displaystyle\int_\Omega Du\, D\phi\, dx = 0 \\ \text{for every } \phi \in H_0^1(\Omega), \ \phi = 0 \text{ q.e. on } E \end{cases}$$

and this gives a variational sense to (4.4). It should be noted that a solution u of (4.5) will not be in general a distribution solution of (4.4), for $\Omega - E$ may not be open.

Let us now describe the main results from [15].

Let ψ_1, ψ_2 and a point $x_0 \in \mathbf{R}^N$ be fixed. By

$$\{u\}_{\psi_1, \psi_2, x_0}$$

we denote the set of all functions $u(x)$ that satisfy (4.1) in some arbitrary open neighborhood Ω of x_0, possibly depending on u.

Definition 4.1. We say that x_0 is a (variational) *regular point* of $\{\psi_1, \psi_2\}$ if the set $\{u\}_{\psi_1, \psi_2, x_0}$ is not empty and every solution $u \in \{u\}_{\psi_1, \psi_2, x_0}$ is finite and continuous at x_0 in $\mathbf{R}^N$.

For every $\varepsilon > 0$ let

(4.6) $$L_1(\varepsilon) = \left\{ x \in \mathbf{R}^N : \psi_1(x) \geq \overline{\psi}_1(x_0) - \varepsilon \right\}$$

(4.7) $$L_2(\varepsilon) = \left\{ x \in \mathbf{R}^N : \psi_2(x) \leq \underline{\psi}_2(x_0) + \varepsilon \right\}$$

where

$$\overline{\psi}_1(x_0) = \inf_{\rho>0} \sup_{B_\rho(x_0)} \psi_1 , \quad \underline{\psi}_2(x_0) = \sup_{\rho>0} \inf_{B_\rho(x_0)} \psi_2 ,$$

the sup and inf over $B_\rho(x_0)$ being in the q.e. sense.

Definition 4.2. We say that $x_0 \in \mathbf{R}^N$ is a *Wiener point* of $\{\psi_1, \psi_2\}$ if for every $\varepsilon > 0$ condition (2.1) of Section 2 is satisfied by both $E = L_1(\varepsilon)$ and $E = L_2(\varepsilon)$, given by (4.6) and (4.7) respectively.

The following *Wiener criterion* holds

Theorem 4.1. *A point* $x_0 \in \mathbf{R}^N$ *is a regular point of* $\{\psi_1, \psi_2\}$ *according to Definition* 4.1 *if and only if all the conditions* (j), (jj) *and* (jjj) *below hold:*

(j) *There exists* $R > 0$ *and* $w \in H^1(B_R(x_0))$, *such that* $\psi_1(x) \leq w(x) \leq \psi_2(x)$ q.e. in $B_R(x_0)$

(jj) $\overline{\psi}_1(x_0) < +\infty$, $\underline{\psi}_2(x_0) > -\infty$ *and* $\overline{\psi}_1(x_0) \leq \underline{\psi}_2(x_0)$

(jjj) x_0 *is a Wiener point of* $\{\psi_1, \psi_2\}$ *according to Definition* (4.2).

At a regular point x_0 we have $\overline{\psi}_1(x_0) \leq u(x_0) \leq \underline{\psi}_2(x_0)$ for every $u \in \{u\}_{\psi_1, \psi_2, x_0}$.

The theorem shows that the regularity of u at x_0 is affected by each one of the two obstacles separately, even in the case when they touch one each other at x_0 and $\overline{\psi}_1(x_0) = u(x_0) = \underline{\psi}_2(x_0)$.

Let us point out that $\psi_1(x)$ and $\psi_2(x)$ may both oscillate very much in arbitrary small neighborhoods of x_0, interpenetrating one each other, see Fig. 1. Condition (j), requiring the existence of a separating function $w(x)$ of finite energy, may be violated if these oscillations are too wild, see Fig. 2. If this is the case, no variational solution exists. However a separating $w(x)$ may still exist, which is continuous in the q.e. sense. Then an appropriate notion of generalized solution can be introduced, as a uniform limit of variational solutions, and Theorem 4.1 can be extended to this more general setting. A useful tool, in dealing with generalized solutions is the estimate of the following Theorem 4.2, which is stable under uniform convergence of obstacles and solutions.

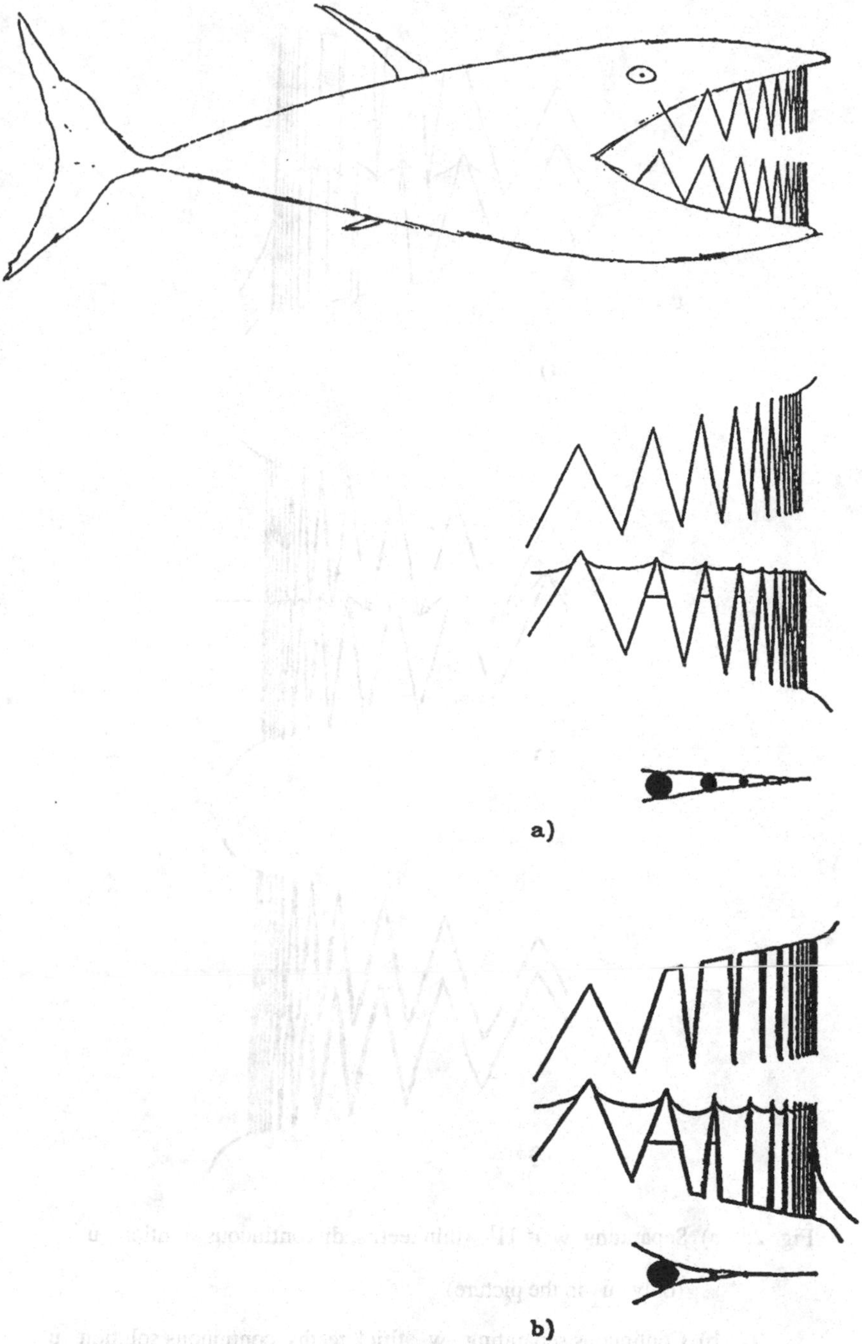

Fig. 1. a) Thick teeths, continuous solution

b) Thin teeths, discontinuous solution

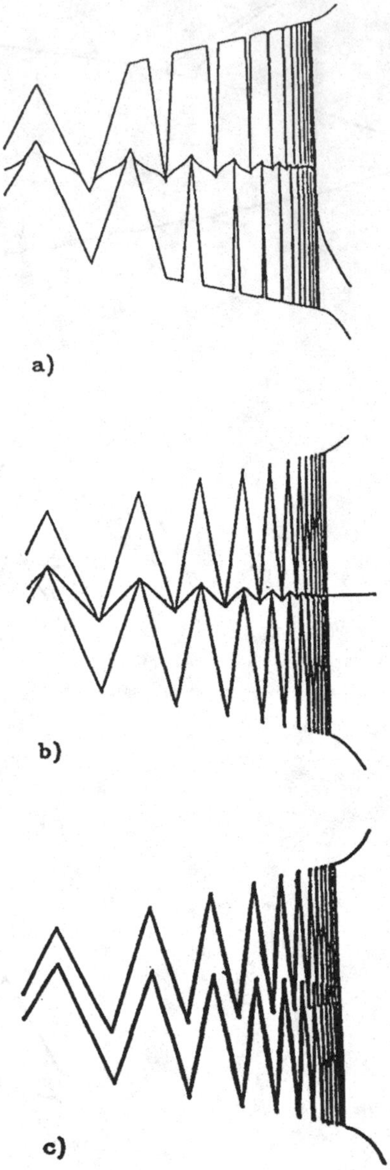

Fig. 2. a) Separating $w \in H^1$, thin teeths, discontinuous solution u (only u in the picture)

b) Continuous separating w, thick teeths, continuous solution u (only w in the picture)

c) No good separating w exists.

Let us now come back to the variational solutions $u \in \{u\}_{\psi_1, \psi_2, x_0}$. We shall first describe a Maz'ja-type estimate for the oscillation of u, then a structural estimate for the potential seminorm $V(r)$ associated with u by (3.4). For simplicity, we shall confine ourselves to the most interesting case

$$-\infty < \overline{\psi}_1(x_0) = \underline{\psi}_2(x_0) < +\infty \tag{4.8}$$

and we refer to [15] for the general result.

The Wiener modulus of ψ_1 at x_0 is defined by (3.5), (3.6) and (3.7), where $\psi = \psi_1$. The Wiener modulus of ψ_2 at x_0 is again defined according to (3.5), (3.6), where now

$$E(\varepsilon,\rho) = \left\{ x \in B_\rho(x_0) : \psi_2(x) \leq \inf_{B_\rho(x_0)} \psi_2 + \varepsilon \right\}$$

for every $\varepsilon > 0$ and $\rho > 0$.

Theorem 4.2. *There exist constants* c *and* $\beta > 0$ *such that if* x_0 *is a point of* $\mathbf{R}^N$ *at which condition* (4.8) *is satisfied, then for every* $u \in \{u\}_{\psi_1, \psi_2, x_0}$ *with domain* $\Omega \ni x_0$ *the following estimate holds*

$$(4.9)\quad \begin{cases} \operatorname*{osc}_{B_r(x_0)} u \leq \psi_{\sigma_1,\sigma_2}(r,R) + c\left[\operatorname*{osc}_{B_R(x_0)} u + \psi_{\sigma_1,\sigma_2}(r,R)\right]\left[\omega_{\psi_1,\sigma_1}(r,R) + \omega_{\psi_2,\sigma_2}(r,R)\right]^\beta \\ \text{where} \\ \psi_{\sigma_1,\sigma_2}(r,R) = \sup_{B_R(x_0)} \psi_1 - \inf_{B_R(x_0)} \psi_2 + \sigma_1\omega_{\psi_1,\sigma_1}(r,R) + \sigma_2\omega_{\psi_2,\sigma_2}(r,R) \end{cases}$$

for arbitrary $\sigma_1 > 0$, $\sigma_2 > 0$ and $0 < r \leq 1/2\, R \leq 1/2 \operatorname{dist}(x, \partial\Omega)$.

The constants c and β depends only on N and on the ellipticity constants of L as in (2.4), if such an operator takes the place of the Laplacian.

Maz'ja's estimate (2.3), with D^C replacing ∂D, is obtained as a special case of (4.9), under the assumption (4.2), by taking ψ_1 and ψ_2 as in (4.3) and by choosing

$$\sigma_1 = \sigma_2 = \operatorname*{osc}_{B_R(x_0)\cap E} h \, \exp\left(\int_r^R \frac{\operatorname{cap}\left(E \cap B_\rho(x_0), B_{2\rho}(x_0)\right)}{\operatorname{cap}\left(B_\rho(x_0), B_{2\rho}(x_0)\right)} \frac{d\rho}{\rho} \right), \quad E = D^C ,$$

(u is extended to be equal to h in D^C) if $\operatorname{osc} h > 0$, and arbitrary $\sigma_1 = \sigma_2 > 0$ otherwise.

In addition, we obtain that (2.3) also holds for general Dirichlet problems of the type (4.5).

Let us now state the structural estimates for the potential seminorm (3.4). They are obtained under the further assumption that a *separating function* $w \in H^1$ exists, such that Δw, or more generally Lw, belongs to the *Kato class* K_N.

By $K_N(\Omega)$, where Ω is a bounded open subset of $\mathbf{R}^N$, we denote the set of all (signed) Radon measures ν in Ω such that

$$\|\nu\|_{K^N(\Omega)} < +\infty ,$$

where we define

$$(4.10)\qquad \begin{cases} \|\nu\|_{K^N(\Omega)} = \sup\limits_{x \in \Omega} \displaystyle\int_\Omega |x-y|^{2-N}\, d|\nu|(y) \quad \text{if } N \geq 3 , \\[2ex] \|\nu\|_{K^N(\Omega)} = \sup\limits_{x \in \Omega} \displaystyle\int_\Omega \log \frac{\operatorname{diam}(\Omega)}{|x-y|}\, d|\nu|(y) + |\nu|(\Omega) \quad \text{if } N = 2 , \end{cases}$$

where $\operatorname{diam}(\Omega)$ denotes the diameter of Ω, see [3].

Theorem 4.3. *Let* ψ_1, ψ_2, x_0 *be such that* (4.8) *holds. Moreover, let* $R_0 > 0$ *and* $w \in H^1\big(B_{R_0}(x_0)\big)$ *be such that* $\psi_1(x) \leq w(x) \leq \psi_2(x)$ *for* x q.e. *in* $B_{R_0}(x_0)$ *and* $\Delta w \in K_N\big(B_{R_0}(x_0)\big)$. *Then for every* $u \in \{u\}_{\psi_1,\psi_2,x_0}$, *with domain* $\Omega \ni x_0$ *in* $\mathbf{R}^N$, *the potential seminorm* (3.4) *satisfies*

$$(4.11)\qquad \begin{cases} V(r) \leq k\, V(R) \Big[\omega_{\psi_1,\sigma_1}(r,R) + \omega_{\psi_2,\sigma_2}(r,R)\Big]^{\beta} + c \Big[\sigma_1\, \omega_{\psi_1,\sigma_1}(r,R) + \sigma_2\, \omega_{\psi_2,\sigma_2}(r,R)\Big] + \\[1ex] + c\, \|\Delta w\|_{K_N(B_R(x_0))} + c\, R^{-\frac{N}{2}}\, \|w - w_R\|_{L^2(B_R(x_0))} \end{cases}$$

for every $\sigma_1 > 0$, $\sigma_2 > 0$, $0 < r \leq R \leq \min\{R_0, \operatorname{dist}(x_0, \partial\Omega)\}$, *where* $w_R = |B_R|^{-1} \displaystyle\int_{B_R(x_0)} w\, dx$.

Moreover

(4.12) $$V(R) \le c_0 \left[R^{-\frac{N}{2}} \left(\|u\|_{L^2(B_{2R}(x_0))} + \|w\|_{L^2(B_{2R}(x_0))} \right) + \|\Delta w\|_{K_N(B_{2R}(x_0))} \right]$$

for every $0 < r \le R \le 1/2 \min \{R_0, \mathrm{dist}(x_0, \partial\Omega)\}$.

The constant k depends only on N, the constants β, c and c_0 on N and on the ellipticity constants of L, if such an operator replaces $-\Delta$.

If $\psi_1 \equiv \psi$ and $\psi_2 \equiv +\infty$, then (4.11), (4.12) reduce to (3.8), (3.9) (note that if $\psi_2 \equiv +\infty$, then

$$\omega_{\psi_2,\sigma_2}(r,R) \equiv \frac{r}{R} \ge \omega_{\psi_1,\sigma_1}(r,R)$$

for arbitrary $\sigma_1 > 0$, $\sigma_2 > 0$ and $0 < r \le R$).

By applying (4.11) to a solution u of the Dirichlet problem (4.5) in $\Omega \ni x_0$, under the assumption that the function $h(x)$ is the restriction to E of a function $h \in H^1(\mathbb{R}^N)$ with $\Delta h \in K_N\left(B_{R_0}(x_0)\right)$ for some large $R_0 > 0$, we obtain

(4.13) $$V(r) \le k\, V(R) \exp\left(-\beta \int_r^R \frac{\mathrm{cap}\left(E \cap B_\rho(x_0), B_{2\rho}(x_0)\right)}{\mathrm{cap}\left(B_\rho(x_0), B_{2\rho}(x_0)\right)} \frac{d\rho}{\rho} \right) + c\, \|\Delta h\|_{K_N\left(B_R(x_0)\right)}$$

that generalizes the energy's decay formula (3.3), valid at an interior point.

The proof of Theorem 4.2 is achieved by comparison of the solution u with solutions of suitable one-sided obstacle problems. The proof of Theorem 4.3 uses estimation techniques from [48] and [13] and special choices of text functions. The main steps of the proof, once we ignore all the technical points due to the presence of the constraints, are similar to those of the proof sketched at the end of the next section for the relaxed Dirichlet problems.

5. Wiener points and γ-convergence of Borel measures

We shall now describe the main lines of the pointwise regularity theory for the *relaxed Dirichlet problems* studied by G. Dal Maso - U. Mosco in [12], [13] and [14] (see also [49] for a summary).

The relaxed Dirichlet problems appear naturally in the limit of Dirichlet problems in highly perturbed domains, as well as in the limit of stationary Schrödinger equations with perturbed singular potentials.

Our purpose here is to show that Wiener's criterion and Maz'ja's estimate of Section 2 keep to hold for such relaxed problems, provided they are suitably formulated in terms of *intrinsic* notions of energy and capacity.

From the point of view of the general two-obstacle theory of Section 4, we are now confining ourselves to the special problems of the type (4.5), with $h \equiv 0$, and to their limits when the set E is perturbed.

Let D be a fixed bounded open domain of $\mathbf{R}^N$ and let E_j be a sequence of closed subsets of D, with $\cup E_j \subset\subset D$.

For each j, let u_j be the weak solution of the Dirichlet problem

$$(5.1) \qquad \begin{cases} -\Delta u_j = 0 \ \text{ in } \ D - E_j, \ \ u_j = 0 \ \text{ on } \ E_j \\ \\ u_j = g \ \text{ on } \ \partial D \end{cases}$$

where $g \in H^1(D)$ is some fixed function, independent of j.

The sequence u_j is bounded in $H^1(D)$, therefore a subsequence, still denoted by u_j, converges strongly in $L^2(D)$ and weakly in $H^1(D)$ to some function $u \in H^1(D)$.

Let us point out that no assumption is made on the behaviour of the E_j's as $j \to +\infty$ and therefore the perturbed u_j's may develop wild oscillations in D as $j \to +\infty$.

We would like to answer the following questions:

1) Does the limit function u satisfy an equation in D?

2) Does u necessarily vanish at some points of D, independently of the prescribed boundary value g?

The answer to 1) is that a *limit equation* does actually exist in D. This equation can be formally written as

$$(5.2) \qquad -\Delta u + \mu u = 0$$

and a *measure* μ appears in it.

The measure μ in (5.2) is a *non-negative Borel measure* in $\mathbf{R}^N$, that vanishes on all polar subsets of $\mathbf{R}^N$ (i.e., sets of null capacity in $\mathbf{R}^N$), but may be $+\infty$ on "large" Borel subsets of $\mathbf{R}^N$. We denote the set of all such μ 's by $\mathcal{M}_0$.

The precise sense in which (5.2) is satisfied by u is the following variational one:

$$
(5.3) \quad \left\{
\begin{array}{l}
u \in H^1_{loc}(D) \cap L^2_{loc}(D,\mu) \\[1em]
\displaystyle\int_\Omega Du\, Dv\, dx + \int_\Omega u\, v\, d\mu = 0 \\[1em]
\text{for every } \; v \in H^1(D) \cap L^2(D,\mu) \; \text{ with compact support} .
\end{array}
\right.
$$

We say that u satisfying (5.3) is a *local (variational) solution* of (5.2) in D.

In addition to (5.3), the limit u of the u_j's satisfies the fixed boundary condition on ∂D, that is

$$
(5.4) \qquad u - g \in H^1_0(D) \; .
$$

By comparing (5.3), (5.4) with (5.1), we see that the zero-order term involving μ in (5.2) appears to be the "variational limit" of the Dirichlet conditions $u_j = 0$ on the "holes" E_j of D, as $j \to +\infty$.

This term contributes a zero-order term also to the *energy* associated with (5.2),

$$
\mathcal{E}_\mu = \int_D |Du|^2 dx + \int_D u^2 d\mu \; .
$$

Since $\mathcal{E}_\mu$ is locally minimized in D by any solution u of (5.3), the nature of the term involving μ is therefore that of a *penalization*, that forces the solution u to be "small" on the support of μ.

Let us point out that in general u is *not* a solution of (5.2) in the distribution sense, since we cannot replace v in (5.3) by an arbitrary $\phi \in C^\infty_0(D)$.

Problem (4.4) with $h = 0$, that was seen in Section 3 as a special two-obstacle problem, can be equivalently considered as a problem of the form (5.2), by defining

$$
(5.5) \qquad \mu = \infty_E
$$

to be the measure $\infty_E(F) = +\infty$ if $\text{cap}(F \cap E) > 0$, $\infty_E(F) = 0$ if $\text{cap}(F \cap E) = 0$.

In [12] a convergence of variational type is introduced in the set $\mathcal{M}_0$, called

γ*-convergence,* that makes $\mathcal{M}_0$ sequentially compact. Moreover, if μ_j γ-converges to μ in $\mathcal{M}_0$ as $j \to \infty$, then the corresponding solutions u_j of (5.3), (5.4), where $\mu = \mu_j$, converge to u strongly in $L^2(D)$ and weakly in $H^1(D)$ as $j \to \infty$.

The answer to the question 1) mentioned before is obtained by applying these results, with $\mu_j = \infty_{E_j}$ for every j.

Let us point out that the limit measure μ in (5.2), while depending on the sequence E_j, does *not* depend on the specific sequence of solutions u_j in (5.1).

Let us now come to question 2). The fact that the measure μ acts as a penalization on the size of $|u|$ suggests that u might be forced to vanish at special points of the domain, where such a penalization becomes "very strong", independently of the boundary condition (5.4). Therefore, we give the following:

Definition 5.1. Let μ be a measure of the class $\mathcal{M}_0$ and let x_0 be an arbitrary point of $\mathbf{R}^N$. We say that x_0 is a *regular Dirichlet point* of μ if every local solution of the equation (5.3) in some neighborhood Ω of x_0 is continuous at x_0 and $u(x_0) = 0$.

As just remarked, we expect these points to be those at which the "strength" of μ as a penalization to $|u|$ is very high.

Let us consider the case that u has been obtained as the limit of a sequence of perturbed solutions, like the u_j's above. Then, we can first estimate the size of $|u_j|$ in a ball $B_r(x_0)$ by using Maz'ja estimate (2.3) on the domain $D - E_j$, see also (4.11). We know indeed from Section 4 that we can apply (2.3) at an arbitrary point x_0 of D, disregarding whether $x_0 \in \partial E_j$ (we extend u_j to be $= 0$ an E_j). Therefore,

$$\sup_{B_r(x_0)} |u_j| \leq c\, R^{-\frac{N}{2}} \, \|u_j\|_{L^2(B_{2R}(x_0))} \, \omega_j(r,R)^{\beta} \tag{5.6}$$

for every $0 < r \leq R \leq 1/2 \operatorname{dist}(x_0, \partial D)$, where

$$\omega_j(r,R) = \exp\left(- \int_r^R \frac{\operatorname{cap}\left(E_j \cap B_\rho(x_0), B_{2\rho}(x_0)\right)}{\operatorname{cap}\left(B_\rho(x_0), B_{2\rho}(x_0)\right)} \frac{d\rho}{\rho} \right) \tag{5.7}$$

and c and $\beta > 0$ are constants depending only on N.

Now we pass to the limit in (5.6) as $j \to +\infty$. The crucial point is taking the limit of the capacities $\mathrm{cap}(E_j \cap B_\rho(x_0), B_{2\rho}(x_0))$ as $j \to +\infty$. Once again, we note that no assumption has been made on the E_j's and on their asymptotic behavior as $j \to +\infty$.

Here we apply another result from [12], namely:

If the sequence of measures ∞_{E_j} γ-converge to μ in $\mathcal{M}_0$ as $j \to +\infty$, then for every $\rho > 0$, except possibly a countable set, we have

$$(5.8) \qquad \lim_{j\to+\infty} \mathrm{cap}\Big(E_j \cap B_\rho(x_0), B_{2\rho}(x_0)\Big) = \mathrm{cap}_\mu\Big(B_\rho(x_0), B_{2\rho}(x_0)\Big) ,$$

where the μ-capacities at the right hand side of (5.8) are defined by

$$(5.9) \qquad \mathrm{cap}_\mu(B,A) = \inf\left\{ \int_A |Dv|^2\, dx + \int_B v^2\, d\mu : v - 1 \in H^1_0(A) \right\}$$

for arbitrary $B \subset$ open $A \subset \mathbf{R}^N$.

Note that if $\mu = \infty_E$, then $\mathrm{cap}_\mu(B,A) = \mathrm{cap}(B,A)$, the usual capacity in $\mathbf{R}^N$, and for arbitrary $\mu \in \mathcal{M}_0$

$$(5.10) \qquad 0 \le \mathrm{cap}_\mu(B,A) \le \mathrm{cap}(B,A)$$

for every $B \subset A$.

By going to the limit in (5.6) as $j \to +\infty$, we thus obtain

$$(5.11) \qquad \sup_{B_r(x_0)} |u| \le c\, R^{-\frac{N}{2}}\, \|u\|_{L^2(B_{2R}(x_0))}\, \omega_\mu(r,R)^\beta ,$$

once we define $\omega_\mu(r,R)$ to be

$$(5.12) \qquad \omega_\mu(r,R) = \exp\left(- \int_r^R \frac{\mathrm{cap}_\mu\big(B_\rho(x_0), B_{2\rho}(x_0)\big)}{\mathrm{cap}\big(B_\rho(x_0), B_{2\rho}(x_0)\big)} \frac{d\rho}{\rho} \right) .$$

We call (5.12) the *Wiener modulus* of μ at x_0.

Definition 5.2. Given an arbitrary $\mu \in \mathcal{M}_0$ we say that $x_0 \in \mathbf{R}^N$ is a *Wiener point* of μ if

$$(5.13) \qquad \lim_{r\to0} \omega_\mu(r,R) = 0$$

for some (hence all) $R > 0$.

As a consequence of (5.11), if x_0 is a Wiener point of u, then u is continuous at x_0 and $u(x_0) = 0$. As we shall see in a moment, (5.11) can be proved for an arbitrary local solution u of (5.2) in a neighborhood of x_0, therefore we also find that if x_0 is a Wiener point of μ, then x_0 is a regular Dirichlet point of μ, according to our previous definition.

Since the converse is also true, we can state the following *Wiener criterion* for relaxed Dirichlet problems.

Theorem 5.1. *Given an arbitrary measure* $\mu \in \mathcal{M}_0$, *a point* $x_0 \in \mathbf{R}^N$ *is a regular Dirichlet point of* μ *according to Definition* 5.1, *if and only if* x_0 *is a Wiener point of* μ *according to Definition* 5.2.

Let us now describe from [13] some more general structural estimates satisfied by an arbitrary local solution u of (5.2) in the interior of its domain Ω.

Let $x_0 \in \Omega$ be fixed and let $R_0 > 0$ be such that $B_{R_0}(x_0) \subset \Omega$. For every $r < R_0$ we consider the following μ*-potential norm*

$$\text{(5.14)}\qquad V_\mu(r) = \left\{ \sup_{B_r(x_0)} |u|^2 + \int_{B_r(x_0)} |Du|^2 |x - x_0|^{2-N} dx + \int_{B_r(x_0)} u^2 |x - x_0|^{2-N} d\mu \right\}^{\frac{1}{2}}$$

if $N \geq 3$ and

$$\text{(5.15)}\qquad V_\mu(r) = \left\{ \sup_{B_r(x_0)} |u|^2 + \int_{B_r(x_0)} |Du|^2 \log\left(\frac{2r}{|x - x_0|^2}\right) dx + \int_{B_r(x_0)} u^2 \log\left(\frac{2r}{|x - x_0|^2}\right) dx \right\}^{\frac{1}{2}}$$

if $N = 2$.

Theorem 5.2. *There exist constants* c *and* $\beta > 0$, *such that for every solution* u *of* (5.3) *in the domain* Ω *we have*

$$\text{(5.16)}\qquad V_\mu(r) \leq c V_\mu(R)\, \omega_\mu(r,R)^\beta$$

for every $0 < r \leq R \leq R_0$. *Moreover,*

$$\text{(5.17)}\qquad V_\mu(R) \leq c_0 R^{-\frac{N}{2}} \| u \|_{L^2(B_{2R}(x_0))}$$

for every $0 < r \leq R \leq 1/2\, R_0$.

Corollary 1. *If* x_0 *is a Wiener point of* μ, *then*

(5.18) $$\lim_{r\to 0} V_\mu(r) = \lim_{x\to x_0} u(x) = u(x_0) = 0 .$$

Corollary 2. *If we define*

$$\mathcal{E}_\mu(r) = \int_{B_r(x_0)} |Du|^2 dx + \int_{B_r(x_0)} u^2 d\mu ,$$

then

(5.19) $$\mathcal{E}_\mu(r) \le k \mathcal{E}_\mu(2R) \frac{r^{N-2}}{\mathrm{cap}_\mu\left(B_{2R}(x_0), B_{4R}(x_0)\right)} \omega_\mu(r,R)^{2\beta}$$

for every $0 < r \le R \le 1/2 \operatorname{dist}(x_0, \Omega)$.

These results show that the modulus of continuity of u, as well as the decay of the energy, at a regular Dirichlet point of the measure μ depends on the rate of divergence of the integral in (5.12) as $r \to 0$. In [14] several examples are given of (radial) μ's in the class $\mathcal{M}_0$ whose Wiener modulus $\omega_\mu(r,R)$ has a decay of logarithmic type as $r \to 0$.

Let us also remark that the estimates above are *stable* under the γ-convergence of μ in $\mathcal{M}_0$. In fact, the continuity property (5.8) is proved in [12] to hold for an arbitrary sequence of measures μ_j that γ-converge to μ in $\mathcal{M}_0$. This property was exploited in our previous derivation of (5.11), when the measure μ was the γ limit of a sequence $\mu_j = \infty_{E_j}$. Indeed, it is also shown in [12] that *every* $\mu \in \mathcal{M}_0$ can be obtained as the γ limit of a sequence of measures of this special form, for a suitable choice of the subsets E_j's.

In [9] a kind of converse of property (5.8) is studied. From the knowledge of the limit of the μ_j-capacities $\mathrm{cap}_{\mu_j}\left(B_\rho(x_0), B_{2\rho}(x_0)\right)$, $\rho > 0, x_0 \in \Omega$, as $j \to +\infty$ it is possible to construct the γ-limit μ of the μ_j's, namely, its *density* $\frac{du}{dv}$ with respect to some reference measure v. This is accomplished by relying on an analogue of the Radon-Nykodim derivation theorem on balls, in which the measures $\mu(B_\rho(x_0))$ are replaced by the capacities $\mathrm{cap}_\mu(B_\rho(x_0))$, see [9] for more details and applications.

Finally, let us mention that the γ-convergence of measures of $\mathcal{M}_0$ can be also described in a probabilistic context of stochastic Wiener diffusions, where it becomes the *stable convergence* of the randomized stopping measures associated with the μ's of the class $\mathcal{M}_0$, see J. Baxter, G. Dal

Maso, U. Mosco [2]. Moreover, it is also shown in [2] that the γ-convergence of μ_j to μ in $\mathcal{M}_0$ is equivalent to the strong convergence of the *resolvent operators* $(-\Delta + \mu_j + \lambda)^{-1}$ to $(-\Delta + \mu + \lambda)^{-1}$ as $j \to \infty$ and this provides a link with the convergence properties of eigenvalues and eigenfunctions of the corresponding relaxed Dirichlet problems.

We conclude by giving the main lines of the proof of Theorem 5.2. A main role in the proof is played by intrinsec notions of energy and capacity and by the related Poincare's inequality.

Proof of Theorem 5.2.

1st step. A Caccioppoli-De Giorgi-Moser inequality.

For fixed $0 < r \leq \rho \leq R$ and $s \in (0,1)$ the μ-potential norm $V = V_\mu$ on the ball $B_{s\rho}(x_0)$ is estimated by the L^2-norm of u in the *annulus* $B_\rho(x_0) - B_{s\rho}(x_0)$:

$$V(s\rho)^2 \leq k\,\rho^{-N} \int_{B_\rho(x_0) - B_{s\rho}(x_0)} u^2\,dx\,. \tag{5.20}$$

2nd step. A Poincare' inequality for the μ-capacity and the μ-energy:

$$\int_{B_\rho(x_0) - B_{s\rho}(x_0)} u^2\,dx \leq k \frac{\rho^N}{\mathrm{cap}_\mu\big(B_\rho(x_0) - B_{s\rho}(x_0), B_{2\rho}(x_0)\big)} \left[\int_{B_\rho(x_0) - B_{s\rho}(x_0)} |Du|^2\,dx + \int_{B_\rho(x_0) - B_{s\rho}(x_0)} u^2\,d\mu \right] \tag{5.21}$$

$k = k(N,s)$, which is shown to hold for every $u \in H^1(B_\rho(x_0))$ and every $\mu \in \mathcal{M}_0$.

By estimating the right-hand side of (5.20) by (5.21) and by introducing the relative capacities

$$\delta_s(\rho) = \frac{\mathrm{cap}_\mu\big(B_\rho(x_0) - B_{2\rho}(x_0), B_{2\rho}(x_0)\big)}{\mathrm{cap}\big(B_\rho(x_0), B_{2\rho}(x_0)\big)}\,, \tag{5.22}$$

we obtain

$$k\delta_s(\rho)\,V(s\rho)^2 \leq \left[\int_{B_\rho(x_0) - B_{s\rho}(x_0)} |Du|^2\,dx + \int_{B_\rho(x_0) - B_{s\rho}(x_0)} u^2\,d\mu \right]. \tag{5.23}$$

3rd step. Hildebrandt-Widman "hole-filling".

By adding $V(s\rho)^2$ to both sides of (5.23) we get

$$V(s\rho)^2 \le \frac{1}{1+k\,\delta_s(\rho)}\,V(\rho)^2 \tag{5.24}$$

for every $0 < r \le \rho \le R$.

4th step. Maz'ja's integration.

Both sides of (5.24) are integrated in $\frac{d\rho}{\rho}$ over the interval $[r,R]$. This leads to

$$V(r)^2 \le c\,V(R)^2 \exp\left(-\alpha \int_r^R \delta_s(\rho)\,\frac{d\rho}{\rho}\right), \tag{5.24}$$

where c and $\alpha > 0$ are constants that depend only on n and s.

5th step. Kellogg-Vasilesco's argument for replacing the anuli $B_\rho(x_0) - B_{s\rho}(x_0)$ with the balls $B_\rho(x_0)$.

By relying on properties of the μ-capacity analogue to those of the usual capacity, one shows that

$$\int_r^R \delta_s(\rho)\,\frac{d\rho}{\rho} \ge c \int_r^R \delta(\rho)\,\frac{d\rho}{\rho} - c \tag{5.25}$$

where $\delta(\rho)$ is now given by

$$\delta(\rho) = \frac{\mathrm{cap}_\mu\left(B_\rho(x_0), B_{2\rho}(x_0)\right)}{\mathrm{cap}\left(B_\rho(x_0), B_{2\rho}(x_0)\right)},$$

and $c = c(s)$.

By using (5.25) to the right hand side of (5.24) and choosing a value of $s \in (0,1)$, the proof of (5.16) is accomplished. As to (5.17), this inequality is already an immediate consequence of (5.20) proved in the 1st step.

This concludes the sketch of the proof of Theorem 5.2. Corollary 1 is immediate from definition and Corollary 2 follows from (5.16), (5.17) by recalling the definition of V_μ and by estimating the L^2-norm in (5.17) by the Poincare' inequality (5.21), with the ball $B_{2R}(x_0)$ replacing $B_\rho(x_0) - B_{s\rho}(x_0)$.

Acknowledgements. The author is grateful for hospitality and support from the Institut für Angewandte Mathematik and SFB 72 at the University of Bonn where this paper was written.

References

[1] **D.R. Adams,** Capacity and the obstacle problem, Appl. Math. Optim. 8 (1981), 39-57.

[2] **J. Baxter, G. Dal Maso and U. Mosco,** Stopping times and Γ -convergence, IMA Preprint Series # 229 Univ. of Minnesota (1986), 1-54; to appear in Trans. AMS.

[3] **H. Beirão da Veiga,** Sulla Hölderianità delle soluzioni di alcune disuguaglianze variazionali con condizioni unilaterali al bordo, Ann. Mat. Pura Appl. (4) 83 (1969), 73-112.

[4] **H. Beirão da Veiga,** Punti regolari per una classe di operatori ellittici non-lineari, Ricerche Mat. 21 (1972), 3-16.

[5] **H. Beirão da Veiga and F. Conti,** Equazioni ellittiche non lineari con ostacoli sottili, Applicazione allo studio dei punti regolari, Ann. Scuola Norm. Sup. Pisa (3) 26 (1972), 533-562.

[6] **M. Biroli,** A De Giorgi-Nash-Moser result for a variational inequality, Boll. UMI (5) 16-A (1979), 598-605.

[7] **M. Brelot,** Norbert Wiener and potential theory, Bull. AMS 72 (II) (1966), 39-41.

[8] **H. Brezis and G. Stampacchia,** Sur la regularité de la solution d'inéquations elliptiques, Bull. Soc. Math. France 96 (1968), 153-180.

[9] **G. Buttazzo, G. Dal Maso and U. Mosco,** A derivation theorem for capacities with respect to a Radon measure, IMA Preprint Series # 203 (1985), 1-17; J. Funct. Anal. 71, No. 2 (1987), 263-278.

[10] **L.A. Caffarelli and D. Kinderlehrer,** Potential methods in variational inequalities, J. Analyse Math. 37 (1980), 285-295.

[11] **P. Charrier,** Continuité de la solution d'un problème unilateral elliptique et application, C.R. Acad. Sci. Paris., T. 290 (1980), 9-12.

[12] **G. Dal Maso and U. Mosco,** Wiener's criterion an Γ -convergence, IMA Preprint Series # 173 Univ. of Minnesota (1985), 1-76; J. Appl. Math. Optim. 15 (1987), 15-63.

[13] **G. Dal Maso and U. Mosco,** Wiener's criteria and energy decay for relaxed Dirichlet problems, IMA Preprint Series # 197 Univ. of Minnesota (1985), 1-64; Arch. Rational Mech. Anal. 95 (1986), 345-387.

[14] **G. Dal Maso and U. Mosco,** Wiener's modulus of a radial measure, IMA Preprint Series # 194, Univ. of Minnesota (1985), 1-29, to appear in Houston J. Math.

[15] **G. Dal Maso, U. Mosco and M.A. Vivaldi,** A pointwise regularity theory for the two-obstacle problem, to appear.

[16] **G. Fichera,** Problemi elastostatici con vincoli unilaterali: Il problema di Signorini con ambigue condizioni al contorno, Atti Accad. Naz. Lincei Mem. Cl. Sci. Fis. Mat. Natur. Sez. I (8) 7 (1963/64), 91-140.

[17] **G. Fichera,** Sul problema elastostatico di Signorini con ambigue condizioni al contorno, Atti Accad. Naz. Lincei Rend. Cl. Sci. Fis. Mat. Natur. (8) 34 (1963), 138-142.

[18] **J. Frehse,** Two dimensional variational problems with thin obstacles, Math. Z. 143 (1975) n. 3, 278-288.

[19] **J. Frehse,** On Signorini's problem and variational problems with thin obstacles, Ann. Scuola Norm. Sup. Pisa Cl. Scienze (4) 4 (1977), 343-362.

[20] **J. Frehse,** On the smoothness of solutions of variational inequalities with obstacles, Proceedings of the Banach Center Semester on PDE, 1978, vol. 10 edited by B. Bojarski, Partial Differential Equations, PWN Polski Sci. Publ. Warsaw (1983), 87-128.

[21] **J. Frehse,** Capacity methods in the theory of partial differential equations, Jber d. Dt. Math. Verein 84 (1982), 1-44.

[22] **J. Frehse and U. Mosco,** Variational inequalities with one-sided irregular obstacles, Manuscripta Math. 28 (1979), 219-233.

[23] **J. Frehse and U. Mosco,** Sur la regularité de certaines inequations variationnelles et quasi-variationnelles, C. R. Acad. Sci. Paris, Serie A. 289 (1979), 627-630.

[24] **J. Frehse and U. Mosco,** Irregular obstacles and quasi-variational inequalities of stochastic impulse control, Ann. Scuola Norm. Sup. Pisa 9 (1982), 105-157.

[25] **J. Frehse and U. Mosco,** Sur la continuité ponctuelle des solutions locales faibles du problème d'obstacle, C. R. Acad. Sci. Paris, Serie A. 295 (1982), 571-574.

[26] **J. Frehse and U. Mosco,** Wiener obstacles, Seminaires Collège de France, Paris, ed. H. Brezis and J.L. Lions, Vol. 6, Pitman Publ. 1984.

[27] **R. Gariepy and W.P. Ziemer,** Behaviour at the boundary of solutions of quasilinear elliptic equations, Arch. Rational Mech. Anal. 56 (1974/75), 372-384.

[28] **R. Gariepy and W.P. Ziemer,** A regularity condition at the boundary for solutions of quasilinear elliptic equations, Arch. Rational Mech. Anal. 67 (1977), 25-39.

[29] **M. Giaquinta,** Remarks on the regularity of weak solutions to some variational inequalities, Math. Z. 177 (1981), no. 1, 15-31.

[30] **V.P. Kahvin and V.G. Maz'ja,** Nonlinear potential theory, Russ. Math. Surv. 27 (1972), 71-148.

[31] **O.D. Kellogg and F. Vasilesco,** A contribution to the theory of capacity, Amer. J. Math. 51 (1929), 515-526.

[32] **D. Kinderlehrer,** Variational inequalities with lower dimensional obstacles, Israel J. Math. 10, 3 (1971), 339-348.

[33] **D. Kinderlehrer,** The smoothness of the solution of the boundary obstacle problem, J. Math. pures et appl. 60 (1981), 193-212.

[34] **D. Kinderlehrer,** Remarks about Signorini's Problem in Linear Elasticity, Ann. Scuola Norm. Sup. Pisa, IV, vol. III, 4 (1981), 605-645.

[35] **H. Lewy,** On a variational problem with inequalities on the boundary, J. Math. Mech. 17 (1968), 861-844.

[36] **H. Lewy,** On a refinement of Evans' law in potential theory, Atti Accad. Naz. Lincei VIII Sez. Rend. 48 (1970), 1-9.

[37] **H. Lewy,** On the coincidence set in variational inequalities, J. Diff. Geom. 6 (1972), 497-501.

[38] **H. Lewy and G. Stampacchia,** On the regularity of the solution of a variational inequality, Comm. Pure Appl. Math. 22 (1969), 153-188.

[39] **H. Lewy and G. Stampacchia,** On the smoothness of superharmonics which solve a minimum problem, J. Analyse Math. 23 (1970), 227-236.

[40] **J.L. Lions and G. Stampacchia,** Variational inequalities, Comm. Pure Appl. Math. 20 (1967), 493-519.

[41] **W. Littman, G. Stampacchia and H. Weinberger,** Regular points for elliptic equations with discontinuous coefficients, Ann. Scuola Norm. Sup. Pisa 17 (1963), 45-79.

[42] **V.G. Maz'ja,** Regularity at the boundary of solutions of elliptic equations and conformal mappings, Dokl. Akad. Nauk. SSSR 152 (1963), 1297-1300; Soviet Math. Dokl. 4 (1963), 1547-1551.

[43] **V.G. Maz'ja,** Behaviour near the boundary of solutions of the Dirichlet problem for a second order elliptic equation in divergence form, Mat. Zametki 2 (1967), 209-220; Math. Notes 2 (1967), 610-617.

[44] **V.G. Maz'ja,** On the continuity at a boundary point of solutions of quasi-linear elliptic equations, Vestnik Leningrad Univ. Math. 3 (1976), 225-242.

[45] **V.G. Maz'ja,** Sobolev spaces (Translated from Russian), Springer Series in Soviet Math., Springer Verlag Berlin 1985.

[46] **J.H. Michael and W.P. Ziemer,** Interior regularity for solutions to obstacle problems, Nonlinear Analysis, TMA, 10 (1987), 1427-1448.

[47] **U. Mosco,** Pointwise potential estimates for elliptic obstacle problems, Proc. Symposia in Pure Math. 45 Part 2 (1986), 207-217.

[48] **U. Mosco,** Wiener criterion and potential estimates for the obstacle problem, IMA Preprint Series # 135, Univ. of Minn. (1985), 1-56; to appear in Indiana U. Math. J..

[49] **U. Mosco,** Variational stability and relaxed Dirichlet problems, in "Metastability and incompletely posed problems", Ed. S. Antman, J.L. Ericksen, D. Kinderlehrer, I. Müller, The IMA Volumes in Math. and its Appl. 3, Springer Verlag, 1987.

[50] **U. Mosco and G.M. Troianiello,** On the smoothness of solutions of unilateral Dirichlet problems, Boll. Un. Mat. Ital. 8 (1973), 57-67.

[51] **J.C.C. Nitsche,** Variational problems with inequalities as boundary conditions or how to fashion a cheap hat for Giacometti's brother, Arch. Rational Mech. Anal. 35 (1969), 83-113.

[52] **G. Stampacchia,** Formes bilinéaires coercitives sur les ensembles convexes, C.R. Acad. Sci. Paris 258 (1964), 4413-4416.

[53] **G. Stampacchia,** Le problème de Dirichlet pour les équations elliptiques du second ordre à coefficients discontinus, Ann. Inst. Fourier 15 (1965), 189-258.

[54] **N. Wiener,** Certain notions in potential theory, J. Math. and Phys. 3 (1924), 24-51.

[55] **N. Wiener,** The Dirichlet problem, J. Math. and Phys. 3 (1924), 127-146.

[56] **W.P. Ziemer,** Preprint.

FULLY NONLINEAR SECOND ORDER ELLIPTIC EQUATIONS

Louis Nirenberg
Courant Institute of Mathematical Sciences
New York University, 251 Mercer Street
New York, N.Y. 10012 (U.S.A.)

Dedicated to Hans Lewy

It's a particular pleasure to participate in this conference dedicated to Hans Lewy. Quite a number of the problems on which many of us have worked are direct outgrowths of ideas originated by him.

1.

In this talk I wish to describe some of the recent existence results for general nonlinear elliptic equations for a single real function u of the form

$$F(x,u,Du,D^2u) = 0 \ .$$

Here u is a real function defined in a bounded domain Ω in R^n (with smooth boundary) or a manifold, $x \in \Omega$, $u_i = \frac{\partial u}{\partial x_i}$ etc., F is a smooth function. The equation is called elliptic at the function u if

$$\frac{\partial F}{\partial u_{jk}}\left(x,u(x),Du(x),D^2u(x)\right)$$

is a definite (say positive definite) matrix. It may be elliptic at some u and not at others, so in treating the equation as an elliptic one we may be forced to restrict the functions u to lie in some special class.

The Dirichlet problem is to find a solution u (at which the equation is elliptic) with prescribed smooth boundary values ϕ on $\partial\Omega$. This is a very old subject and traditional approaches are via the continuity method or the use of degree theory. To make these work one has to establish a priori estimates for the solutions and their derivatives. The traditional a priori estimates are for the $C^{2,\mu}$ norm of the solutions, i.e. the C^2 norm of u plus the μ-Hölder continuity of D^2u, for some positive $\mu < 1$. The regularity theory of elliptic machinery then implies that the solution u is $C^\infty(\Omega)$ or analytic - in case the data and $\partial\Omega$ are C^∞ or analytic. Progress in this subject, as in many branches of mathematics, has come when some new technical lemma is discovered -- a lemma improving the elliptic machinery. It's long been known that if for solutions u one can estimate $|u|_{C^2}$ and the modulus of continuity of D^2u in $\overline{\Omega}$ then an estimate for the $C^{2,\mu}$ norm and the rest follows.

The following is a long standing important

Open Problem. If one has an a priori estimate for $|u|_{C^2}$, can one estimate a modulus of continuity of D^2u? and so $|u|_{C^{2,\mu}}$?

For $n = 2$ the answer is yes - due mainly to techniques of C.B. Morrey. The recent progress in the subject for $n > 2$ is the following: the answer is yes in case F is also assumed to be concave (or convex) in its dependence on the variables D^2u, i.e. on the symmetric matrix $\{u_{jk}\}$. The proof of this is involved and technical and many people have contributed to it. I will just mention a few names: L.C. Evans [7], N.V. Krylov [11], see also [12], N.S. Trudinger [19] (he, and then F. Schulz [15], simplified Evans' arguments, see also D. Gilbarg, N.S. Trudinger [9], L. Caffarelli, J.J. Kohn, L. Nirenberg, J. Spruck [3]. Further references may be found in [2-4].

Here I will give a report of some of the results of Caffarelli, Spruck and myself [2-6]. The problems we treat are primarily of two kinds.

A. $$F = f(\lambda(u_{jk}(x))) - \psi(x) = 0 \quad \text{in } \Omega \qquad (1)$$

$$u = \phi \quad \text{on } \partial\Omega \qquad (2)$$

Here $\lambda = (\lambda_1, \ldots, \lambda_n)$ are the eigenvalues of the Hessian matrix $\{u_{jk}\}$ and f is a symmetric function (under permutation) of the λ_i - satisfying suitable conditions.

B. $$F = f(\kappa(\text{graph})) - \psi(x) = 0 \qquad (3)$$

and (2) is imposed. Here $\kappa = (\kappa_1, \ldots, \kappa_n)$ are the principal curvatures of the graph $(x, u(x))$ and f is again symmetric.

Examples.

Ex. 1. $f(\lambda) = \sum \lambda_i = \Delta u$

Ex. 1'. $f(\kappa) = \sum \kappa_i = n \cdot$ mean curvature of graph

$$= \frac{n}{\sqrt{1+|\nabla u|^2}}\left(\Delta u - \frac{u_j u_k u_{jk}}{1+|\nabla u|^2}\right)$$

Ex. 2. $f(\lambda) = [\Pi\ \lambda_i]^{1/n} = \det(u_{jk})^{1/n}$

Ex. 2'. $f(\kappa) = (\Pi\ \kappa_i)^{1/n}$ = (Gauss curvature)$^{1/n}$

$$= \left|\frac{\det(u_{jk})}{\left(1+|\nabla u|^2\right)^{\frac{n+2}{2}}}\right|^{1/n}$$

Ex. 3. (H.B. Lawson and F. Reese Harvey [13]):

$$f(\lambda) = \mathrm{Im}\left(\delta_{jk} + i\, u_{jk}\right) = \sum_{k=0}^{\left[\frac{n-1}{2}\right]} (-1)^k\, \sigma^{(2k+1)}\left(\lambda(u_{ij})\right) = 0 \tag{4}$$

Here $\sigma^{(j)}$ is the j – th elementary symmetric function

$$\sigma^{(j)}(\lambda) = \sum_{i_1<\ldots<i_j} \lambda_{i_1}\lambda_{i_2}\ldots\lambda_{i_j} \quad .$$

For example 1 the equation (1) is simply the inhomogeneous Laplace equation, and the Dirichlet problem may be solved for any domain Ω with smooth boundary. Example 1' is elliptic at any (smooth) u. However the Dirichlet problem (2) for the minimal surface equation $\Sigma\ \kappa_i = 0$ is solvable for all ϕ only if $\partial\Omega$ satisfies a certain condition: the mean curvature of $\partial\Omega$ is nonnegative at every point (see H. Jenkins, J. Serrin [10]).

In example 2 and 2' the equations (1), (3) are of Monge-Ampère type. They are not elliptic at all u. They *are* elliptic at u if u is strictly convex, i.e. $\{u_{jk}(x)\}$ is positive definite if $x \in \Omega$. Thus it is natural to restrict oneself to that class of functions. Then one has to necessarily require ψ to be positive in Ω. Furthermore, if say $\phi = 0$, $\partial\Omega$ is to be the level surface of a strictly convex function, and hence it has also to be strictly convex. The Dirichlet problem (2) for example 2, in a strictly convex bounded domain Ω is then solvable, see [11] and [2]. However, for example 2', this will not be the case if Ω is large. Essentially necessary and sufficient conditions have been presented in [1] and [20]. As shown in [2] a sufficient condition is the existence of a subsolution satisfying (2).

Example 3 arises in the study of minimal submanifolds with high codimension. In [13] the authors show that the equation (4) is elliptic *at any solution* u. Furthermore for any solution u, the graph of the gradient, $(x, \nabla u)$, an n-dimensional manifold in R^{2n}, is absolutely area minimizing. They posed the question: can the Dirichlet problem be solved?

2.

In Example 2 and 2' we consider the function f only in the positive cone Γ^+ in R^n. In [4] we treat general f defined in an open convex cone Γ in λ–space with vertex at the origin containing Γ^+, $\Gamma \neq R^n$. It is assumed that:

(i) f is symmetric in the λ_i (under permutation), $f_{\lambda_i} > 0$ and f is concave. Also $f > 0$ in Γ, $f = 0$ on $\partial\Gamma$ (this is less general than in [4]).

(ii) For any $\lambda \in \Gamma, f(R\lambda) \to +\infty$ as $R \to \infty$.

A smooth function u in Ω with $u = \phi$ on $\partial\Omega$ is called admissible if $\lambda(u_{jk}(x)) \in \Gamma \; \forall x \in \Omega$. The condition $f_{\lambda_i} > 0$ in (i) implies that $F = f(\lambda(u_{jk})) - \psi(x)$ is elliptic at admissible u. The condition that f is concave implies that F is concave.

We wish to find an admissible u satisfying the equation

$$f(\lambda(u_{jk}(x))) = \psi(x) > 0 \quad \text{in} \quad \overline{\Omega} \; . \tag{5}$$

Here is an example,

Ex. 4. $$f(\lambda) = \left[\sigma^{(k)}(\lambda)\right]^{1/k} . \tag{6}$$

Γ is then the component containing Γ^+ in which $\sigma^{(k)} > 0$. That this is a cone and that f satisfies the condition above is a consequence of the results of L. Gårding [8]. This is a very special case of his inequality and it would be nice to have a completely elementary proof of (i) for this f.

It turns out that the nature of the results depend very much on whether the positive axes lie on $\partial\Gamma$ or inside Γ. Such cones are called respectively type 1 or 2.

Theorem 1 ([4]). *With* f *satisfying* (i), (ii) *and* Γ *of type 2, there exists a unique admissible solution of* (5) $\forall \; \phi \in C^\infty(\partial\Omega)$ *and* $\forall \; \Omega$ *(with* $\partial\Omega$ *smooth*).

The theorem is proved using the continuity method and is based on a priori estimates for the C^2 norm of solutions.

At this point let me make a side remark about cones of type 2. Consider the minimal surface equation

$$\sum \kappa_i = 0 \; . \tag{7}$$

One has the well known (extended)

Bernstein Theorem. *If* u *is a solution of* (7) *in all of* R^n *and* $n \le 7$ *then* u *is an affine function.*

The assertion is false for $n > 7$. Our methods for estimating the second derivatives of solutions of (5) give a weak extension of Bernstein's theorem for all n - under suitable growth conditions at infinity.

Generalized Bernstein Theorem. *If* u *is a solution of* (7) *in all of* R^n *and*

$$\nabla u(x) = o\left(|x|^{1/2}\right) \quad \textit{at infinity} \tag{8}$$

then u *is an affine function.*

This is an improvement of the corresponding result by Moser who assumed ∇u bounded. It is somewhat surprising that the result (proved by the same method) applies to more general equations:

Theorem A ([6]). *Let* Γ *be a cone of type* 2 *with* $\partial\Gamma \setminus \{0\}$ *smooth. If* u *is a function on all of* R^n *satisfying*

$$\kappa\,(\text{graph of } u(x)) \in \partial\Gamma \ \ \forall\, x \in R^n\,, \tag{7$'$}$$

and (8), *then* u *is an affine function.*

For the equation (7) the cone Γ is the half space $\sum \kappa_i > 0$. The result need not hold for cones of type 1.

The theorem suggests a natural

Question. Can one drop the condition (8) for low n?

In case $n = 2$ the answer is yes, by a result of L. Simon (Theorem 4.1 in [16]).

Let us return to equation (5) and suppose now the cone Γ is of type 1.

Theorem 2 ([4]). *Assume* (i) *and* (ii) *and also (here* $e_n = (0,\dots,0,1)$ *)*

(iii) $$f\left(\lambda + Re_n\right) \to \infty \quad \textit{as} \ R \to \infty \quad \forall\, \lambda \in \Gamma\,.$$

For any $\phi \in C^\infty(\partial\Omega)$ *there exists an admissible solution* u *of* (7) $\Leftrightarrow$

(iv) $\partial\Omega$ is connected and $\forall\, x \in \partial\Omega$, if $\mu_1(x),\dots,\mu_{n-1}(x)$ are the principal curvatures of $\partial\Omega$ (relative to the interior normal) then for some large R, $(\mu_1(x),\dots,\mu_{n-1}(x),R) \in \Gamma$. Moreover in the case of Example 4 this last condition means that $\partial\Omega$ is connected and $\sigma^{(k-1)}(\mu_1(x),\dots,\mu_{n-1}(x)) > 0 \quad \forall\, x \in \partial\Omega$. Some of the results described are of interest also in the complex case. The complex Monge-Ampère equation for the complex Hessian $u_{z_j\bar z_k}$ was treated in

[3]. Theorems 1 and 2 have been extended to corresponding complex cases by A. Vinacua [21].

In case ψ = constant in (7) Theorems 2 also holds if condition (iii) is omitted and this fact has interesting consequences. We describe one:

Let G be an open unbounded convex region in λ-space with smooth boundary Σ satisfying

(a) G is symmetric in the λ_i (under permutation), and the interior normal at every point of Σ lies in Γ^+.

(b) $0 \notin G$ and 0 lies on the opposite side from G of every tangent hyperplane to Σ.

These conditions imply that the cone Γ with vertex at the origin generated by all points of Σ contains Γ^+.

The problem we pose is the following: Let Ω be a bounded domain in R^n with $\partial\Omega$ smooth. Find $u \in C^\infty(\overline{\Omega})$ with $u = \phi \in C^\infty(\partial\Omega)$ on $\partial\Omega$ satisfying

$\lambda(u_{jk}(x)) \in \Sigma \ \forall x \in \overline{\Omega}$. (9)

Theorem 3 ([4]). *If* Γ *is of type* 2 *then this problem always has a solution. If* Γ *is of type* 1 *then the problem is solvable for all* $\phi \Leftrightarrow$ *condition* (iv) *holds.*

Remark. We have not written (9) in the usual form of a differential equation. In fact there are many ways of doing so. The way we choose is to write it in the form (7):

$f(\lambda\ (u_{jk}(x))) = 1$ (7)"

where f is the function in Γ homogeneous of degree one which is identically equal to one on Σ. Condition (a) then corresponds to ellipticity and the convexity of G corresponds to concavity of f.

Theorem 3 is applied in [4] (see Theorem 5) to yield at least two solutions of the Dirichlet problem for equation (4) in a strictly convex (and even more general) domain Ω. A component of the set $f(\lambda) = 0$ given in (4) has all the properties of Σ in Theorem 3 and so the theorem applies. I wish to stress that in treating that component we don't work with the specific equation (4) but rather with (7)". We believe more solutions should exist: n or $(n-1)$ if n is odd or even. For example in case $n = 3$ equation (4) takes the form

$$\prod \lambda_i - \sum \lambda_i = \det(u_{jk}) - \Delta u = 0 \quad .$$

We believe there should be at least three solutions of the Dirichlet problem. If $\phi = 0$, the function $u = 0$ is a solution but it is neither of the two solutions we have found.

P.L. Lions, N.S. Trudinger and J. Urbas [14] have treated Monge-Ampère equations with Neumann boundary conditions.

3.

We turn finally to problems of type B, i.e., equations of the form (3). In [5] we have extended results of A. Treibergs [17] and Treibergs and S. Wei [18] who prescribed mean curvature for a compact star-shaped hypersurface in R^{n+1} (see [5] for other references). Given a positive function $\psi(y)$ in R^{n+1} satisfying $(\rho = |y|)$

$$(\rho\,\psi(y))_\rho \le 0$$

we wish to find a compact star-shaped (about the origin) hypersurface lying in $r_1 \le |y| \le r_2$ of

$$f(-\kappa) = \psi(y) \quad . \qquad (10)$$

Here κ represents the principal curvatures of the hypersurface and $f(\lambda) > 0$ is a symmetric function of the λ_i defined in an open convex symmetric (under permutation of the λ_i) cone $\Gamma \subset R^n$ with vertex at the origin containing Γ^+. f is assumed to satisfy (i) above. In addition we assume $f(1,...,1) = 1$ and

(ii)' $$\sum f_{\lambda_i} \, , \; \sum \lambda_i f_{\lambda_i} \ge \phi(f)$$

with ϕ a positive increasing function on R^+.

We also assume

$$\psi(y) \ge f\left(\frac{1}{r_1}, ..., \frac{1}{r_1}\right) \quad \text{for } |y| = r_1$$

$$\psi(y) \le f\left(\frac{1}{r_2}, ..., \frac{1}{r_2}\right) \quad \text{for } |y| = r_2 \; .$$

Theorem 4. *Under the conditions above there is a smooth starshaped solution of* (10). *Furthermore any two solutions such that at every point their principal curvatures belong to* $-\Gamma$ *are end points of a one-parameter family of homothetic dilations, all of which are solutions.*

Theorem 4 includes the case $f(\lambda) = \left[\dfrac{\sigma^{(k)}(\lambda)}{\left(\dfrac{n}{k}\right)} \right]^{1/k}$.

The Dirichlet problem for (3) has turned out to be more difficult. We have only been able to treat successfully the case of Ω strictly convex and ϕ = constant.

Under suitable technical hypotheses on f we have the following result:

Assume that there exists a smooth subsolution u *of* (3) *with* $u = \phi$ = constant *on* $\partial\Omega$ *(strictly convex), i.e.*

$$f\big(\kappa(\text{graph of } u(x))\big) \geq \psi(x) \ \text{ in } \ \overline{\Omega} \ .$$

Then (3) *has an admissible solution* $u \in C^\infty(\Omega)$.

The paper with this result is now being prepared for publication. The derivation of the C^2 estimates is quite elaborate.

The work was supported by ARO-DAA629-84-K-0150.

References

[1]   **I. Bakelman,** Generalized elliptic solutions of the Dirichlet problem for the Monge-Ampère equations, Proc. Symp. Pure Math., AMS $\underline{44}$ (1986), 1-30.

[2]   **L. Caffarelli, L. Nirenberg and J. Spruck,** The Dirichlet problem for nonlinear second order elliptic equations. I: Monge-Ampère equation, Comm. Pure Appl. Math. $\underline{37}$ (1984), 369-402.

[3]   **L. Caffarelli, J.J. Kohn, L. Nirenberg and J. Spruck,** The Dirichlet problem for nonlinear second order elliptic equations. II: Complex Monge-Ampère, and uniformly elliptic equations, Comm. Pure Appl. Math. $\underline{38}$ (1985), 209-252.

[4]   **L. Caffarelli, L. Nirenberg and J. Spruck,** The Dirichlet problem for nonlinear second order elliptic equations, III: Functions of eigenvalues of the Hessian, Acta Math. $\underline{155}$ (1985), 261-301.

[5]   **L. Caffarelli, L. Nirenberg and J. Spruck,** Nonlinear second order elliptic equations. IV: Starshaped compact Weingarten surfaces, Current topics in partial differential equations, ed. Y. Ohya, K. Kasahara, N. Shimakura. Kinokuniya, Tokyo (1985), 1-26.

[6]   **L. Caffarelli, L. Nirenberg and J. Spruck,** On a form of Bernstein's theorem, to appear.

[7]   **L.C. Evans,** Classical solutions of fully nonlinear, convex, second order elliptic equations, Comm. Pure Appl. Math. $\underline{35}$ (1982), 333-363.

[8] **L. Gårding,** An inequality for hyperbolic polynomials, J. Math. Mech. 8 (1959), 957-965.

[9] **D. Gilbarg and N.S. Trudinger,** Elliptic partial differential equations of second order, Springer-Verlag, Berlin-Heidelberg-New York, Second edition, 1983.

[10] **H. Jenkins and J. Serrin,** The Dirichlet problem for the minimal surface equation in higher dimensions, J. für die reine und angew. Math. 229 (1968), 170-187.

[11] **N.V. Krylov,** Boundedly nonhomogeneous elliptic and parabolic equations in a domain, Izvestia Math. Ser. 47 (1983), 75-108.

[12] **N.V. Krylov,** Nonlinear elliptic and parabolic equations of second order, Moscow Nauk Glavnaya Red. Fisico-Math. Lit., 1983.

[13] **H.B. Lawson and F. Reese Harvey,** Calibrated geometries, Acta Math. 148 (1982), 47-157.

[14] **P.L. Lions, N.S. Trudinger and J.E. Urbas,** The Neumann problem for equations of Monge-Ampère type, preprint.

[15] **F. Schulz,** Über nichtlineare konkave elliptische Differentialgleichungen 11 Math., Gött Schriftenreihe der Sonderforschungsbereichs Geom. und Anal. 21 (1985).

[16] **L. Simon,** A Hölder estimate for maps between surfaces in Euclidean space, Acta Math. 139 (1977), 19-51.

[17] **A.E. Treibergs,** Existence and convexity for hyperspheres of prescribed mean curvature, preprint.

[18] **A.E. Treibergs and S.W. Wei,** Embedded hyperspheres with prescribed mean curvature, J. Diff'l. Geometry 18 (1983), 513-521.

[19] **N.S. Trudinger,** Fully nonlinear, uniformly elliptic equations under natural structure conditions, Trans. Amer. Math. Soc. 278 (1983), 751-769.

[20] **N.S. Trudinger and J.I.E. Urbas,** The Dirichlet problem for the equation of prescribed Gauss curvature, Bull. Austral. Math. Soc. 28 (1983), 217-231.

[21] **A. Vinacua,** Nonlinear elliptic equations written in terms of functions of the eigenvalues of the complex Hessian, Ph.D. thesis, Courant Institute, June 1986.

POSITIVE SOLUTIONS OF A PRESCRIBED MEAN CURVATURE PROBLEM

James Serrin
School of Mathematics - University of Minnesota
206 Church Street S.E.
Minneapolis - Minnesota 55455 (U.S.A.)

Dedicated to Hans Lewy

In this paper we shall discuss the existence of positive radial solutions of the following Dirichlet problem

$$(I)\qquad \begin{cases} \operatorname{div} \dfrac{Du}{\sqrt{1+|Du|^2}} + f(u) = 0 & \text{in } \Omega \\ u(x) = 0 & \text{for } x \in \partial\Omega , \end{cases}$$

where

$$f(u) = -\lambda u + u^q$$

and Ω is a ball of given radius R in the n-dimensional Euclidean space R^n. Here $n > 1, q > 1$, λ is a real parameter and Du denotes the gradient of u.

In view of the interest in the analogous problem for the Laplace operator as well as the importance of the minimal surface equation, it is surprising that Problem (I) has barely been studied. Indeed, except for certain non-existence results [3], [5] and an existence theorem [4] for the case $\lambda > 0$ and $\Omega \equiv R^n$, the author knows of no earlier treatments.

The situation when $\lambda \le 0$ appears fairly complicated and accordingly throughout this paper we shall restrict consideration to the case $\lambda > 0$ (Theorem 2 is however also of interest when $\lambda \le 0$). When $\lambda \ge 0$ and $q \ge (n+2)/(n-2)$ it was shown in [5] that problem (I) has no non-trivial solutions (radial or otherwise). Hence we can confine our attention to the case $q < (n+2)/(n-2)$, this being no restriction at all if $n = 2$. It is, of course, interesting that the exponent $(n+2)/(n-2)$ plays an equally important role in problem (I) as in the corresponding problem for the Laplace operator.

In view of the results in [2] it can be anticipated that solutions will exist only when λ is not too large, an expectation which is indeed confirmed by Theorem 1. The result of Theorem 3 shows, moreover, that solutions can exist only if the radius R of Ω is not too small. The proof here involves an interesting comparison argument which uses in a crucial way the properties of the mean curvature operator.

In contrast to the above non-existence theorems, numerical calculations of Evers and Levine [6] together with the result of [4] noted above suggest that under appropriate conditions on λ, q and R a solution can exist. In Theorem 4 we given a result which confirms this expectation and complements the earlier non-existence theorems.

1. Non-existence theorems

If $u = u(r)$, $r = |x|$, is a radial solution of the differential equation

$$\operatorname{div} \frac{Du}{\sqrt{1 + |Du|^2}} + f(u) = 0$$

then we have, obviously,

$$E(|u'|)\,u'' + \frac{n-1}{r} A(|u'|)\,u' + f(u) = 0 \tag{1}$$

where $u' = du/dr$ and

$$A(p) = \frac{1}{\left(1+p^2\right)^{1/2}} \qquad E(p) = \frac{1}{\left(1+p^2\right)^{3/2}} .$$

Multiplying (1) by u' and integrating from $r = r_0$ to $r = r_1$ yields the identity

$$H(p_1) - H(p_0) + (n-1) \int_{r_0}^{r_1} A u'^2 \frac{dr}{r} + F(u_1) - F(u_0) = 0 , \tag{2}$$

where $u_1 = u(r)$, $u_0 = u(r_0)$, $p_1 = |u'(r_1)|$, $p_0 = |u'(r_0)|$, and

$$H(p) = \int_0^p \rho\, E(\rho) d\rho = 1 - \frac{1}{\left(1+p^2\right)^{1/2}}$$

$$F(u) = \int_0^u f(s) ds .$$

Taking $r_0 = 0$ and $r_1 = R$ and using the boundary conditions

$$u'(0) = 0\,, \qquad u(R) = 0$$

then gives $F(u(0)) > 0$. Next take $r_0 = r$ in $(0,R)$ and $r_1 = R$. This gives

$$H(p_R) - H(p) + (n-1)\int_r^R Au'^2\,\frac{dr}{r} - F(u) = 0\,,$$

so in turn

(3) $$-F(u) \le H(p) < 1\,.$$

We can now state our first result.

Theorem 1. Let $u = u(r)$ *be a positive radial solution of Problem (I). Then*

$$F(u(0)) > 0\,.$$

Moreover no solution can exist when

$$\lambda > \left(2\,\frac{q+1}{q-1}\right)^{(q-1)/(q+1)}.$$

Proof. Only the second part of the result remains to be shown. Let α be the positive root of $f(u) = 0$. Clearly the range of u includes the interval $(0,u(0)]$, and since $F(u(0)) > 0$ it particularly includes the value $u = \alpha$. An easy calculation shows that

$$-F(\alpha) = \frac{1}{2}\,\frac{q-1}{q+1}\,\lambda^{(q+1)/(q-1)}\,,$$

and the theorem now follows directly from (3). (The method used here is taken from [2]).

Theorem 2. Consider the initial value problem

(4) $$E\,v'' + \frac{n-1}{r}\,Av' + g(v) = 0$$

(5) $$v(0) = \xi, \quad v'(0) = 0$$

where $g(v)$ *is a given function on* $[0,\infty)$ *with* $g(0) = 0$, $g''(v) \ge 0$. *If* $\xi > 0$ *and*

(6) $$\frac{1}{\xi^2} \le g'(\xi) \le \frac{1}{4n^2}\,g(\xi)^2\,, \quad g(\xi) > 0$$

then the solution cannot be continued to the line $v = 0$.

Proof. Consider the auxiliary function

$$w(r) = \xi - b + \sqrt{b^2 - r^2}\,, \qquad 0 \le r < b\,,$$

whose graph is a quarter circle of radius b centered at $(0, \xi - b)$ and passing through the initial point $(0, \xi)$ of the solution $v(r)$.

We suppose b is chosen so that

$$\frac{n}{g(\xi)} < b \leq \xi \ . \tag{7}$$

Since $v''(0) = -g(\xi)/n$ and $w''(0) = -1/b$ it follows in particular that $v''(0) < w''(0)$. Consequently the graph of v is, at least for sufficiently small values of r, beneath the graph of w.

Let us assume for contradiction that v can be continued to the line $v = 0$. Then there exists a value $\eta \leq b$ such that

$$0 < v < w \quad \text{for} \quad 0 < r < \eta \ ,$$

with either

$$v(\eta) = w(\eta) \quad \text{or} \quad v(\eta) = \xi - b \ .$$

An examination of the different cases which can occur shows that there are now values r_1 and r_2, with $0 < r_1 < \eta$, $r_1 < r_2 < b$, such that

$$v(r_1) = w(r_2)$$

$$v'(r_1) = w'(r_2)$$

$$v''(r_1) \geq w''(r_2) \ .$$

It follows that

$$E\, v''(r_1) \geq E\, w''(r_2) = -\frac{1}{b}$$

and consequently, from the differential equation,

$$\frac{n-1}{r_1} Av'(r_1) + g(v(r_1)) \leq \frac{1}{b} \ .$$

The quantity $Av(r_1)$ can be estimated by writing the differential equation in the form

$$\left(r^{n-1} Av'\right)' + r^{n-1} g(v) = 0$$

and integrating from 0 to r_1. Thus we get

$$r_1^{n-1} Av'(r_1) = -\int_0^{r_1} r^{n-1} g(v)dr \geq -\frac{r_1^n}{n} g(\xi) \ .$$

(note that $g(v) \leq g(\xi)$ since $v(r) < \xi$ for $0 < r < \eta$ and $g'' \geq 0$). It follows that

$$-\frac{n-1}{n} g(\xi) + g(v) \leq \frac{1}{b}$$

or in turn

$$\frac{g(\xi)}{n} \le \frac{1}{b} + g(\xi) - g(v) < \frac{1}{b} + b\, g'(\xi) \ , \tag{8}$$

using the fact that $g''(v) \ge 0$ and $g'(\xi) > 0$.

Now choose b explicitly as $1/\sqrt{g'(\xi)}$. That this choice is allowable in (7) follows from condition (6). Then (8) becomes

$$\frac{g(\xi)}{n} < 2\sqrt{g'(\xi)} \ ,$$

violating (6) and completing the proof.

Corollary. No solution of Problem (I) can exist unless the initial value $\xi = u(0)$ *satisfies*

$$\xi^{q+1} - 2\lambda\,\xi^2 < 4n^2 q \ . \tag{9}$$

Proof. Suppose the initial value ξ is such that

$$\xi^{q+1} - 2\lambda\,\xi^2 \ge 4n^2 q \ .$$

Choosing $g(v) = f(v) = -\lambda v + v^q$ in Theorem 2, it is easily verified that condition (6) holds. Hence the solution $u(r)$ cannot be continued to the line $u = 0$ as is necessary to satisfy the boundary condition $u = 0$ when $r = R$.

Theorem 3. There exists a positive number R_1, *depending only on* q *and* n, *such that Problem (I) has no positive radial solution when* $R < R_1$.

Proof. Consider the function w introduced in Theorem 2, but now with the explicit choice

$$b = \frac{1}{g(\xi)} \ , \qquad g(\xi) > 0 \ .$$

We assert that the solution $v(r)$ of the initial value problem (4), (5) lies above $w(r)$.

To this end, note first that $v''(0) > w''(0)$ so that at least initially the graph of v lies above the graph of w. Moreover since $v''(0) < 0$ and $v'(0) = 0$, the function $v(r)$ is initially decreasing. Now, for contradiction, suppose the graphs of v and w cross at some positive value of r, say for the first time at $r = \eta > 0$. There are two cases to consider.

1. v *is monotone decreasing for* $0 < r < \eta$. Then the construction in Theorem 2 provides the inequality

$$\frac{n-1}{r_2} A v'(r_2) + g(v(r_2)) \ge \frac{1}{b} \ .$$

Since $v'(r_2) < 0$ it follows that $g(v(r_2)) > 1/b$, and so in turn $g(\xi) > 1/b$, a contradiction.

2. v *is not monotone on* $0<r<\eta$. In this case there must be numbers r_0 and r_1, with $0<r_0<r_1<\eta$, such that

$$v'(r_0)=0, \quad v(r_1)=v(r_0) .$$

This, however, contradicts the identity (2), which (provided $F(u)$ is replaced by the corresponding function $G(u)$) of course holds equally well for solutions $v(r)$ of (4). The assertion is thus proved.

We now turn explicitly to the consideration of positive radial solutions of Problem (I). Putting $u(0)=\xi$, there are again two cases to consider.

Case 1. $1/f(\xi)\le\xi$. The graph of w then lies completely above the r axis. Accordingly, since the graph of u lies above the graph of w by the previous argument, we have

$$u(r) > 0 \quad \text{for} \quad 0\le r\le b ,$$

where $b=1/f(\xi)$. Consequently $R>1/f(\xi)$.

By the preceding corollary the initial value ξ must satisfy (9) if a solution is to exist at all. Hence $\xi\le y$ and

$$f(\xi)\le f(y) ,$$

where y is defined by the condition $y^{q+1}-2\lambda y^2 = 4n^2 q$. Thus in this case $1/f(y)$ provides a lower bound for the radius R for which Problem (I) has a positive radial solution.

Case 2. $1/f(\xi)>\xi$. The graph of w now cuts the r axis at the point $\bar{r}$, where

$$\bar{r}^{-2}=\xi(2b-\xi)=\xi\left(\frac{2}{f(\xi)}-\xi\right)>\frac{\xi}{f(\xi)} .$$

This is a monotonically decreasing function of ξ so obviously $\bar{r}\ge\sqrt{y/f(y)}$, which again provides a lower bound for the radius R.

To complete the proof we note that (for fixed q and n) λ is bounded for any eventual solution and so obviously y is also bounded. Hence R is bounded from zero for fixed q and n.

2. Existence

In this section we obtain an existence theorem for Problem (I) in the case $\lambda>0$. In view of the results of the previous section, and the condition $q<(n+2)/(n-2)$ noted in the Introduction, it is clearly necessary to restrict λ, q and R. The specific result is as follows.

Theorem 4. Suppose $\lambda > 0$, *and* $q < (n+2)/(n-2)$ *if* $n > 2$. *There exists a positive constant* λ_0, *depending only on* q *and* n, *and a positive radius* R_0, *depending on* λ, q *and* n, *such that Problem (I) has a positive radial solution for each* $\lambda < \lambda_0$.

Remark. The proof gives R_0 in the form

$$R_0 = a(q,n)\,\lambda^{-(q-1)/4q},$$

an expression which tends to ∞ as $\lambda \to 0$. In fact, an alternate but rather more delicate method of proof shows that $R_0(\lambda, q, n)$ can be taken to depend only on q and n.

Proof of Theorem 4. We follow a method which was used in [4] to prove existence of positive radial solutions of Problem (I) in the case $\Omega \equiv \mathbf{R}^n$. Assuming familiarity with that proof, a continuity argument shows that there is a constant ε_0, suitably small and depending only on q and n, such that for $0 < \varepsilon < \varepsilon_0$ the solution of the initial value problem

$$\left(\frac{v'}{\sqrt{1+\varepsilon v'^2}}\right)' + \frac{n-1}{r}\,\frac{v'}{\sqrt{1+\varepsilon v'^2}} - \varepsilon v + v^q = 0 \tag{10}$$

$$v(0) = 1, \qquad v'(0) = 0 \tag{11}$$

first crosses the r-axis at a point $\bar{r}$ satisfying

$$|C(q,n) - \bar{r}| < 1.$$

Here $C(q,n)$ is the first positive root of the initial value problem

$$v'' + \frac{n-1}{r}v' + v^q = 0$$

$$v(0) = 1, \quad v'(0) = 0,$$

known to exist by the work of Fowler [1]. (It is here that the condition $q < (n+2)/(n-2)$ is crucial.)

Now fix $s \in (C(q,n)+1, \infty)$. Using the methods of [4] we see that for each $\varepsilon < \varepsilon_0$ there exists a value $\xi_s \in (\alpha, 1)$ such that the solution of (10) with initial data

$$v(0) = \xi_s, \qquad v'(0) = 0$$

has its first zero at $r = s$.

Since solutions of (10) are transformed into solutions of (1) by the change of variables

$$v(r) = \lambda^{-1/2q}\,u\!\left(\lambda^{-(q-1)/4q}\,r\right), \quad \varepsilon = \lambda^{(q+1)/2q},$$

it follows that Problem (I) is solvable for any $\lambda < \varepsilon_0^{2q/(q+1)}$ and any

$$R > \left(C(q,n) + 1\right)\lambda^{-(q-1)/4q} .$$

This completes the proof.

References

[1] **R.H. Fowler,** Further studies of Emden's and similar differential equations, Quart. J. Math. 2 (1931), 259-288.

[2] **B. Franchi, E. Lanconelli and J. Serrin,** Esistenza e unicità degli stati fondamentali per equazione ellittica quasilineare, Rend. Accad. Naz. Lincei, in press.

[3] **W.-M. Ni and J. Serrin,** Non-existence theorems for quasilinear partial differential equations, Rend. Circ. Mat. Palermo, Suppl. 5 (1985), 171-185.

[4] **L.A. Peletier and J. Serrin,** Ground states for the prescribed mean curvature equation, Proc. AMS (1987).

[5] **P. Pucci and J. Serrin,** A general variational identity, Indiana U. Math. J. 35 (1986), 681-703.

[6] **T.K. Evers,** Numerical search for ground state solutions of a modified capillary equation, M. Sci. Thesis, Iowa State Univ. 1985.

ON THE CONVERGENCE AT INFINITY OF SOLUTIONS WITH FINITE DIRICHLET INTEGRAL TO THE EXTERIOR DIRICHLET PROBLEM FOR THE STEADY PLANE NAVIER-STOKES SYSTEM OF EQUATIONS

Dan Socolescu
Institut für Angew. Mathematik - Universität Karlsruhe
Englerstraße 2 - 7500 Karlsruhe (W. Germany)

Dedicated to Hans Lewy

1. Introduction

Let Ω be a two-dimensional domain exterior to a compact set Δ with smooth boundary $\partial\Delta$. The steady flow in Ω of a viscous incompressible fluid past the obstacle Δ with uniform velocity $\underset{\sim}{v}_\infty$ at infinity is described by the Navier-Stokes equations and the continuity equation

$$(1) \qquad \Omega : \begin{cases} \nu\nabla^2 \underset{\sim}{v} - (\underset{\sim}{v} \cdot \nabla)\underset{\sim}{v} - \rho_\infty^{-1} \nabla p = \underset{\sim}{0} \ , \\ \nabla \cdot \underset{\sim}{v} = 0 \ , \end{cases}$$

with the boundary conditions

$$(2) \qquad \partial\Delta : \underset{\sim}{v} = \underset{\sim}{0} \ ,$$

$$(3) \qquad [0,2\pi] : \lim_{r\to\infty} \underset{\sim}{v}(r,\theta) = \underset{\sim}{v}_\infty \ .$$

Here ν is the coefficient of viscosity, $\nabla := (\partial_x,\partial_y)$ is the Nabla differential operator, $\underset{\sim}{v} = (u,v)$ is the velocity vector, ρ_∞ is the fluid density, p is the pressure, r is the radius vector, i.e. the distance from the given point to the origin, taken interior to Δ, and θ is the polar angle.

In his study on this exterior Dirichlet problem in 1933, Leray [8] constructed a certain solution $(\underset{\sim}{v}_L,p_L)$ satisfying (1) and (2) and having a velocity with finite Dirichlet integral

$$(4) \qquad \int_\Omega |\nabla \underset{\sim}{v}|^2 dxdy < \infty \ .$$

Whether this solution had the desired limit behaviour (3) was left open. Leray's construction went as follows. Let Ω_R be the set of points in Ω of radius vector $r < R$. He first proved that for every $R > \max_{\Delta} r =: r_{\Delta}$ and for every constant vector $\underset{\sim}{v}_{\infty}$ there is at least one solution $(\underset{\sim}{v}_R, p_R)$ of (1) in Ω_R satisfying the boundary conditions

(5) $$\partial\Omega_R : \underset{\sim}{v}_R = \begin{cases} \underset{\sim}{0} & \text{on } \partial\Delta , \\ \underset{\sim}{v}_{\infty} & \text{for } r = R . \end{cases}$$

Concerning all such solutions $\tilde{v}_R$ Leray proved the existence of a uniform bound for the Dirichlet integral, namely for some positive constant C independent of R and ν

(6) $$\int_{\Omega_R} |\nabla \underset{\sim}{v}_R|^2 dxdy \le C^2(1+\nu^{-1})^2 .$$

He then showed that a sequence $R_i \to \infty$ exists, such that the solutions $(\underset{\sim}{v}_{R_i}, p_{R_i})$ converge uniformly together with all their first order derivatives in any compact subset of Ω to a solution $(\underset{\sim}{v}_L, p_L)$ satisfying (1), (2) and (4) - cf. also [2], [5], [7] -. The further behaviour of $\underset{\sim}{v}_D$ and p_D as $r \to \infty$ was left unsettled and remained so for more than four decades.

In 1974 Gilbarg and Weinberger [5] proved that

(i) the Leray solution $(\underset{\sim}{v}_L, p_L)$ is bounded,

(ii) the velocity $\underset{\sim}{v}_L$ has a limit in the mean at infinity

(7) $$\begin{cases} \lim\limits_{r\to\infty} \int_0^{2\pi} |\underset{\sim}{v}_L(r,\theta) - \underset{\sim}{v}_0|^2 d\theta = 0 , \\ \lim\limits_{r\to\infty} \overline{\underset{\sim}{v}}_L(r) = \underset{\sim}{v}_0 \end{cases}$$

where $\overline{\underset{\sim}{v}}_L(r) := \frac{1}{2\pi}\int_0^{2\pi} \underset{\sim}{v}_L(r,\theta)d\theta$, $|\underset{\sim}{v}_0| := \lim\limits_{r\to\infty} \max\limits_{\theta \in [0,2\pi]} |\underset{\sim}{v}_L(r,\theta)|$,

(iii) the pressure p_L converges pointwise and in the mean to the same limit at infinity

(8_1) $$[0,2\pi) : \lim_{r\to\infty} p_L(r,\theta) = p_{\infty} ,$$

$$(8_2) \qquad \lim_{r\to\infty} \int_0^{2\pi} |p_L(r,\theta) - p_\infty|^2 d\theta = 0 \ .$$

The questions of whether $\underset{\sim}{v}_L$ tends pointwise to its asymptotic value $\underset{\sim}{v}_0$ and whether $\underset{\sim}{v}_0$ is equal to the prescribed asymptotic value $\underset{\sim}{v}_\infty$ were however left open.

Four years later the same authors [6] investigated the asymptotic behaviour of an arbitrary solution with finite Dirichlet integral $(\underset{\sim}{v}_D, p_D)$ of (1) and (2) and showed that

(i) the velocity $\underset{\sim}{v}_D$ has at infinity the behaviour

$$(9) \qquad \lim_{r\to\infty} |\underset{\sim}{v}_D(r,\theta)|^2/\ln r = 0 \ , \quad \text{uniformly in } \theta \ ,$$

$$(10) \qquad \begin{cases} \lim_{r\to\infty} \int_0^{2\pi} |\underset{\sim}{v}_D(r,\theta) - \underset{\sim}{v}_D(r)|^2 d\theta = 0 \ , \\ \lim_{r\to\infty} |\underset{\sim}{v}_D(r)| = \lim_{r\to\infty} \max_{\theta\in[0,2\pi]} |\underset{\sim}{v}_D(r,\theta)| \ . \end{cases}$$

If, furthermore, $0 < \lim_{r\to\infty} |\underset{\sim}{v}_D(r)| < \infty$, then there exists a constant vector $\underset{\sim}{v}_0$, such that

$$(11) \qquad \begin{cases} \lim_{r\to\infty} |\underset{\sim}{v}_D(r)| = |\underset{\sim}{v}_0| \ , \\ \lim_{r\to\infty} \arg(u_D(r) + iv_D(r)) = \arg(u_0 + iv_0) \ , \end{cases}$$

$$(12) \qquad \lim_{r\to\infty} \int_0^{2\pi} |\underset{\sim}{v}_D(r,\theta) - \underset{\sim}{v}_0|^2 d\theta = 0 \ .$$

(ii) the pressure p_D converges pointwise and in the mean to the same limit at infinity

$$(13_1) \qquad [0,2\pi) : \lim_{r\to\infty} p_D(r,\theta) = p_\infty \ ,$$

$$(13_2) \qquad \lim_{r\to\infty} \int_0^{2\pi} |p_D(r,\theta) - p_\infty|^2 d\theta = 0 \ .$$

The questions of whether every bounded $\underset{\sim}{v}_D$ converges pointwise to its asymptotic mean value $\underset{\sim}{v}_0$ and whether the boundedness condition can be dropped were also left open.

The present paper is concerned with these open problems. In fact we prove that

(i) every $\underset{\sim}{v}_D$ is bounded and converges pointwise to its asymptotic mean value $\underset{\sim}{v}_0$; in the particular case of the Leray solution $\underset{\sim}{v}_L$, $\underset{\sim}{v}_0$ is equal to the prescribed asymptotic value $\underset{\sim}{v}_\infty$ - Theorem 1 and respectively Corollary 1-,

(ii) the Leray sequence of solutions $(\underset{\sim}{v}_{R_i}, p_{R_i})$, $i \in N$, of (1) and (5) in Ω_{R_i} converges quasi-uniformly on Ω to $(\underset{\sim}{v}_L, p_L)$ - Corollary 1-. A subsequent paper [17] will be concerned with the rate of convergence and uniqueness of solutions with finite Dirichlet integral, namely (iii) every $\underset{\sim}{v}_D$ is physically reasonable [3], [12], $\underset{\sim}{v}_D = \underset{\sim}{v}_{PR}$, e.g.

(14) $$\Omega : \underset{\sim}{v}_{PR} - \underset{\sim}{v}_0 = O(r^{-1/4-\varepsilon}),\ 0 < \varepsilon \leq \frac{1}{4},$$

with $\varepsilon = 1/4$, provided $\underset{\sim}{v}_0 \neq 0$, this estimate being sharp,

(iv) every p_D converges at infinity to p_∞ as $O(r^{-1/2})$, provided $\underset{\sim}{v}_0 \neq 0$, and this estimate is sharp,

(v) the Leray solution $(\underset{\sim}{v}_L, p_L)$ is unique in the Leray class of solutions $(\underset{\sim}{v}_D, p_D)$, provided $v > v_{cr}$. Moreover, the Leray class coincides with Finn's class of solutions $(\underset{\sim}{v}_{PR}, p_{PR})$.

These results were announced in [13], [14], [15] and [16]. However, since some proofs presented there were only sketched or incomplete, we give here and respectively in [17] all the details of the revised proofs.

2. Convergence at infinity of a solution with finite Dirichlet integral of the velocity

Theorem 1 [16]. *Every solution* $(\underset{\sim}{v}_D, p_D)$ *of* (1), (2) *and* (4) *is bounded and tends pointwise to its asymptotic mean value* $(\underset{\sim}{v}_0, p_\infty)$.

For the proof of this theorem we need the following

Theorem of Gilbarg and Weinberger [6, pp. 384, 396, 399, 400]. *The vorticity* $\omega_D := \partial_y u_D - \partial_x v_D$ *of the velocity* $\underset{\sim}{v}_D$, *its gradient* $\nabla\omega_D$ *as well as* $r^{1/2}\ln^{-1/4} r\nabla\omega_D$ *are square integrable in* Ω

(15) $$\int_\Omega r\ln^{-1/2} r\, |\nabla\omega_D|^2 dxdy \leq C \int_\Omega \omega_D^2 dxdy \leq 2C \int_\Omega |\nabla \underset{\sim}{v}_D|^2 dxdy < \infty,$$

where C *is a positive constant. Moreover*

(16) $$\lim_{r\to\infty} r^{3/4} \ln^{-1/8} r\, |\omega_D(r,\theta)| = 0,\ \text{uniformly in } \theta,$$

$$(17)\qquad |\omega_D(z_1,z_1) - \omega_D(z_2,z_2)| \le C\mu(R)\,|z_1 - z_2|^{1/2} ,$$
$$|z_1|,\ |z_2| > R + 2 ,\quad |z_1 - z_2| \le 1 ,$$

where C *is a positive constant independent of* R *and*

$$(18)\qquad \lim_{R\to\infty} R^{3/4}\ln^{-3/8}R\,\mu(R) = 0 .$$

Proof of Theorem 1. Let $(\underset{\sim}{v}_D, p_D)$ be an arbitrary solution of (1), (2) and (4). According to (13_1) and (13_2) p_D converges pointwise to its asymptotic mean value p_∞. Hence it remains only to show that $\underset{\sim}{v}_D$ is bounded, has an asymptotic mean value $\underset{\sim}{v}_0$ and tends pointwise to it. To this end we introduce the Bernoulli-Helmholtz function

$$H_D := \frac{1}{2}|\underset{\sim}{v}_D|^2 + \frac{p_D}{\rho_\infty}$$

and give the Navier-Stokes equations (1_1) the equivalent form

$$(19)\qquad \Omega : \begin{cases} \dfrac{\partial H_D}{\partial x} = \nu\dfrac{\partial \omega_D}{\partial y} - v_D\omega_D , \\[2mm] \dfrac{\partial H_D}{\partial y} = -\nu\dfrac{\partial \omega_D}{\partial x} + u_D\omega_D . \end{cases}$$

From (19) it follows immediately that H_D and ω_D are solutions of

$$(20)\qquad \Omega : \nu\nabla^2 H - (\underset{\sim}{v}_D \cdot \nabla)H = \nu\omega_D^2 ,$$

and respectively

$$(21)\qquad \Omega : \nu\nabla^2\omega - (\underset{\sim}{v}_D \cdot \nabla)\omega = 0 .$$

Next we write the continuity equation (1_2), the vorticity function ω_D and the equations (19), (20) and (21) in polar coordinates

$$(1_2')\qquad \Omega : \frac{1}{r}\frac{\partial}{\partial r}\Big[r(u_D\cos\theta + v_D\sin\theta)\Big] - \frac{1}{r}\frac{\partial}{\partial\theta}(u_D\sin\theta - v_D\cos\theta) = 0 ,$$

$$(22)\qquad \Omega : \frac{1}{r}\frac{\partial}{\partial r}\Big[r(u_D\sin\theta + v_D\cos\theta)\Big] - \frac{1}{r}\frac{\partial}{\partial\theta}(u_D\cos\theta - v_D\sin\theta) = \omega_D ,$$

$$(19')\qquad \Omega:\begin{cases} \dfrac{\partial H_D}{\partial r} = \dfrac{\nu}{r}\dfrac{\partial \omega_D}{\partial \theta} + (u_D\sin\theta - v_D\cos\theta)\omega_D \ , \\ \dfrac{1}{r}\dfrac{\partial H_D}{\partial \theta} = -\nu\dfrac{\partial \omega_D}{\partial r} + (u_D\cos\theta + v_D\sin\theta)\omega_D \ , \end{cases}$$

$$(20')\qquad \Omega:\nu\left(\frac{\partial^2 H_D}{\partial r^2} + \frac{1}{r}\frac{\partial H_D}{\partial r} + \frac{1}{r^2}\frac{\partial^2 H_D}{\partial \theta^2}\right) - \frac{1}{r}\frac{\partial}{\partial r}\Big[r(u_D\cos\theta + v_D\sin\theta)H_D\Big] +$$
$$+\frac{1}{r}\frac{\partial}{\partial \theta}\Big[(u_D\sin\theta - v_D\cos\theta)H_D\Big] = \nu\omega_D^2 \ ,$$

$$(21')\qquad \Omega:\nu\left(\frac{\partial^2 \omega_D}{\partial r^2} + \frac{1}{r}\frac{\partial \omega_D}{\partial r} + \frac{1}{r^2}\frac{\partial^2 \omega_D}{\partial \theta^2}\right) - \frac{1}{r}\frac{\partial}{\partial r}\Big[r(u_D\cos\theta + v_D\sin\theta)\omega_D\Big] +$$
$$+\frac{1}{r}\frac{\partial}{\partial \theta}\Big[(u_D\sin\theta - v_D\cos\theta)\omega_D\Big] = 0 \ .$$

Multiplying (1') and (22) by $\sin\theta$ and $\cos\theta$ and integrating with respect to θ we get

$$(23)\qquad [r_\Delta,\infty):\ \frac{d\hat{u}_D}{dr} = -\frac{d\hat{u}_D^{c2}}{dr} - \frac{d\hat{v}_D^{s2}}{dr} - \frac{2}{r}\left(u_D^{c2} + v_D^{s2}\right) \ ,$$

$$(24)\qquad [r_\Delta,\infty):\ \frac{d\hat{v}_D}{dr} = -\frac{d\hat{u}_D^{s2}}{dr} + \frac{d\hat{v}_D^{c2}}{dr} - \frac{2}{r}\left(\hat{u}_D^{s2} - \hat{v}_D^{c2}\right) \ ,$$

$$(25)\qquad [r_\Delta,\infty):\ \frac{d\hat{u}_D}{dr} = \hat{\omega}_D^{s1} \ ,$$

$$(26)\qquad [r_\Delta,\infty):\ \frac{d\hat{v}_D}{dr} = -\hat{\omega}_D^{c1} \ ,$$

where $\hat{f}^{cn(sn)} = \dfrac{1}{2\pi}\displaystyle\int_0^{2\pi} f(r,\theta)\cos n\theta(\sin n\theta)d\theta$, $n \in \mathbf{N} \cup \{0\}$, $\hat{f} := \hat{f}^{c0}$, are the Fourier coefficients of $f(r, \theta)$, i.e.

$$(27)\qquad \Omega:f(r,\theta) = \hat{f}(r) + 2\sum_{n=1}^{\infty}\left[\hat{f}^{cn}(r)\cos n\theta + \hat{f}^{sn}(r)\sin n\theta\right] \ .$$

Using the Parseval equality, from (10_1) we infer

$$(28)\qquad \lim_{r\to\infty} \hat{u}_D^{cn(sn)}(r) = \lim_{r\to\infty} \hat{v}^{cn(sn)}(r) = 0 \ , \ n \in \mathbf{N} \ .$$

From (23), (24), (25) and (26) it follows then

(29) $\underset{\sim}{v}_D$ is bounded in Ω iff $\frac{1}{r}(\hat{u}_D^{c2} + \hat{v}_D^{s2})$ and $\frac{1}{r}(\hat{u}_D^{s2} - \hat{v}_D^{c2})$ are integrable on $[r_\Delta, \infty)$

[respectively $\hat{\omega}_D^{c1}$ and $\hat{\omega}_D^{s1}$ are integrable on $[r_\Delta, \infty)$].

Noting now that by the integral theorem of the mean $\underset{\sim}{v}_D(r, \theta_1(r)) = \hat{\underset{\sim}{v}}_D(r)$, $H_D(r, \theta_2(r)) = \hat{H}_D(r)$, and applying the Cauchy-Schwarz inequality we obtain

$$(30_1) \quad [0,2\pi) : |\underset{\sim}{v}_D(r,\theta) - \hat{\underset{\sim}{v}}_D(r)|^2 = |\int_{\theta_1}^{\theta} \partial_\phi \underset{\sim}{v}_D(r,\theta)d\phi|^2 \le 2\pi \int_0^{2\pi} |\partial_\phi \underset{\sim}{v}_D|^2 d\phi \ ,$$

$$(30_2) \quad [0,2\pi) : |\underset{\sim}{v}_D(r,\theta_2^*) - \underset{\sim}{v}_D(r,\theta_1^*)|^2 \le 2\pi \int_0^{2\pi} |\partial_\phi \underset{\sim}{v}_D|^2 d\phi \ ,$$

and similarly for H_D, and hence, by taking account of (4), (9), (15) and (19),

$$(31_1) \quad [0,2\pi) : \int_{r_\Delta}^{\infty} r^{-1} |\underset{\sim}{v}_D(r,\theta) - \hat{\underset{\sim}{v}}_D(r)|^2 dr \le 2\pi \int_\Omega |\nabla \underset{\sim}{v}_D|^2 dxdy < \infty \ ,$$

$$(31_2) \quad [0,2\pi) : \int_{r_\Delta}^{\infty} r^{-1} |\underset{\sim}{v}_D(r,\theta_2^*) - \underset{\sim}{v}_D(r,\theta_1^*)|^2 dr < \infty \ ,$$

$$(32_1) \quad [0,2\pi) : \int_{r_\Delta}^{\infty} r^{-1} \ln^{-1} r |H_D(r,\theta) - \hat{H}_D(r)|^2 dr \le 2\pi \int_\Omega \ln^{-1} r |\nabla H_D|^2 dxdy \le$$

$$\le (v^2 \int_\Omega \ln^{-1} r |\nabla \omega_D|^2 dxdy + \int_\Omega \ln^{-1} r |\underset{\sim}{v}_D|^2 \omega_D^2 dxdy < \infty \ ,$$

$$(32_2) \quad [0,2\pi) : \int_{r_\Delta} r^{-1} \ln^{-1} r |H_D(r,\theta_2^*) - H_D(r,\theta_1^*)|^2 dr < \infty \ .$$

Integrating the left side integral in (31_1) and (32_1) with respect to θ and taking into account the Parseval equality we get

$$(33) \qquad \int_{r_\Delta}^{\infty} r^{-1} |\hat{\underset{\sim}{v}}_D^{cn(sn)}(r)|^2 dr < \infty, \ n \in N \ ,$$

$$(34) \qquad \int_{r_\Delta}^{\infty} r^{-1} \ln^{-1} r |\hat{H}_D^{cn(sn)}(r)|^2 dr < \infty, \ n \in N \ .$$

We need now more information about the asymptotic behaviour of the Fourier coefficients of $\underset{\sim}{v}_D$. To this end we integrate (19') and (20') with respect to θ and using (1') and (22) we obtain

$$(35) \qquad [r_\Delta,\theta): \frac{d\hat{H}_D}{dr} = \frac{1}{2\pi}\int_0^{2\pi} (u_D\sin\theta - v_D\cos\theta)\omega_D d\theta ,$$

$$(36) \qquad [r_\Delta,\infty): \frac{d\hat{\omega}_D}{dr} = \frac{1}{2\pi\nu r^2}\frac{d}{dr}\left[r^2 \int_0^{2\pi} (u_D\cos\theta + v_D\sin\theta)(u_D\sin\theta - v_D\cos\theta)d\theta \right] ,$$

$$(37) \qquad [r_\Delta,\infty): \frac{d}{dr}\left[r\frac{d\hat{H}_D}{dr} - \frac{r}{2\pi\nu}\int_0^{2\pi} (u_D\cos\theta + v_D\sin\theta)H_D d\theta \right] = \frac{1}{2\pi\nu}\int_0^{2\pi} r\omega_D^2 d\theta .$$

Integrating next (37) with respect to r and using the square integrability of ω_D in Ω we get

$$(38) \qquad [r_\Delta,\infty): \frac{d\hat{H}_D}{dr} = \frac{1}{2\pi\nu}\int_0^{2\pi} (u_D\cos\theta + v_D\sin\theta)H_D d\theta + \frac{C}{r} - \frac{1}{2\pi\nu r}\int_r^{\infty}\int_0^{2\pi} \omega_D^2 dxdy .$$

For the study of the asymptotic behaviour of the left side term in (38) we write (19) as an inhomogeneous Cauchy-Riemann equation

$$(19'') \qquad \Omega: \frac{\partial(H_D + i\nu\omega_D)}{\partial\bar{z}} = \frac{i}{2}\bar{w}_D\omega_D , \quad \frac{\partial}{\partial\bar{z}} = \frac{1}{2}\left(\frac{\partial}{\partial x} + i\frac{\partial}{\partial y}\right) ,$$

where $z = x + iy$ and $\bar{z} = x - iy$ are the complex variables, $w_D = u_D - iv_D$ is the complex velocity. We use (19") to prove that

$$(39) \qquad \lim_{r\to\infty} r^{3/4}\ln^{-13/8} r\,|\nabla H_D(r,\theta)| = 0 \ , \ \underline{\text{uniformly in}}\ \theta .$$

Indeed, according to the Pompeiu formula [11], [18, p. 22], the solution of (19") in the disc $D(z,R) = \{\,\zeta \in R^2 \mid |\zeta - z| < R,\ |z| > 2R > 2r_\Delta\,\}$ is given by

$$(40) \quad D(z,R): H_D(\bar{z}) + i\nu\omega_D(\bar{z}) = \frac{1}{2\pi i}\left\{ \int_{\partial D} \frac{H_D(\zeta) + i\nu\omega_D(\zeta)}{\zeta - \bar{z}} d\zeta + PV\int_D (\zeta - \bar{z})^{-1} \times \bar{w}_D(\zeta)\omega_D(\zeta)d\xi d\eta \right\}$$

where $\zeta = \xi + i\eta$ and PV denotes the Cauchy principal value. By differentiating (40) with respect to z we obtain

$$(41)\quad D(z,R): \frac{\partial(H_D + i v\omega_D)}{\partial z}(\bar{z}) = \frac{1}{2\pi i}\Big\{\int_{\partial D} \frac{H_D(\zeta) + iv\omega_D(\zeta)}{(\zeta - \bar{z})^2} d\zeta + PV\int_D (\zeta - \bar{z})^{-2} \times$$

$$\times\Big[\overline{w}_D(\zeta)\omega_D(\zeta) - \overline{w}_D(\bar{z})\omega_D(\bar{z})\Big] d\xi d\eta\Big\},\quad \frac{\partial}{\partial z} = \frac{1}{2}(\partial_x - i\partial_y),$$

where we have used on one side the equality

$$(42)\qquad D(z,R): \bar{\bar{z}} - \bar{z} = -\frac{1}{\pi} PV\int_D \frac{d\xi d\eta}{\zeta - \bar{z}},$$

on the other side the analyticity in $z \in D(z,R)$ of the line integral in (40) and respectively of the modified area integral, i.e.

$$(43)\qquad D(z,R): PV\int_D \frac{\overline{w}_D(\zeta)\omega_D(\zeta)}{\zeta - z} d\zeta d\eta + \pi\Big[\overline{w}_D(\bar{z})\omega_D(\bar{z})\Big]\bar{z}\ .$$

We estimate now the two integrals in (41), taking without loss of generality $\tilde{z} = z$. From (9), (13_1) and (16) it follows then

$$(44)\qquad \lim_{R\to\infty} R\ln^{-1}R\,\Big|\int_{\partial D} (\zeta - z)^2[H_D(\zeta) + iv\omega_D(\zeta)]d\zeta\Big| = 0\ .$$

On the other side using (17) and the theorem of the mean we get

$$(45)\quad \Big|PV\int_D \frac{\overline{w}_D(\zeta)\omega_D(\zeta) - \overline{w}_D(z)\omega_D(z)}{(\zeta - z)^2} d\xi d\eta\Big| \le \Big(PV\int_{|\zeta - z|\le 1} + \int_{1<|\zeta - z|<R}\Big)\frac{|\overline{w}_D(\zeta)\omega_D(\zeta) - \overline{w}_D(z)\omega_D(z)|}{|\zeta - z|^2} d\xi d\eta \le$$

$$\le C\mu(R)\max_{\overline{D}}|\overline{w}_D|\times\int_0^{2\pi}\int_0^1 r^{-1/2} dr d\theta + 2\max_{\overline{D}}|\omega_D|\max_{\overline{D}}\Big[\Big|\frac{\partial\overline{w}_D}{\partial z}\Big| + \Big|\frac{\partial\overline{w}_D}{\partial\bar{z}}\Big|\Big]$$

$$\int_0^{2\pi}\int_0^1 dr d\theta + 2\sup_{r\ge R}|\overline{w}_D(r,\theta)\omega_D(r,\theta)|\int_0^{2\pi}\int_1^R r^{-1} dr d\theta \le C_1\Big\{\mu(R)\max_{\overline{D}}|w_D| +$$

$$+\max_{\overline{D}}|\omega_D|\max_{\overline{D}}\Big[\Big|\frac{\partial\overline{w}_D}{\partial z}\Big| + \Big|\frac{\partial\overline{w}_D}{\partial\bar{z}}\Big|\Big] + \sup_{r\ge R}|\overline{w}_D(r,\theta)\omega_D(r,\theta)|\ln R\Big\},$$

where $C_1 = \max\{4\pi C, 4\pi\}$. Taking now account of (9), (16) and (18) as well as of the estimate [6, p. 402]

$$(46)\qquad \lim_{r\to\infty} r^{3/4}\ln^{-9/8} r\,|\nabla w_D(r,\theta)| = 0\ ,\ \text{uniformly in}\ \theta\ ,$$

from (19") and (45) we infer

$$\lim_{r\to\infty} r^{3/4}\ln^{-5/8} r \left| \frac{\partial(H_D + i\nu\omega_D)}{\partial\bar{z}} \right| = 0 \ , \text{ uniformly in } \theta \ , \tag{47}$$

$$\lim_{r\to\infty} r^{3/4}\ln^{-13/8} r \left| \frac{\partial(H_D + i\nu\omega_D)}{\partial z} \right| = 0 \ , \text{ uniformly in } \theta \ . \tag{48}$$

Using now the identities

$$\partial_x = \partial_z + \partial_{\bar{z}} \ , \ \partial_y = i(\partial_z - \partial_{\bar{z}}) \ , \tag{49}$$

(39) then follows. Hence, from (38) we obtain

$$\lim_{r\to\infty} r^{3/4}\ln^{-13/8} r \left| \int_0^{2\pi} (u_D\cos\theta + v_D\sin\theta)H_D d\theta \right| = 0 \ . \tag{50}$$

On the other side, taking account of the l'Hospital rule [10, p. 319], from (38) we get

$$\lim_{r\to\infty} \frac{1}{\ln r} \left\{ \hat{H}_D(r) - \hat{H}_D(r_\Delta) - \frac{1}{2\pi\nu} \int_{r_\Delta}^{r}\int_0^{2\pi} (u_D\cos\theta + v_D\sin\theta)H_D d\theta d\rho \right\} = \tag{51}$$

$$= \lim_{r\to\infty} \frac{1}{r} \left\{ \frac{d\hat{H}_D}{dr} - \frac{1}{2\pi\nu} \int_0^{2\pi} (u_D\cos\theta + v_D\sin\theta)H_D d\theta \right\} = C \ .$$

Using (9), from (51) it follows

$$\lim_{r\to\infty} \ln^{-1} r \left| \int_{r_\Delta}^{r}\int_0^{2\pi} (u_D\cos\theta + v_D\sin\theta)H_D d\theta d\rho \right| = 0 \ . \tag{52}$$

Integrating now (1_2) and applying the Green theorem we find, $\forall R > r_\Delta$,

$$\int_{\Omega_R} \nabla \cdot \underset{\sim}{v}_D dxdy = R \int_0^{2\pi} \left[u_D(R,\theta)\cos\theta + v_D(R,\theta)\sin\theta \right] d\theta = 0 = 2\pi R \left[\hat{u}_D^{c1}(R) + \hat{v}_D^{s1}(R) \right] \ , \tag{53}$$

where we have used the non-slip condition (3). Taking into account the Fourier expansions of $\underset{\sim}{v}_D$, p_D and ω_D, as well as (53) we obtain

$$(54)\quad [r_\Delta,\infty):\ \frac{1}{2\pi}\int_0^{2\pi}(u_D\cos\theta+v_D\sin\theta)H_D\,d\theta=\left(\hat{u}_D^2-\hat{v}_D^2\right)\hat{u}_D^{c1}+\hat{u}_D\hat{v}_D\left(\hat{u}_D^{s1}+\hat{v}_D^{c1}\right)+$$

$$+\hat{u}_D\left\{\frac{1}{2\pi}\int_0^{2\pi}\left[(u_D-\hat{u}_D)\cos\theta+(v_D-\hat{v}_D)\sin\theta\right](u_D-\hat{u}_D)\,d\theta+\frac{\hat{p}_D^{c1}}{\rho_\infty}\right\}+$$

$$\hat{v}_D\left\{\frac{1}{2\pi}\int_0^{2\pi}\left[(u_D-\hat{u}_D)\cos\theta+(v_D-\hat{v}_D)\sin\theta\right](v_D-\hat{v}_D)\,d\theta+\frac{\hat{p}_D^{s1}}{\rho_\infty}\right\}+$$

$$+\frac{1}{2\pi}\int_0^{2\pi}\left[(u_D-\hat{u}_D)\cos\theta+(v_D-\hat{v}_D)\sin\theta\right]\frac{p_D-\hat{p}_D}{\rho_\infty}\,d\theta+\frac{1}{2\pi}\int_0^{2\pi}(u_D\cos\theta+v_D\sin\theta)\,\frac{|\underset{\sim}{v}_D-\hat{\underset{\sim}{v}}_D|^2}{2}\,d\theta\ .$$

Denoting the line integral in (36) by $f(r)$, i.e.

$$(55)\quad [r_\Delta,\infty):\ \nu f(r)=\frac{1}{2\pi}\int_0^{2\pi}(u_D\cos\theta+v_D\sin\theta)(u_D\sin\theta-v_D\cos\theta)\,d\theta=\hat{u}_D\times\left(\hat{u}_D^{s2}-\hat{v}_D^{c2}\right)-$$

$$-\hat{v}_D\left(\hat{u}_D^{c2}+\hat{v}_D^{s2}\right)+\frac{1}{2\pi}\int_0^{2\pi}\left[(u_D-\hat{u}_D)\cos\theta+(v_D-\hat{v}_D)\times\sin\theta\right]\left[(u_D-\hat{u}_D)\sin\theta-(v_D-\hat{v}_D)\cos\theta\right]d\theta\ ,$$

and using the l'Hospital rule, from (16) and (36) it follows, $\forall\ \varepsilon>0$,

$$(56)\qquad \lim_{r\to\infty}\frac{\left(\frac{5}{4}+\varepsilon\right)(f-\hat{\omega}_D)}{r^{-3/4+\varepsilon}}=\lim_{r\to\infty}\frac{\frac{d}{dr}[r^2(f-\hat{\omega}_D)]}{r^{1/4+\varepsilon}}=-\lim_{r\to\infty}2r^{\frac{3}{4}-\varepsilon}\,\hat{\omega}_D=0\ .$$

We are able now to prove that $\underset{\sim}{v}_D$ is bounded. To this purpose we show at first that (50) and (56) imply

$$(57)\qquad \lim_{r\to\infty} r^{3/4-\varepsilon}\int_0^{2\pi}|\underset{\sim}{v}_D(r,\theta)-\hat{\underset{\sim}{v}}_D(r)|^2\,d\theta=0\ ,\ \&\forall\ \varepsilon>0\ .$$

Indeed, let us isolate the following terms in (54) and (55), and give them, using the integral theorem of the mean, the equivalent forms

$$(58)\quad [r_\Delta,\infty):\ \frac{\hat{u}_D(r)}{2\pi}\int_0^{2\pi}(u_D-\hat{u}_D)^2\cos\theta\,d\theta+\frac{\hat{v}_D(r)}{2\pi}\int_0^{2\pi}(v_D-\hat{v}_D)^2\sin\theta\,d\theta+$$

$$+\frac{1}{2\pi}\int_0^{2\pi}(u_D-\hat{u}_D)(v_D-\hat{v}_D)(\hat{u}_D\sin\theta+\hat{v}_D\cos\theta)\,d\theta+\frac{1}{4\pi}\int_0^{2\pi}(u_D\cos\theta+v_D\sin\theta)\,|\underset{\sim}{v}_D-\hat{\underset{\sim}{v}}_D|^2\,d\theta=$$

$$= \frac{\hat{u}_D(r)\cos\bar{\theta}(r) - \hat{v}_D(r)\sin\bar{\theta}(r)}{2\pi}\int_0^{2\pi}(u_D - \hat{u}_D)^2 d\theta + \frac{u_D(r)\sin\bar{\bar{\theta}}(r) + v_D(r)\cos\bar{\bar{\theta}}(r)}{4\pi}\int_0^{2\pi}(u_D - \hat{u}_D +$$

$$+ v_D - \hat{v}_D)^2 d\theta + \frac{u_D(r,\theta^*(r))\cos\theta^*(r) - \hat{u}_D(r)\sin\theta^*(r) + (v_D(r,\theta^*(r)) + 2\hat{v}_D(r)) \times \sin\theta^*(r) -}{4\pi}$$

$$\frac{\hat{v}_D(r)\cos\theta^*(r)}{4\pi}\int_0^{2\pi} |\underset{\sim}{v}_D - \underset{\sim}{\hat{v}}_D|^2 d\theta = \frac{\hat{v}_D(r)\sin\bar{\theta}(r) - \hat{u}_D(r)\cos\bar{\theta}(r)}{2\pi}\int_0^{2\pi}(v_D - \hat{v}_D)^2 d\theta +$$

$$+ \frac{\hat{u}_D(r)\sin\bar{\theta}(r) + \hat{v}_D(r)\cos\bar{\theta}(r)}{4\pi} \times \int_0^{2\pi}(u_D - \hat{u}_D + v_D - \hat{v}_D)^2 d\theta + \frac{[(u_D(r,\theta^{**}(r) + 2\hat{u}_D(r))\cos\theta^{**}(r) -}{4\pi}$$

$$\frac{\hat{u}_D(r)\sin\theta^{**}(r) + v_D(r,\theta^{**}(r))\sin\theta^{**}(r)}{4\pi}\int_0^{2\pi} |\underset{\sim}{v}_D - \underset{\sim}{\hat{v}}_D|^2 d\theta \ ,$$

(59) $$[r_\Delta,\infty) : \frac{1}{2\pi}\int_0^{2\pi}\Big[(u_D - \hat{u}_D)\cos\theta + (v_D - \hat{v}_D)\sin\theta\Big]\Big[(u_D - \hat{u}_D)\sin\theta - (v_D - \hat{v}_D) \times \cos\theta\Big] d\theta =$$

$$= \frac{\sin 2\bar{\theta}(r) + \cos 2\bar{\theta}(r)}{4\pi}\int_0^{2\pi}(u_D - \hat{u}_D)^2 d\theta + \frac{\cos 2\bar{\theta}(r) - \sin 2\bar{\theta}(r)}{4\pi}\int_0^{2\pi}(v_D - \hat{v}_D)^2 d\theta -$$

$$- \frac{\cos 2\hat{\hat{\theta}}(r)}{4\pi}\int_0^{2\pi}(u_D - \hat{u}_D + v_D - \hat{v}_D)^2 d\theta \ ,$$

where $\bar{\theta}(r)$, $\bar{\bar{\theta}}(r)$, $\theta^*(r)$, $\theta^{**}(r)$, $\hat{\theta}(r)$ and $\hat{\hat{\theta}}(r) \in [0,2\pi]$.

Taking into account the identity

(60) $$R : \cos^2\theta(r) + \sin^2\theta(r) = 1 \ ,$$

one obtains that the left-side terms in (58) and (59) behave at infinity exactly like

(61_1) $$C_1\left(\hat{u}_D^2 + \hat{v}_D^2\right)^{1/2}\int_0^{2\pi} |\underset{\sim}{v}_D - \underset{\sim}{\hat{v}}_D|^2 d\theta$$

and, respectively, like

(61_2) $$C_2\int_0^{2\pi} |\underset{\sim}{v}_D - \underset{\sim}{\hat{v}}_D|^2 d\theta \ ,$$

where the positive constants C_1 and C_2 are independent of r. Giving now (50) and (56) the equivalent forms

$$(62)\qquad (r_\infty,\infty): \frac{1}{2\pi}\int_0^{2\pi}(u_D\cos\theta + v_D\sin\theta)H_D d\theta = k_1(r)r^{-3/4}\ln^{13/8} r \ ,$$

$$(63)\qquad (r_\infty,\infty): \frac{1}{2\pi}\int_0^{2\pi}(u_D\cos\theta + v_D\sin\theta)(u_D\sin\theta - v_D\cos\theta)d\theta = k_2(r)r^{-3/4+\varepsilon} \ ,$$

where

$$(64)\qquad \lim_{r\to\infty} k_1(r) = \lim_{r\to\infty} k_2(r) = 0 \ ,$$

taking account of (54) and (55) and eliminating $\hat{v}_D$ and respectively $\hat{u}_D$ between (62) and (63), we obtain second order equations in $\hat{u}_D$ and respectively $\hat{v}_D$. Using now (58), (59), (61_1) and (61_2), (57) then follows. Taking account of (57) and the Parseval equality, from (23) and (24) we obtain by integration

$$(65)\qquad [r_\Delta,\infty): \hat{u}_D(r) - u_0 = -\left[\hat{u}_D^{c2}(r) + \hat{v}_D^{s2}(r)\right] + 2\int_r^\infty \rho^{-1}\left[\hat{u}_D^{c2}(\rho) + \hat{v}_D^{s2}(\rho)\right]d\rho \ ,$$

$$(66)\qquad [r_\Delta,\infty): v_D(r) - \hat{v}_0 = -\left[\hat{u}_D^{s2}(r) + \hat{v}_D^{c2}(r)\right] + 2\int_r^\infty \rho^{-1}\left[\hat{u}_D^{s2}(\rho) + \hat{v}_D^{c2}(\rho)\right]d\rho \ ,$$

where $\underset{\sim}{v}_0$, a constant vector, is according to (11) and (12) the asymptotic mean velocity. Moreover

$$(67)\qquad \lim_{r\to\infty} r^{3/8-\varepsilon/2}\,|\hat{\underset{\sim}{v}}_D(r) - \underset{\sim}{v}_0| = 0\ , \quad \forall\, \varepsilon > 0 \ .$$

Let us show now that $\underset{\sim}{v}_D$ converges pointwise to $\underset{\sim}{v}_0$ at infinity. For this purpose we assume at first that $\underset{\sim}{v}_0 = 0$. Then from (10_2) we get

$$(68)\qquad \lim_{r\to\infty} |\hat{\underset{\sim}{v}}_D(r)| = \lim_{r\to\infty}\ \max_{\theta\in[0,2\pi]} |\underset{\sim}{v}_D(r,\theta)| = 0 \ .$$

Using the inequality

$$(69)\qquad \Omega: 0 \le |\underset{\sim}{v}_D(r,\theta)| \le \max_{\theta\in[0,2\pi]} |\underset{\sim}{v}_D(r,\theta)|$$

and letting $r \to \infty$ we infer then

$$(70)\qquad [0,2\pi): \lim_{r\to\infty} \underset{\sim}{v}_D(r,\theta) = \underset{\sim}{0} \ .$$

It remains to consider the case where $\underset{\sim}{v}_0 \ne 0$. Let choose the point $P(R,\theta)$, $R > r_\Delta$, as the origin of a new system of polar coordinates (r', θ'), and give $(19_1')$ the equivalent form

$$(19_1')\quad \Omega: \frac{\partial}{\partial r'}\left[\frac{|\underset{\sim}{v}_D - \underset{\sim}{v}_0|^2}{2} + \frac{p_D}{\rho_\infty}\right] = \frac{\nu}{r'}\frac{\partial \omega_D}{\partial \theta'} + \left[(u_D - u_0)\sin\theta' - (v_D - v_0)\times\cos\theta'\right]\omega_D + \frac{u_0}{r'}\frac{\partial v_D}{\partial \theta'} - \frac{v_0}{r'}\frac{\partial u_D}{\partial \theta'} \,.$$

Integrating first with respect to r' on $[0,R/2]$ and then with respect to θ' on $[0,2\pi]$ and taking the absolute value we obtain

$$(71)\quad \Omega: \frac{1}{2}|\underset{\sim}{v}_D(P) - \underset{\sim}{v}_0|^2 \le \frac{1}{4\pi}\int_0^{2\pi}\left|\underset{\sim}{v}_D\left(\frac{R}{2},\theta\right) - \underset{\sim}{v}_0\right|^2 d\theta + \left|p_D(P) - p_D\left(\frac{R}{2}\right)\right| + \frac{1}{2\pi}\int_0^{R/2}\int_0^{2\pi}\left|\left[(u_D - u_0)\sin\theta - (v_D - v_0)\cos\theta\right]\omega_D\right| d\theta dr \,.$$

Applying the Cauchy-Schwarz inequality we obtain

$$(72)\quad \int_1^{R/2}\int_0^{2\pi}\left|\left[(u_D - u_0)\sin\theta - (v_D - v_0)\cos\theta\right]\omega_D\right| d\theta dr \le \left(4\int_1^{R/2}\int_0^{2\pi} r^{-1}|\underset{\sim}{v}_D - \underset{\sim}{v}_0|^2 d\theta dr \int_1^{R/2}\int_0^{2\pi}\omega_D^2 dxdy\right)^{1/2}$$

$$\le \left(4\int_1^{R/2}\int_0^{2\pi} r^{-1}|\underset{\sim}{v}_D - \underset{\sim}{v}_0|^2 d\theta dr\right)^{1/2}\left(\int_{R/2}^{3R/2}\int_0^{2\pi}\omega_D^2 dxdy\right)^{1/2} ,$$

where we have used the fact that the disc $r' < R/2$ is contained in the annulus $R/2 < r < 3R/2$. On the other side we have

$$(73)\quad \int_0^1\int_0^{2\pi}\left|\left[(u_D - u_0)\sin\theta - (v_D - v_0)\cos\theta\right]\omega_D\right| d\theta dr \le C\times \max_{\overline{\Omega}_{\frac{R}{2}+1}\setminus\Omega_{\frac{R}{2}-1}} |\omega_D| \,,$$

where C depends on the (bounded) velocity $\underset{\sim}{v}_D$. From (71) it follows then

$$(74)\quad \Omega: \frac{1}{2}|\underset{\sim}{v}_D(P) - \underset{\sim}{v}_0|^2 \le \frac{1}{4\pi}\int_0^{2\pi}\left|\underset{\sim}{v}_D\left(\frac{R}{2},\theta\right) - \underset{\sim}{v}_0\right|^2 d\theta + \left|p_D(P) - p_D\left(\frac{R}{2}\right)\right| + \left(8\int_1^{R/2}\int_0^{2\pi} r^{-1}\left[|\underset{\sim}{v}_D - \underset{\sim}{v}_D|^2 + |\underset{\sim}{v}_D - \underset{\sim}{v}_0|^2\right] d\theta dr\right)^{\frac{1}{2}}\left(\int_{R/2}^{3R/2}\int_0^{2\pi}\omega_D^2 dxdy\right)^{\frac{1}{2}} +$$

$$+ C \max_{\bar{\Omega}_{\frac{R}{2}+1} \setminus \Omega_{\frac{R}{2}-1}} |\omega_D| \ .$$

Letting now $R \to \infty$ and using (4), (10), (13_1), (15), (31_1) and the fact that (67) is independent of the choice of the system of coordinates, i.e. of R, from (74) we infer that also for $\underset{\sim}{v}_0 \neq 0$

$$[0,2\pi) : \lim_{r\to\infty} \underset{\sim}{v}_D(r,\theta) = \underset{\sim}{v}_0 \ . \tag{75}$$

Remark 1. The argument used to prove the estimate (39) follows closely the argument employed in [6, p. 402] to estimate the derivatives of the velocity $\underset{\sim}{v}_D$.

Remark 2. The idea to use (19_{11}") in order to show the pointwise convergence at infinity of the velocity $\underset{\sim}{v}_D$, e.g. (75), was suggested by the proof of the pointwise convergence at infinity of the pressure p_D [6, p. 394].

In the particular case, where $(\underset{\sim}{v}_D, p_D)$ is the Leray solution $(\underset{\sim}{v}_L, p_L)$, we obtain the following

Corollary 1 [13]. *The Leray solution* $(\underset{\sim}{v}_L, p_L)$ *of* (1), (2) *and* (3) *tends pointwise at infinity to* $(\underset{\sim}{v}_\infty, p_\infty)$. *Furthermore, the Leray sequence of solutions* $(\underset{\sim}{v}_{R_i}, p_{R_i})$, $i \in N$, *of* (1) *and* (5) *in* Ω_{R_i} *converges quasi-uniformly on* $\bar{\Omega}$ *to* $(\underset{\sim}{v}_L, p_L)$.

For the proof of this Corollary we need the following results:

Definition 1 [9, p. 66]. *Let* X *and* Y *be metric spaces and let* f_n, $n \in N$, *map* X *into* Y. *The sequence* $\{f_n\}_{n\in N}$ *is said to converge quasi-uniformly on* X *to* $f : X \to Y$ *if*

(i) f_n *converges pointwise to* f,

(ii) *for every* $\varepsilon > 0$ *there exist a sequence* $\{n_p\}_{p\in N}$ *and a sequence* $\{D_p\}_{p\in N}$ *of open sets* $D_p \subset X$, $X = \bigcup_{p=1}^{\infty} D_p$, *such that*

$$\mathrm{dist}_y(f(x), f_{n_p}(x)) < \varepsilon \, , \ p \in N \, , \ \forall x \in D_p \ .$$

Theorem of Arzelà, Gagaeff and Alexandrov [1], [4], [9, p. 68]. *Let* X, Y *be metric spaces and let* f_n, $n \in N$, *map* X *into* Y *continuously. The sequence* $\{f_n\}$ *converges on* X *to a continuous mapping* $f : X \to Y$, *if and only if the convergence is quasi-uniform.*

Proof of Corollary 1. Let apply the Pompeiu formula to the Leray sequence

(w_{R_i},p_{R_i}), $i \in N$, of solutions of (1) and (5) in Ω_{R_i}, $(w_{R_i}, p_{R_i}) \to (w_L,p_L)$, uniformly on compact subsets of $\overline{\Omega}$, by noting that w_{R_i} is solution of an inhomogeneous Cauchy-Riemann equation

(76) $$\Omega_{R_i} : \frac{\partial w_{R_i}}{\partial \bar{z}} = \frac{i}{2}\,\omega_{R_i} \;,$$

(77) $$\Omega_{R_i} : w_{R_i}(z) = w_\infty + \frac{1}{2\pi i}\,\mathrm{PV}\int_{\Omega_{R_i}} \frac{\omega_{R_i}(\zeta)}{\zeta - z}\, d\xi d\eta \;.$$

On the other side by the theorem of the mean we have

(78) $$\Omega_{R_i} : w_{R_i}(r,\theta) = \omega_\infty + (r - R_i)\,\frac{\partial w_{R_i}}{\partial r}(r + [R_i - r]\beta(r,\theta;R_i),\theta) \;,$$

where $0 < \beta < 1$. Without loss of generality we can assume that

(79) $$\Omega : \lim_{R_i\to\infty} \beta(r,\theta;R_i) = \beta_\infty(r,\theta) \;,\; 0 \le \beta_\infty \le 1 \;,$$

for a suitable chosen subsequence of $\{R_i\}$ which we denote again by $\{R_i\}$. By letting in (77) and (78) $R_i \to \infty$ we get

(80) $$\Omega : w_L(z) = w_\infty + \frac{1}{2\pi i}\lim_{R_i\to\infty}\mathrm{PV}\int_{\Omega_{R_i}} \frac{\omega_{R_i}(\zeta)}{\zeta - z}\, d\xi d\eta \;,$$

(81) $$\Omega : w_L(r,\theta) = w_\infty + \lim_{R_i\to\infty}\left\{ (r - R_i)\,\frac{\partial w_{R_i}}{\partial r}(r + [R_i - r]\beta(r,\theta;R_i),\theta) \right\} \;.$$

Using the fact that the limits in (80) and (81) exist and are finite and assuming that $w_L \equiv 0$ in the case $w_\infty \neq 0$, from (79) we infer

(82) $$\Omega : \beta_\infty(r,\theta) = 0 \;.$$

Indeed, otherwise from (81) it results that

(83) $$\lim_{r\to\infty} r\frac{\partial w_L}{\partial r}(r,\theta) = C \neq 0 \;,\; \text{uniformly in } \theta \;,$$

in contradiction with (4), (9) and the boundedness of w_L. But (82) implies the existence of a positive constant γ, such that

(84) $$\lim_{r\to\infty} r^{1+\gamma}\frac{\partial w_L}{\partial r}(r,\theta) = 0 \;,\; \text{uniformly in } \theta \;.$$

From (75) and (84) it follows then

(85) $$\underset{\sim}{v}_0 = \underset{\sim}{v}_\infty \; .$$

Let us prove now that $w_\infty \neq 0$ implies $w_L \neq 0$. To this end we write (23) and (24) for $\underset{\sim}{v}_{R_i}$ and integrating with respect to r we get

(86) $$[r_\Delta, R_i] : \hat{u}_{R_i}(r) - u_\infty = -\left[\hat{u}^{c2}_{R_i}(r) + \hat{v}^{s2}_{R_i}(r)\right] + 2\int_r^{R_i} \frac{1}{\rho}\left[\hat{u}^{c2}_{R_i}(\rho) + \hat{v}^{s2}_{R_i}(\rho)\right] d\rho \; ,$$

(87) $$[r_\Delta, R_i] : \hat{v}_{R_i}(r) - v_\infty = -[\hat{u}^{s2}(r) - \hat{v}^{c2}(r)] - v^{c2}(r)] + 2\int_r^{R_i} \frac{1}{\rho}\left[\hat{u}^{s2}_{R_i}(\rho) - \hat{v}^{c2}_{R_i}(\rho)\right] d\rho \; .$$

Using (31_1) and (57), both written for $\underset{\sim}{v}_{R_i}$, from (86) and (87) we obtain

(88) $$[r_\Delta, R_i] : \int_r^{R_i} \rho^{-1} |\underset{\sim}{v}_{R_i}(\rho) - \underset{\sim}{v}_\infty|^2 d\rho \le C \; ,$$

where, according to the construction, C is independent of R_i, and hence, letting $R_i \to \infty$ we infer

(89) $$\int_{r_\Delta}^{\infty} r^{-1} |\underset{\sim}{v}_L(r) - \underset{\sim}{v}_\infty|^2 dr < \infty \; .$$

But (89) implies $\underset{\sim}{v}_L \neq 0$. For the sake of completeness we give another proof of (85), which uses directly (89). Indeed, taking in account the fact that $\underset{\sim}{v}_L$ has finite Dirichlet integral and using (67) we get

(90) $$\int_{r_\Delta}^{\infty} r^{-1} |\underset{\sim}{v}_L(r) - \underset{\sim}{v}_0|^2 dr < \infty \; .$$

From (89) and (90) it follows then (85). It remains now to show that the Leray sequence of solutions $(\underset{\sim}{v}_{R_i}, p_{R_i})$, $i \in \mathbf{N}$, of (1) and (5) in Ω_{R_i} converges quasi-uniformly on $\overline{\Omega}$ to the Leray solution $(\underset{\sim}{v}_L, p_L)$. To this end we define $\underset{\sim}{v}^e_{R_i}$ and $\underset{\sim}{v}^e_L$ as follows

$$(91) \qquad \overline{\Omega} : \underset{\sim}{v}^e_{R_i} := \begin{cases} \underset{\sim}{v}_{R_i} \text{ in } \Omega_{R_i} , \\ & i \in \mathbf{N} , \\ \underset{\sim}{v}_\infty \text{ in } \overline{\Omega} \backslash \Omega_{R_i} , \end{cases}$$

$$(92) \qquad \overline{\Omega} : \underset{\sim}{v}^e_L := \begin{cases} \underset{\sim}{v}_L \text{ in } \Omega , \\ \underset{\sim}{v}_\infty \text{ at infinity} . \end{cases}$$

It is easy to see on one side that $\underset{\sim}{v}^e_{R_i}$, $i \in \mathbf{N}$, and $\underset{\sim}{v}^e_L$ are continuous on $\overline{\Omega}$. On the other side by using the stereographic projection the extended plane $\mathbf{R}^2$ becomes a metric space. By Definition 1 and the Theorem of Arzela, Gagaeff and Alexandrov the result then follows.

Remark 3. The relations (81) and (89) can be used in order to prove the pointwise convergence of w_L at infinity.

References

[1] **P.S. Alexandrov,** On the so-called quasi-uniform convergence, Uspekhi Mat. Nauk 3 (1948), 213-215.

[2] **R. Finn,** On the steady-state solutions of the Navier-Stokes equations III, Acta Math. 105 (1961), 197-244.

[3] **R. Finn,** Mathematical questions relating to viscous fluid flow in an exterior domain, Rocky Mountain J. Math. 3 (1973), 107-140.

[4] **B. Gagaeff,** Sur les suites convergentes des fonctions mesurables B, Fund. Math. 18 (1932), 182-188.

[5] **D. Gilbarg and H. Weinberger,** Asymptotic properties of Leray's solution of the stationary two-dimensional Navier-Stokes equations, Uspekhi Mat. Nauk 29 (1974), 109-122.

[6] **D. Gilbarg and H. Weinberger,** Asymptotic properties of steady plane solutions of Navier-Stokes equations with bounded Dirichlet integral, Ann. Scuola Norm. Sup. Pisa 5 (1978), 381-404.

[7] **O.A. Ladyzhenskaya,** The mathematical theory of viscous incompressible flow, Second edition, Gordon & Breach, New York 1969.

[8] **J. Leray,** Etudes de diverses équations intégrales non linéaires et de quelques problèmes que pose l'hydrodynamique, J. Math. Pures Appl. 12 (1933), 1-82.

[9] **M. Nicolescu,** Analiza matematica, Vol. II, Editura tehnica, Bucuresti 1958.

[10] **M. Nicolescu, N. Dinculeanu and S. Marcus,** Manual de analiza matematica, Vol. 1, Editura didactica si pedagogica, Bucuresti 1963.

[11] **D. Pompeiu,** Sur une classe de fonctions d'une variable complexe, Rend. Circ. Mat. Palermo 33 (1912), 108-113; Sur une classe de fonctions d'une variable complexe et sur certaines équations intégrales, Rend. Circ. Mat. Palermo 35 (1913), 277-281.

[12] **D.R. Smith,** Estimates at infinity for stationary solutions of the Navier-Stokes equations in two dimensions, Arch. Rational Mech. Anal. 20 (1965), 341-372.

[13] **D. Socolescu,** On the convergence at infinity of the steady plane Leray solution of the Dirichlet problem for the Navier-Stokes equations, Mathematical methods in Fluid Mechanics, Editors E. Meister, K. Nickel and J. Polasek, Verlag P. Lang, Frankfurt am Main (1982), 211-222.

[14] **D. Socolescu,** On the asymptotic decay order and uniqueness of the Leray solution of the Dirichlet problem for the Navier-Stokes equations, Analyse non linéaire et applications à la Mécanique, Editeurs R. Faure and G. Hecquet, Publications I.R.M.A. de l'U.E.R. de Mathématiques pures et appliquées de l'Université de Lille 5 (2), Villeneuve d'Ascq 1983.

[15] **D. Socolescu,** On the relation between the Leray class of solutions with finite Dirichlet integral and Finn's class of physically reasonable solutions of the Dirichlet problem for the steady plane Navier-Stokes equations, Bericht Nr. 23, Fakultät für Mathematik, Karlsruhe (1984).

[16] **D. Socolescu,** On the convergence at infinity of solutions with finite Dirichlet integral of the exterior Dirichlet problem for the steady plane Navier-Stokes equations, Bericht Nr. 27, Fakultät für Mathematik, Karlsruhe (1986).

[17] **D. Socolescu,** On the convergence at infinity, rate of convergence and uniqueness of solutions with finite Dirichlet integral to the exterior Dirichlet problem for the steady plane Navier-Stokes system of equations, Manuscript.

[18] **I.N. Vekua,** Verallgemeinerte analytische Funktionen, Akademie-Verlag, Berlin 1963.

THE ELLIPTIC SINH GORDON EQUATION AND THE CONSTRUCTION OF TOROIDAL SOAP BUBBLES

Joel Spruck*
Dept. of Mathematics and Statistics
University of Massachusetts at Amherst
Amherst, MA 01003 (U.S.A.)

Dedicated to Hans Lewy

In this paper we will study the semilinear elliptic equation

$$\Delta u + \lambda \sinh u = 0$$

in bounded domains $D \subset R^2$. This equation has sometimes been called the elliptic Sinh-Gordon equation. Of particular interest is the study of the following boundary value problem of "nonlinear eigenvalue" type:

$$(1) \qquad \begin{aligned} \Delta u + \lambda \sinh u &= 0 \quad \text{in } R \\ u &= 0 \quad \text{on } \partial R \\ u &\geq 0 \end{aligned}$$

where R is a rectangle in R^2. This problem arises in plasma physics and also statistical mechanics as a way of modeling point vortices. However, it arises in a surprising and central way in the construction of compact surfaces of constant mean curvature. This will be explained in the following section.

The basic questions to be answered concern the uniqueness of solutions (the existence is very easy) and the behavior of solutions as λ tends to zero. Moreover, we would like a complete classification of all the positive solutions. Generally speaking, such an ambitious goal has never been accomplished for nonlinear eigenvalue problems except in very special cases. Because of the special geometric significance of problem (1) we shall show that a very beautiful and elegant description of all the solutions is possible.

Our main results are as follows.

* Research partially supported by NSF Grant DMS-8501952.

Theorem 1. *Let* (λ_k, u_k) *be a sequence of nontrivial solutions to* (1) *with* $\lambda_k \downarrow 0$. *Then the* u_k *tend to the Green's function*

$$-2 \log |g(z)|^2$$

where $g(z)$ *is the symmetric conformal map of* R *onto the unit disk. The convergence is uniform (in any topology) on compact subsets of* $R - \{0,0\}$ *and in* $W^{1,p}(R), p < 2$.

This theorem easily extends to nonlinearities $f(t)$ with f asymptotic to ce^t as $t \to \infty$. It also holds for a class of domains D which have reflective symmetry about the origin.

Theorem 2. *For each* λ, $0 < \lambda < \lambda_1(R)$, *there is a unique nontrivial solution of* (1). *The solutions can be described in terms of the classical elliptic functions.*

The proof of Theorem 1 takes up the majority of the paper and illustrates the use of apriori bounds and symmetry arguments in proving global results. Section 1 develops the connection between problem (1) and toroidal soap bubbles. In Section 2 we prove an apriori bound for all solutions of (1) that is central to the proof of Theorem 1. In Section 3 we use this bound and symmetry arguments to complete the proof of Theorem 1.

In Section 4 we briefly discuss an iterative procedure [ES] that is very useful in computing the solutions of (1). Section 5 discusses the representation of solutions of (1) and contains the proof of Theorem 2.

1. The Construction of Toroidal Soap Bubbles

A classical problem in differential geometry, made famous by H. Hopf [H] is the following: Does there exist a compact surface in R^3 of constant mean curvature (without boundary) other than the standard sphere? Hopf himself showed that such a surface could not be of genus zero. Later, A.D. Alexandrov [A] showed that such a surface could not be embedded, that is, it must have self-intersections if it exists. In fact, Alexandrov invented the method of moving planes to prove that an embedded compact surface of constant mean curvature is a sphere. Finally, in 1984 [W] proved the existence of infinitely many non-congruent compact surfaces of constant mean curvature of genus one. A key point in his construction is the analysis of problem (1) for λ near zero.

To see how problem (1) arises, let $F : D \subset R^2 \to R^3$ be a conformal representation of a surface, that is

$$F_x^2 = F_y^2 = E > 0$$

$$F_x \bullet F_y = 0 \quad .$$

Here, $z = x + iy$ are coordinates in R^2. The metric on the surface is given by $ds^2 = E|dz|^2$ with normal field

$$\xi = \frac{F_x \wedge F_y}{|F_x \wedge F_y|} \quad .$$

The second fundamental form of the surface is given by $l\,dx^2 + 2m\,dxdy + n\,dy^2$ where

$$l = F_{xx} \bullet \xi \quad m = F_{xy} \bullet \xi \quad n = F_{yy} \bullet \xi \quad .$$

The mean curvature H and Gauss curvature K are given by

$$H = \frac{l+n}{2E}, \qquad K = \frac{ln - m^2}{E^2} \quad .$$

The first fundamental equation of surface theory is the Gauss equation

(2) $$-K = \frac{\Delta \log E}{2E} \quad .$$

The other fundamental equations are the Codazzi-Mainardi equations. These can be expressed as follows (see [H]).

Let $\phi = l - n - 2im$. Then

(3) $$|\phi| = 2E\sqrt{H^2 - K} = E(k_2 - k_1)$$

(where $k_2 \geq k_1$ are the principal curvatures).

The Codazzi-Mainardi equations are given (in complex notation) by

(4) $$\phi_{\bar{z}} = EH_z \quad .$$

Now suppose F is a doubly periodic immersion of constant mean curvature (genus 1). Then from (4), ϕ is an analytic function which is doubly periodic. Hence

(5) $$|\phi| = \lambda \text{ constant} \quad .$$

By making a rotation of the x,y coordinates we can arrange that ϕ is real, that is, $m = 0$. Geometrically, this means that the coordinate lines are lines of curvature.

From (3) and (5) we find

$$H^2 - K = \frac{\lambda^2}{4E^2} \quad .$$

Substitution into equation (2) gives

$$\frac{\lambda^2}{4E^2} = \frac{\Delta \log E}{2E} + H^2$$

or

$$\Delta \log E + 2EH^2 - \frac{\lambda^2}{2E} = 0 \quad .$$

Finally, we set $E = \frac{\lambda}{2H} e^u$. Then

(6) $$\Delta u + 2\lambda H \sinh u = 0 \quad .$$

For convenience we may assume $H = 1/2$. Then (6) is just the elliptic Sinh-Gordon equation. We can also easily compute

(7) $$k_2 = \frac{1+e^{-u}}{2} \qquad l = k_1 E \qquad E = \lambda e^u$$
$$m = 0$$
$$k_1 = \frac{1-e^{-u}}{2} \qquad n = k_2 E \quad .$$

Now we want to reverse this procedure and start with a doubly periodic solution of

$$\Delta u + \lambda \sinh u = 0 \quad \text{in} \quad R^2$$

and construct the map F with first and second fundamental forms given by (7). The simplest way to do this is to solve

(8) $$\Delta u + \lambda \sinh u = 0 \quad \text{on} \quad R$$
$$u = 0 \quad \text{on} \quad \partial R$$

where $R = [0,a] \times [0,b]$. Then we can extend u by odd reflection to a doubly periodic solution on R^2. The parameter λ is a scale parameter which can be removed by the transformation

$$\omega(x,y) = \frac{1}{2} u \left(\frac{x}{\sqrt{\lambda}}, \frac{y}{\sqrt{\lambda}} \right) \quad .$$

Then,

(9) $$\Delta \omega + \sinh \omega \cosh \omega = 0 \quad \text{in} \quad R^\lambda$$
$$\omega = 0 \quad \text{on} \quad \partial R^\lambda$$

where $R^\lambda = [0,\sqrt{\lambda}\ a] \times [0,\sqrt{\lambda}\ b]$.

For later reference, we have $E = e^{2\omega}$ and

$$k_2 = e^{-\omega}\cosh\omega \quad l = e^{\omega}\sinh\omega$$
$$k_1 = e^{-\omega}\sinh\omega \quad m = 0$$
$$n = e^{\omega}\cosh\omega\ .$$

The fundamental equations of surface theory for the triple $F_x,\ F_y,\ \xi$ are

$$\begin{aligned} F_{xx} &= \omega_x F_x - \omega_y F_y + e^{\omega}\sinh\omega\ \xi \\ F_{xy} &= \omega_y F_x + \omega_x F_y \\ F_{yy} &= -\omega_x F_x + \omega_y F_y + e^{\omega}\cosh\omega\ \xi \\ \xi_x &= -e^{-\omega}\sinh\omega\ F_x \\ \xi_y &= -e^{-\omega}\cosh\omega\ F_y\ . \end{aligned} \tag{10}$$

This is an overdetermined system which can be integrated if u satisfies (8) or equivalently ω satisfies (9). The surface $F(x,y)$ is unique up to a Euclidean motion (see [E]).

The difficulty is that F need not be doubly periodic; there is a problem of periods. In order to overcome this difficulty Wente required u to be non-negative, that is, u solves (1). Then by Theorem 3.2 of [GNN], u inherits the symmetry of the rectangle about its center. By analyzing the system (10) it is then easy to analyze the inherited symmetries of the surface $F(x,y)$. Wente then shows that it is possible to choose the parameters of the problem $(\lambda, b/a)$ so that "F closes up" (see [W]).

2. An Apriori Bound for Positive Solutions

In this section we will prove an apriori sup norm estimate for solutions of (1) that will be central to our analysis.

Let $u \geq 0$ satisfy

$$\Delta u + \lambda\ f(u) = 0 \quad \text{in } D \subset R^2$$

and let $w = e^{-\alpha u}$ where $1/2 < \alpha < 1$. Then

$$\Delta w = \alpha\lambda w f(u) + \frac{|\nabla w|^2}{w}\ . \tag{11}$$

Define a vector field $V(x)$ by

$$V^j = \frac{1}{w}\left(w_i w_{ij} - 1/2 w_j \Delta w\right)\ .$$

We note that V can be expressed as a gradient:

(12) $$V^j = \left(\frac{|\nabla w|^2}{2w} - \lambda\, G(w) \right)_j \quad \text{where} \quad G'(t) = tf\left(\ln \frac{1}{t}\right) .$$

Using (11), a short computation gives

(13) $$\operatorname{div} V = J + \big(2\alpha f(u) - f'(u)\big) \cdot \frac{\lambda\, |\nabla w|^2}{2w}$$

where $J = \frac{1}{w}\left(\sum_{i,j} w_{ij}^2 - \frac{1}{2}(\Delta w)^2 \right) \geq 0$.

If $f(u) = \sinh u$ then

(14) $$\operatorname{div} V \geq \frac{\lambda\alpha^2}{2}\left((2\alpha - 1)e^{(1-\alpha)u} - (2\alpha + 1)e^{-(1+\alpha)u} \right) |\nabla u|^2 .$$

Let η be a non-negative $C_0^\infty(D)$ function. Then

$$0 = \int_D \operatorname{div} \eta V \, dxdy = \int_D \big(\eta \operatorname{div} V + \nabla\eta \cdot V\big) dx\, dy .$$

Using (12) this gives

(15) $$\int_D \eta \operatorname{div} V \, dx\, dy = \int_D \Delta\eta \left(\frac{|\nabla w|^2}{2w} - \lambda\, G(w) \right) dx\, dy .$$

Using (14) for $f(u) = \sinh u$ (15) leads to the estimate

(16) $$\lambda \int_D \eta\, e^{(1-\alpha)u} |\nabla u|^2 dx\, dy \leq c_1 + c_2 \int_{\operatorname{supp} \eta} e^{-\alpha u} |\nabla u|^2 dx\, dy .$$

Now let D satisfy the conditions of [GNN, p. 223].

Then the maximum of u is achieved in a compact subdomain $D_{4\varepsilon} = \{(x,y) \in D : \operatorname{dist}((x,y),\partial D) \geq 4\varepsilon\}$, where ε is independent of u. Choose η such that $\eta \equiv 1$ on $D_{2\varepsilon}$, $\eta \in C_0^\infty(D_\varepsilon)$. Then we can estimate

(17) $$\int_{D_\varepsilon} e^{-\alpha u} |\nabla u|^2 dx\, dy \leq C(\varepsilon)$$

$$\lambda \int_{D_\varepsilon} e^{(1-\alpha)u} dx\, dy \leq C(\varepsilon) .$$

From (16) and (17) it follows that

(18) $$\int_D |\nabla \eta e^{\frac{1-\alpha}{2}u}|^2 dx\,dy \;.$$

Using the Sobolev inequality, estimate (18) gives the following L^p estimate for e^u :

(19) $$|e^u|_{L^p(D_{2\varepsilon})} \le c_1 + \frac{c_2}{\lambda^{\frac{1}{1-\alpha}}} \;.$$

We can now easily prove

Theorem 2.1. *Let* $u \ge 0$ *satisfy*

$$\Delta u + \lambda \sinh u = 0 \text{ in } D \subset R^2$$

$$u = 0 \text{ on } \partial D \;.$$

Then $u \le \log\left(c_1 + \frac{c_2}{\lambda}\right)$ *where* c_1, c_2 *depend only on* D.

Proof. We compute

$$\begin{aligned}\Delta\zeta^2 e^u &= \zeta^2\Delta e^u + e^u\Delta\zeta^2 + 4\zeta e^u\nabla\zeta\nabla u\\ &= \zeta^2 e^u\left(\Delta u + |\nabla u|^2\right) + e^u\left(\Delta\zeta^2 + 4\zeta\nabla u \bullet \nabla\zeta\right)\\ &\ge -c\left(\lambda\zeta^2 e^{2u} + \left(|\nabla\zeta|^2 + \zeta|\Delta\zeta|\right)e^u\right) \;.\end{aligned}$$

Choosing $\zeta \in C_0^\infty(D_{2\varepsilon})$, $\zeta \equiv 1$ on $D_{4\varepsilon}$, we have from elliptic regularity theory

$$\begin{aligned}\sup_{D_{4\varepsilon}} e^u \le \sup_D \zeta^2 e^u &\le C\left(\lambda\,||e^{2u}||_{L^p(D_{2\varepsilon})} + ||e^u||_{L^p(D_{2\varepsilon})}\right)_{P>1}\\ &\le c_1 + \frac{c_2}{\lambda^{\frac{1+\alpha}{1-\alpha}}}\end{aligned}$$

by (19). This proves the theorem.

3. The Singular Limit as $\lambda \to 0$

In this section we will discuss the asymptotic behavior of solutions as $\lambda \to 0$. In order to understand what happens consider a very classical example.

Example. Let $u \geq 0$ satisfy $\Delta u + \lambda e^u = 0$ in $B_1 \subset R^2$. By Theorem 1 of [GNN] u is radial and it is not difficult to show that all solutions are given by

$$u = -2 \log \frac{r^2 + \beta}{1 + \beta} \quad .$$

Then,

$$\Delta u = \frac{-8\beta}{(r + \beta)^2} = \frac{-8\beta}{(1 + \beta)^2} e^u \quad .$$

Thus β must satisfy $\dfrac{8\beta}{(1 + \beta)^2} = \lambda$, i.e.

$$\beta^{\pm} = \frac{4}{\lambda} - 1 \pm \sqrt{\left(\frac{4}{\lambda} - 1\right)^2 - 1}$$

and the picture is

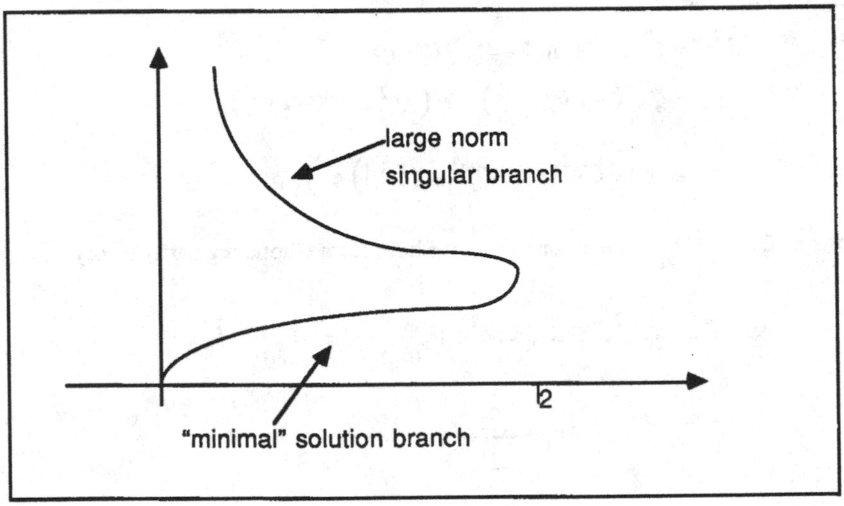

The choice $\beta_- \approx \frac{\lambda}{8}$ for small λ gives the singular branch of solutions which converge to the Green's function $-4 \log r$.

In our situation, the minimal solution is replaced by the trivial solution $u \equiv 0$ and there is a bifurcation from the trivial solution at the first Dirichlet eigenvalue $\lambda_1 = \lambda_1(D)$ of the domain of u

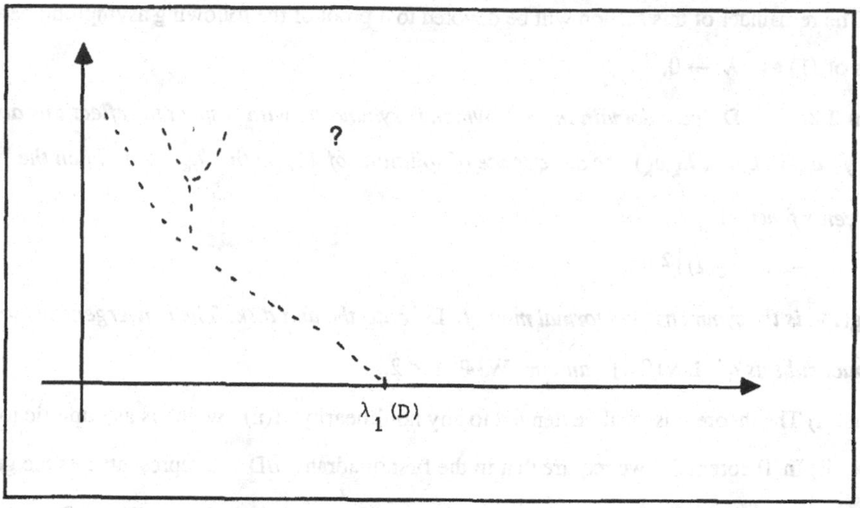

Before we proceed to the study of the singular limit of solutions as $\lambda \to 0$ we summarize the existence and uniqueness situation for a general domain $D \subset R^2$.

Theorem 3.1. *For* $\lambda > \lambda_1(D)$ *there are no nontrivial positive solutions of* (1) *in* D. *There exists a maximally connected closed branch of solutions* C *bifurcating from the trivial solution at* $\lambda = \lambda_1$. *The projection of* C *on* λ *space covers the interval* $(0, \lambda_1)$. *For* λ *near* λ_1 *there is uniqueness.*

Proof. Since $\frac{\sinh u}{u} \geq 1$ the first Dirichlet eigenvalue μ_1 of $\Delta\phi + \mu \cdot \frac{\sinh u}{u}\phi = 0$ satisfies $\mu_1 \leq \lambda_1(D)$ (with equality if and only if $u \equiv 0$) by the variational characterization of μ_1 and λ_1. Hence since $\phi = u$, $\mu = \lambda$ there is no solution of (1) if $\lambda > \lambda_1$.

The existence of C near $\lambda = \lambda_1$ ($\lambda < \lambda_1$) is well known (see [CR]). The global existence of C follows from our apriori bound Theorem 2.1 and a theorem of Rabinowitz [R].

Finally, the asserted uniqueness of solutions follows from known results once we show that for λ near λ_1 all solutions of (1) are close to the trivial solution. This again follows from Theorem 2.1. For if (λ_k, u_k) are a sequence of solutions of (1) then by Theorem 2.1 the u_k are uniformly bounded so by elliptic regularity theory we can choose a subsequence which we still call u_k which converge uniformly to a solution of $\Delta u + \lambda_1 \sinh u = 0$. But we have seen that $u \equiv 0$ is the unique solution for $\lambda = \lambda_1$. It follows that the u_k cannot be bounded away from zero on any compact subset of D, and the uniqueness follows.

The remainder of this section will be devoted to a proof of the following asymptotic behavior of solutions of (1) as $\lambda \to 0$.

Theorem 3.2. *Let* D *be a domain in* R^2 *which is symmetric with respect to reflections about the* x *and* y *axes. Let* (λ_k, u_k) *be a sequence of solutions of* (1) *with* $\lambda_k \to 0$. *Then the* u_k *tend to the Green's function*

$$-2 \log |g(z)|^2$$

where g(z) *is the symmetric conformal map of* D *onto the unit disk. The convergence is uniform on compact subsets of* $D \setminus \{0,0\}$ *and in* $W^{1,p}$ $p < 2$.

Remarks. i) The theorem is easily extended to any nonlinearity $f(u)$ which is asymptotic to ce^u as $u \to +\infty$. ii) In Theorem 3.2 we require that in the first quadrant ∂D be represented as the graph $y = g(x)$ of a monotone decreasing function $g(x)$. This forces any solution of (1) in D to have the required symmetry properties.

For simplicity in presentation, we present the proof for the case $D = R$, a rectangle. The proof proceeds in several steps.

Step 1. Let (λ, u) be a solution of (1) in R. Then u is uniformly bounded (independent of λ!) on compact subsets of $R \setminus \{0,0\}$.

We use the estimate $\int_R \lambda \sinh u \, dx\, dy \le C$ where C is independent of λ. To see this, we write equation (1) in the weak form

$$\int_R \nabla u \cdot \nabla \zeta dx\, dy = \lambda \int_R \zeta f(u) dx\, dy \qquad f(u) = \sinh u$$

and set $\zeta = 1 - e^{-\alpha u}$. Then

$$\int_R \lambda f(u) dx\, dy = \lambda \int_R e^{-\alpha u} f(u) dx\, dy + \alpha \int_R e^{-\alpha u} |\nabla u|^2 dx\, dy \ . \tag{20}$$

Using the symmetry of u, the right hand side of (20) is dominated by

$$C \left\{ \lambda \int_{R_\varepsilon} e^{-\alpha u} f(u) dx\, dy + \alpha \int_{R_\varepsilon} e^{-\alpha u} |\nabla u|^2 dx\, dy \right\}$$

where (as in section 2) R_ε is the subset of R whose points are at least distance ε from R.

Hence from estimate (17) we have shown

$$\int_R |\Delta u| dx\, dy = \lambda \int_R \sinh u \, dx\, dy \le C$$

for a uniform constant C.

To show the uniform boundedness of u at points (x,y) away from the origin consider a point (x_0,y_0) as in the diagram

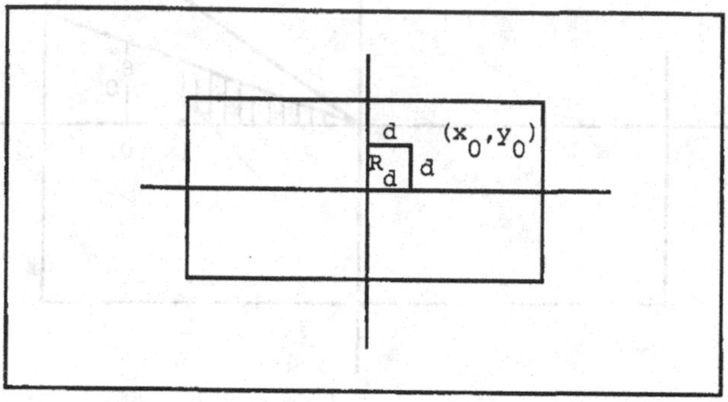

Then by the symmetry properties of u (i.e. $u(x,y) = u(-x,y) = u(x,-y)$, and $u_x \le 0$, $u_y \le 0$ in $\{x \ge 0, y \ge 0\} \cap R$) $u(x,y) \ge u(x_0,y_0)$ for all points (x,y) in R_d. Hence

$$C \ge \int_R \lambda \sinh u \, dx\, dy \ge \int_{R_d} \lambda \sinh u \, dx\, dy \ge |R_d| \lambda \sinh u(x_0,y_0)$$

so that

$$|\Delta u\,(x_0,y_0)| \le \frac{c}{|R_d|} = \frac{c}{d^2} \,. \tag{21}$$

Assuming we have a bound for $|\Delta u|$ on compact subsets of $R \setminus \{0,0\}$, then since we already know $\Delta u \in L^1(R)$, it follows by standard elliptic estimates that $u \in L^p(R)$ for all p and the uniform bound for sup u away from (0,0) follows from (21) (see [GT]).

Of course, the worst case occurs when (x_0,y_0) is on the x or y axes. Since the argument in both cases is the same, assume (x_0,y_0) lies on the positive x axis. We will show using a modification of the method of moving planes that $u(x,y) \ge u(x_0,0)$ for all points (x,y) in the triangle $T_d : d = x_0 > 0$

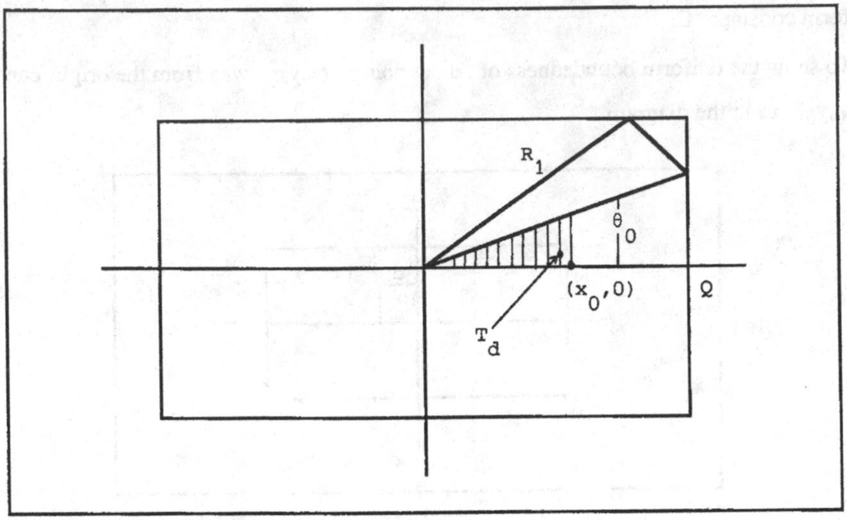

The area of T_d is $d^2/4 \sin 2\theta_0$. The angle θ_0 is determined as follows: Let R_1 be the quarter of R in the first quadrant and consider lines through the origin of slope $\tan\theta$, $\theta > 0$. These lines cut off a "triangular cap" from R_1 (see diagram) and for θ small, the reflection of this triangle about the line is contained in R_1 while for θ too large the reflection leaves R_1. Then θ_0 is the sup of all θ with the good reflection property. Clearly θ_0 depends only on R.

To prove that $u(x,y) \geq u(x_0,0)$ for all (x,y) in T_d we consider the family of "planes" $P_\lambda : y = (\tan\theta_0)x - \lambda$, when θ_0 is defined above and use the method of moving planes for this family with respect to the domain R_1. Note that P_0 cuts off the maximal triangular cap Σ_0 by our previous discussion. What makes our argument nonstandard is the fact that we do not have the boundary condition $u = 0$ on the bottom boundary of R_1; instead we use that on this boundary $u_y = 0$, $u_x \leq 0$ with $u_x < 0$ if $x > 0$.

Let $n = (-\cos\theta_0, \sin\theta_0)$ denote the normal directions to the planes P_λ and let λ_0 be the value of λ so that P_λ passes through Q (see diagram). We will show that for $0 < \lambda < \lambda_0$, $\partial_n u = -\cos\theta_0 u_x + \sin\theta_0 u_y > 0$ (note that n points into R_1 at ∂R_1) and $u(X) < u(X^\lambda)$ for $X \in \Sigma_\lambda$ (here $X = (x,y)$ and Σ_λ is the triangular cap cut off by P_λ from R_1).

To get started we observe that for points on R_1 near Q where $u = 0$ then $\partial_n u > 0$ by the Hopf lemma since $\Delta u \leq 0$. On the other hand, for points on ∂R_1 near Q on the x axis,

$$\partial_n u = -\cos\theta_0 u_x + \sin\theta u_y = -\cos\theta_0 u_x > 0 \quad .$$

It follows that for λ close to λ_0, $\lambda < \lambda_0$ we have

(22) $$\partial_n u > 0,\ u(X) < u\left(X^\lambda\right) \quad X \in \Sigma_\lambda .$$

We decrease λ until a critical value $\mu \geq 0$ is reached beyond which (22) no longer holds. We show that $\mu = 0$. Suppose $\mu > 0$. Then as in "the old argument" we may apply lemma 2.2 of [GNN] to conclude that

$$u(X) < u(X^\mu) \text{ in } \Sigma_\mu \text{ and } \partial_n u > 0 \text{ on } R_1 \cap P_\mu .$$

Thus (22) holds for $\lambda = \mu$ and for some $\varepsilon > 0$

(23) $$\partial_n u > 0 \text{ in } \Sigma_{\mu-\varepsilon} .$$

From the definition of μ there is a sequence λ^j, $0 < \lambda^j \uparrow \mu$ and points X_j in Σ_{λ^j} such that

$$u(X_j) \geq u\left(X_j^{\lambda^j}\right) .$$

A subsequence of the X_j which we still call X_j converges to a point X_0 in Σ_μ; then

$$X_j^{\lambda_j} \to X_0^\mu \text{ and } u(X_0) \geq u\left(X_0^\mu\right) .$$

Since (22) holds for $\lambda = \mu$ we must have $X_0 \in \partial\Sigma_\mu$. If $X_0 \in P_\mu$ we reach a contradiction using (23) as in the "old argument". Otherwise the subsequence $X_j \in \Sigma_\mu$ (for j large) and thus

(24) $$u\left(X_j^\mu\right) > u(X_j) \geq u\left(X_j^{\lambda_j}\right) .$$

It follows from (24) that

(25) $$u(X_0) = u\left(X_0^\mu\right) \text{ and } \partial_n u\left(X_0^\mu\right) = 0$$

and X_0 is a point on $\partial\Sigma_\mu \setminus P_\mu$ where $u(X_0) > 0$. Let $v(X) = u(X^\mu)$ in Σ_μ and set

$$w(X) = v(X) - u(X) \geq 0 \quad w(X) \not\equiv 0 .$$

Then as before w satisfies an elliptic equation for which the Hopf lemma applies (in fact $\Delta w \leq 0$) and $w(X_0) = u(X_0^\mu) - u(X_0) = 0$ by (25). Hence by the Hopf lemma $\partial_n w(X_0) > 0$. But $\partial_n w(X_0) = \partial_n v(X_0) - \partial_n u(X_0) = \partial_n u(X_0^\mu) - \partial_n u(X_0) = -\partial_n u(X_0) < 0$, a contradiction. The proof of Step 1 is complete.

Step 2. Let (λ_k, u_k) be a sequence of solutions of (1) in R with $\lambda_k \downarrow 0$. Then a subsequence of the u_k converges to a multiple of the Green's function of R with pole at the origin. The convergence is uniform away from $(0,0)$ (and in $W^{1,p}(R)$ $1 \le p < 2$).

To prove this we first note that the u_k cannot converge uniformly to zero (the trivial solution) for this would imply that $\lambda = 0$ is a Dirichlet eigenvalue of the Laplace operator Δ. From this we conclude that

$$\lambda_k \sinh u_k(0,0) \to +\infty \quad \text{as} \quad \lambda_k \to 0 \ .$$

For otherwise $|\Delta u_k| \le C$ in R, $u_k = 0$ on ∂R so that $u_k \to 0$ uniformly in R, a contradiction. Therefore,

(26) $$u_k(0,0) \ge \log \frac{1}{\lambda_k} + 1, \quad k \text{ large.}$$

Later in this section we will show (Proposition 3.3)

$$|\nabla u_k| \le \frac{C_\varepsilon \log \frac{1}{\lambda}}{\lambda^{1/2+\varepsilon}} + C_\varepsilon$$

where $\varepsilon > 0$.

Assuming this, we find that for $|X| = \lambda_k$ small

$$u_k(X) \ge u_k(0,0) - |X| \left(\frac{C \log \frac{1}{\lambda_k}}{\lambda_k^{1/2+\varepsilon}} \right)$$

(27) $$\ge \log \frac{1}{|X|} \ .$$

Also,

(28) $$u_k \ge \log \frac{1}{|X|} - C \quad \text{on } \partial R \ .$$

Since $\Delta u_k \le 0$ and $\log \frac{1}{|x|} - C$ is harmonic in $R \setminus B_{\lambda_k}(0,0)$, it follows from the maximum principle (using (27), (28)) that

(29) $$u_k(X) \ge \log \frac{1}{|X|} - C \quad \text{in } R \setminus B_{\lambda_k}(0,0) \ .$$

Since the u_k are uniformly bounded away from $(0,0)$ we may choose a subsequence such that

$u_k \to u$ uniformly in R away from $(0,0)$. Evidently, $u \geq 0$ in R,

$$\Delta u = 0 \text{ in } R \setminus \{0,0\}, \; u = 0 \text{ on } \partial R \text{ and } u \geq \log \frac{1}{|x|} - C \text{ in } R \setminus \{0,0\}$$

by (29).

It follows that u is a positive harmonic in R with a singularity at $(0,0)$. But classical theorems say that $u = cG$ is the only possible positive harmonic with an isolated singularity and the proof of Step 2 is complete.

Step 3. The limit u constructed in Step 2 is given by $u = -2 \log |g(z)|^2$ where $g(z)$ is the symmetric conformal map of R onto the unit disk.

We proceed as follows. Define

$$\phi(z,\bar{z}) = u_{zz} - \frac{u_z^2}{2}$$

where $z = x + iy$ and $\frac{\partial}{\partial z}, \frac{\partial}{\partial \bar{z}}$ have their usual meaning. In complex notation, if u satisfies (1), then

$$u_{z\bar{z}} + \frac{\lambda}{4} \sinh u = 0 \quad \text{in} \quad R \; .$$

Therefore,

$$\phi_{\bar{z}} = u_{zz\bar{z}} - u_z u_{z\bar{z}} = -\frac{\lambda}{4} (\cosh u - \sinh u) u_z$$

$$= -\frac{\lambda}{4} e^{-u} u_z \; .$$

Appealing once more to Proposition (3.3) below,

$$|\phi_{\bar{z}}| \leq c \, \lambda^{1/2-\varepsilon} \; .$$

It follows from standard potential theory that $\phi = \psi(z) + \eta(z,\bar{z})$, where $\psi(z)$ is analytic in R, and $|\eta| \leq c\lambda^{\frac{1}{2}-\varepsilon}$. In fact ψ is just the Cauchy integral of $\phi = u_{zz} - \frac{u_z^2}{2}$ around a simple closed curve $\Gamma \subset R$ containing $(0,0)$. In particular ψ *is uniformly bounded independent of* λ.

Applying this construction to the (λ_k, u_k) of Step 2, we obtain a sequence ψ_k of analytic functions which are uniformly bounded and form a normal family. Therefore, a subsequence of the ψ_k converge uniformly on compact subsets of R to *a regular analytic function* ψ. Evidently

(30) $$u_{zz} - \frac{u_z^2}{2} = \psi$$

where $u = \lim u_k$.

We know that $u = -a \log|g(z)|^2$. Then

(31) $$u_z = -a\frac{g'(z)}{g(z)}, \quad u_{zz} = -a\frac{g''(z)}{g(z)} + a\frac{g'(z)^2}{g(z)^2} \quad .$$

Substitution of (31) into (30) gives

$$-a\frac{g''(z)}{g(z)} + \left(a - a^2/2\right)\left(\frac{g'(z)}{g(z)}\right)^2 = \psi(z) \text{ in } R \ .$$

Observe that $\frac{g'(z)}{g(z)}$ has a simple pole at $z = 0$ while $\frac{g''(z)}{g(z)}$ is in fact regular at $z = 0$ since $g(z)$ is odd. Since ψ is regular we must have $a = a^2/2$ or $a = 2$. This completes the proof of Step 3.

We have now completed the proof of Theorem 3.2 modulo the technical estimate given by

Proposition 3.3. *Let* $u \geq 0$ *satisfy*

$$\Delta u + \lambda \sinh u = 0 \quad \text{in } D \subset R^2$$

$$u = 0 \quad \text{on } \partial D \ .$$

Then

$$|\nabla u| \leq \frac{c_1}{\lambda^{1/2+\delta}} + c_2 \quad \textit{for any } \delta > 0$$

on compact subsets of D.

Proof. Let $v = D_\gamma u$ be any directional derivative of u. Then v satisfies

$$\Delta v + dv = 0 \quad \text{in } D$$

where $d = \lambda \cosh u$.

By standard elliptic theory,

(32) $$|v|_{L^\infty(D_\varepsilon)} \leq c|dv|_{L^p(D_\varepsilon)} + \sup_{\partial D_\varepsilon}|v| \quad p > 1$$

where D_ε is chosen as in Section 2. Since u is uniformly bounded in a neighborhood of ∂D, it follows (again by regularity theory) that $\sup_{\partial D_\varepsilon}|v| \leq C$, a uniform constant.

By Holder's inequality, for $1 < p < 2$

(33) $$|dv|_p \le |d|_{\frac{2p}{2-p}} |v|_2$$

and by L_p interpolation

(34) $$|d|_q \le |d|_1^{\lambda} |d|_r^{1-\lambda} \quad \text{where} \quad \frac{1}{q} = \frac{\lambda}{1} + \frac{(1-\lambda)}{r} \ .$$

Choosing $q = \frac{2p}{2-p}$, we use the estimates

$$|d|_{L^1(D)} \le c, \quad |d|_{L^r(D)} \le \frac{c_1}{\lambda^{\frac{\alpha}{1-\alpha}}} + c_2 \ .$$

(This last estimate is just (19) from Section 2) which gives

(35) $$|d|_q \le \frac{\bar{c}_1}{\lambda^{\frac{\alpha}{1-\alpha}(1-\lambda)}} + \bar{c}_2 \ .$$

We note that $q > 2$ can be made arbitrarily close to 2 (by taking p close to 1) and that $\lambda < \frac{1}{q} < 1/2$ can be made arbitrarily close to $1/2$ by choosing r large enough. In particular, we may take $1 - \lambda = 1/2 + \delta$. Also, by taking $\alpha > \frac{1}{2}$ close to $1/2$, $\frac{\alpha}{1-\alpha} > 1$ can be made arbitrarily close to 1. Thus

(36) $$\frac{\alpha}{1-\alpha}(1-\lambda) = 1/2 + \delta \qquad \delta > 0 \ .$$

Finally,

$$|v|^2_{L^2(D)} \le \int_D |\nabla u|^2 dx\, dy = \int_D u \lambda \sinh u \, dx\, dy$$

$$\le |u|_{L^\infty(D)} \int_D \lambda \sinh u \, dx\, dy$$

$$\le c_1 \log \frac{1}{\lambda} + c_2$$

which is negligible compared to powers of $\frac{1}{\lambda}$. Thus from (32)-(36) the proposition follows.

4. The Inverse Power Method and the Numerical Computation of Solutions

In this section we will discuss an iterative procedure [ES] that is very useful for the numerical solution of nonlinear eigenvalue problems

$$Lu \equiv \partial_i(a_{ij}\partial_j u) - a(x)u = -\lambda f(x,u) \quad \text{in } D$$
$$u = 0 \qquad \text{on } \partial D$$

where D is a bounded domain in R^n and L is self-adjoint and uniformly elliptic with $a(x) \geq 0$.

The method is a generalization of the famous power method for computing the eigenvalues of a matrix.

Let $B(u,v) = \int_D (a_{ij}\partial_j u \partial_i v + a(x)uv)dx$ be the natural bilinear form associated to the operator L, where u,v belong to the Sobolev space $H_1(D)$, and set

$$F(x,u) = \int_0^u f(x,t)dt \quad .$$

Define $A_R = \{u \in H_1(D) : B(u,u) = R^2\}$. We construct a mapping $T : A_R \to A_R$ as follows: given $v \in A_R$, $u = Tv$ is defined by

$$u = \frac{Rw}{(B(w,w))^{1/2}}$$

where w is the solution of the (inverse) problem

$$\Delta w = -f(x,v) \quad \text{in } D$$
$$w = 0 \qquad \text{on } \partial D \; .$$

Assuming f is monotone non-decreasing in u, the mapping T enjoys the following variational property.

Lemma 4.1. *Let* $v \in A_R$ *and let* $u = Tv$. *Then*

$$\int_D F(v)dx \geq \int_D F(v)dx$$

with equality if and only if $u = v$ *a.e.*

Proof. Note that u satisfies

(37) $$\Delta u = -\lambda f(x,v) \quad \text{in } D$$
$$u = 0 \qquad \text{on } \partial D$$

where $\lambda = R/(B(w,w))^{1/2}$. From (37) we find

$$B(u,u) = \lambda \int uf(x,v)dx$$

(38)

$$B(u,v) = \lambda \int v f(x,v)dx \quad .$$

Since $B(u,v) \le 1/2(B(u,u) + B(v,v)) = B(u,u)$, (38) implies that $\int (u-v)f(x,v)dx \ge 0$ with equality if and only if $u = v$ a.e. Now $f(x,u)$ is monotone non-decreasing in u so that $F(x,u)$ is convex in u. Hence

(39) $$F(x,u) \ge F(x,v) + (u-v)f(x,v) \quad .$$

Integrating (39) gives

$$\int_D F(x,u)dx \ge \int_D F(x,v)dx + \int_D (u-v)f(x,u)dx$$

$$\ge \int_D F(x,v)dx$$

proving the lemma.

Given $u_0 \in A_R$ we define a discrete flow on A_R by inductively defining

$$u_{j+1} = Tu_j \quad .$$

Theorem 4.2. *Let* $f(x,u)$ *be uniformly Lipschitz in* x *and locally uniformly Lipschitz in* u *and satisfy Sobolev growth conditions*

(40)
$$|f(x,u)| \le c_1 + c_2 |u|^p \quad 0 \le p < \frac{n+2}{n-2} \quad n > 2$$

$$\le c_1 + c_2 e^{|u|^p} \quad 0 \le p < 2 \quad n = 2 \quad .$$

Assume f is nondecreasing in u. Given any $u_0 \in A_R$ satisfying $u_0 > 0$, the iteration scheme

$$u_{j+1} = Tu_j$$

has the property that any subsequence has a subsequence converging uniformly to a positive solution of $Lu = -\lambda f(x,u)$.

Proof. Since $u_0 > 0$ in D, it follows that $u_j > 0$ in D. Note that from (38)

$$\lambda_{j+1} = \frac{B(u_{j+1},u_j)}{\int u_j f(u_j)} \le R^2 / \int u_j f(u_j) \quad .$$

by convexity of F in u,

$$\int u_j f(u_j) \geq \int F(u_j) - \int F(u_0) + \int u_0 f(u_j)$$

(41)

$$\geq \int F(u_j) - \int F(u_0) \quad .$$

Using Lemma 4.1, $\int F(u_j) \geq \int F(u_0) = \varepsilon > 0$ so that (41) gives

$$0 < \lambda_{j+1} \leq R^2/\varepsilon \quad .$$

Since $u_j \in A_R$ and $\{\lambda_j\}$ bounded, it follows from the Sobolev growth conditions (40) that

$$\int_D F(x,u_j) \leq M$$

for a uniform constant M (see [GT]).

We choose subsequences $\{u_{j_k+1}\}$ $\{u_{j_k}\}$ of the $\{u_j\}$ converging weakly in H_1 to u and v respectively, and such that $\lambda_{j_k+1} \to \lambda$. Then u and $v \in A_R$ are related by

$$u = Tv \quad .$$

But since $\int F(x,u_j)$ was monotone increasing (Lemma 4.1) and uniformly bounded, the limit exists and

$$\int_D F(x,u) = \int_D F(x,v) \quad .$$

Appealing once more to Lemma 4.1 $u = v$ a.e. and u is a solution. □

Theorem 4.2 does not give any convergence rates. Indeed, unless f is linear in u such rates probabily do not exist because there is nonuniqueness in general. However one can easily obtain the estimate

$$|u_{k+1} - u_k|^2_{H_1} \leq C\left(\int_D F(u_{k+1}) - \int_D F(u_k)\right)$$

which gives a reasonable aposteriori criterion.

5. Explicit Representation of Solutions

The numerical algorithm of the last section can be used to generate computer pictures of the surfaces of constant mean curvature obtained by the procedure described in Section 1. Namely, we compute ω by the algorithm of Section 4 and use this to numerically integrate the over-determined system (10). By adjusting the parameters of the problem one can experimentally find good choices of the parameters so that the resulting surface closes up.

Looking at these computer generated pictures U. Abresch [Ab] observed that the curvature lines for the smaller principal curvature k_1 (these are the image of the horizontal lines $y = \text{constant}$) appeared to be planar curves (figure eights). Geometrically, this means that the torsion τ of the curves

$$F(x,\bullet)$$

should vanish. Using equations (10) we will derive an analytic condition on ω which is equivalent to the vanishing of this torsion.

Lemma 5.1. *The torsion* $= \dfrac{F_x \wedge F_{xx} \bullet F_{xxx}}{|F_x \wedge F_{xx}|^2}$

of the horizontal curvature lines $F(x,\bullet)$ *vanishes if and only if* ω *satisfies*

(42) $$\sinh\omega\ \omega_{xy} - \cosh\omega\ \omega_{xy} = 0 \quad \textit{in}\ \ \mathbb{R}\,.$$

Proof. We use (10) to compute $F_x \wedge F_{xx} \bullet \xi$. From (1) (recalling $|F_x| = |F_y| = e^{\omega}$, $F_x \wedge F_y = e^{2\omega}\xi$, $F_x \bullet F_y = 0$).

$$\begin{aligned} F_x \wedge F_{xx} &= -\omega_y e^{2\omega}\xi - e^{\omega}\sinh\omega\, F_y \\ F_{xxx} &= \omega_x\left(\omega_x F_x - \omega_y F_y + e^{\omega}\sinh\omega\,\xi\right) + \omega_{xx}F_x \\ &\quad - \omega_{xy}F_y - \omega_y\left(\omega_y F_x + \omega_x F_y\right) \\ &\quad - e^{\omega}\sinh\omega\left(e^{-\omega}\sinh\omega\, F_x\right) + \left(e^{\omega}\sinh\omega\right)_x\xi\ . \end{aligned}$$

Therefore,

$$\begin{aligned} F_x \wedge F_{xx} \bullet F_{xxx} &= -\omega_y e^{2\omega}\left(e^{\omega}\sinh\omega + \left(e^{\omega}\sinh\omega\right)_x\right)\omega_x \\ &\quad - e^{\omega}\sinh\omega\, e^{2\omega}\left(-2\omega_x\omega_y - \omega_{xy}\right) \\ &= e^{2\omega}\left(\sinh\omega\ \omega_{xy} - \cosh\omega\ \omega_x\omega_y\right)\ . \end{aligned}$$

Since $|F_x \wedge F_y|$ never vanishes the lemma follows.

With this geometric motivation, Abresch [Ab] now considered the overdetermined system

(43) $$\Delta\,\omega + \sinh\omega\cosh\omega = 0 \quad \text{in}\ \ \mathbb{R}$$

$$\sinh \omega\ \omega_{xy} - \cosh \omega\ \omega_x\, \omega_y = 0 .$$

It is remarkable that the additional equation (42) induces a separation of variables and reduces the system (43) to solving ordinary differential equations.

Theorem 5.2. (Abresch [Ab])

i) *There exists a two-parameter family of real analytic solutions* ω *of the system* (43) *which is defined on*

$$P = \left\{ (\alpha,\beta) \in \mathbf{R}^2 : \alpha,\beta \geq 0, \alpha + \beta \geq 1 \right\}$$

by means of the equations:

$$\cosh \omega = \left(1 + f^2 + g^2\right)^{-1} (f'(x) + g'(y)),\ \ \omega(0,0) \geq 0 \tag{44}$$

$$\begin{aligned} \omega_x &= -f(x)\sinh \omega \\ \omega_y &= -g(y)\cosh \omega \end{aligned} \tag{45}$$

$$\begin{aligned} f'^2 &= f^4 + \left(1 + \alpha^2 - \beta^2\right) f^2 + \alpha^2, \quad f(0) = 0 \quad f'(0) = \alpha \\ g'^2 &= g^4 + \left(1 - \alpha^2 + \beta^2\right) g^2 + \beta^2 \quad g(0) = 0 \quad g'(0) = \beta \ . \end{aligned} \tag{46}$$

ii) *Conversely, if* ω *is any real analytic solution of system* (43), *then* ω *is globally bounded,* $|\omega|$ *achieves its maximum, and up to a translation,* ω *or* $-\omega$ *is contained in the above family.*

Theorem 5.2 describes all solutions of the system which are globally defined in R^2. It is easy to check that the trivial solution $\omega \equiv 0$ corresponds to the whole segment $\alpha + \beta = 1$, $\alpha, \beta \geq 0$ on ∂P. Also, the rays $\alpha = 0$, $\beta \geq 1$ and $\beta = 0$, $\alpha \geq 1$ parametrize solutions ω which depend only on either x or y.

The particular subfamily of interest to us is the family of Dirichlet solutions, which satisfy the additional constraint $\omega > 0$ on R, $\omega = 0$ on ∂R for some rectangle R with sides parallel to the coordinate axes.

Assume $\alpha, \beta > 0$, $\alpha + \beta > 1$. The qualitative behavior of f depends on whether or not the quartic $f^4 + (1 + \alpha^2 - \beta^2)f^2 + \alpha^2$ has real zeroes. Since $\alpha \neq 0$ we can define $a = a(\alpha, \beta)$ to be the smallest positive zero of f (possibly $a = +\infty$). When $\beta < \alpha + 1$ $a(\alpha, \beta)$ is a full period of f (which resembles the function tan) while if $\beta > \alpha + 1$ then $a(\alpha, \beta)$ is a half period of f (which resembles the function sin). On the borderline $\beta = \alpha + 1$, f has no positive zero and $a = +\infty$.

The Dirichlet solutions are characterized by

Proposition 5.3 (Abresch [Ab]). *Suppose* $\alpha, \beta > 0$, $\alpha + \beta > 1$.

i) *If* $|\alpha - \beta| < 1$ *then* the connected components of $\{(x,y) \mid \omega(x,y) \neq 0\}$ *(where* ω *is given by* (44) (45) *are rectangles; in fact,* $\omega > 0$ *on* ∂R.

ii) *The strip* $P^D = \{(\alpha, \beta) \in P : |\alpha - \beta| < 1\}$ *around the diagonal in the parameter space is mapped diffeomorphically onto the set*

$$R^D = \left\{ (a_0, b_0) \in R_+^2 : a_0^{-2} + b_0^{-2} \geq \pi^{-2} \right\} .$$

This set contains the edge lengths of all rectangles R_{a_0,b_0} *which can support Dirichlet solutions of the system* (43).

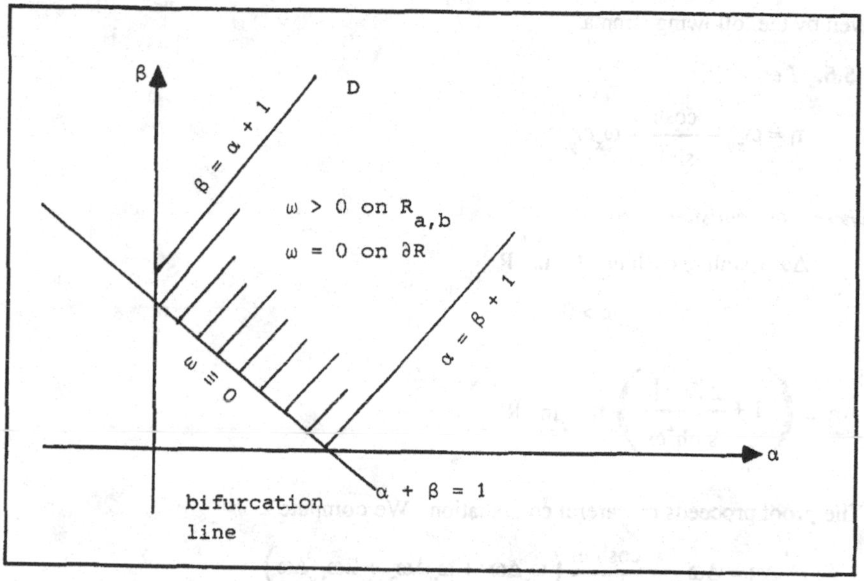

We shall not present the proofs of these results. (The interested reader can refer to [Ab] or carry out the analysis which is not so difficult).

A natural question now arises: are there solutions of the Dirichlet problem

(47)
$$\Delta\omega + \sinh\omega\cosh\omega = 0 \quad \text{in } R$$
$$\omega = 0 \quad \text{on } \partial R$$
$$\omega > 0 \quad \text{in } R$$

(with $R = (-a/2, a/2) \times (-b/2, b/2)$)

which do not arise as members of the family P^D? This is equivalent to asking whether the extra condition (42) in fact holds for all solutions of (47). This is indeed the case!

Theorem 5.4. *Let* ω *be a solution of* (47). *Then* ω *satisfies*

$$\sinh \omega \, \omega_{xy} - \cosh \omega \, \omega_x \omega_y = 0 \quad \text{in } R .$$

An immediate consequence of Theorem 5.4 is that solutions of the nonlinear eigenvalue problem (1) are unique and representable in terms of elliptic functions.

Corollary 5.5. *The solutions of* (1) *are unique, and*

$$\omega = 1/2u \left(\frac{x}{\sqrt{\lambda}}, \frac{y}{\sqrt{\lambda}} \right)$$

is represented by elliptic functions given by Proposition 5.3.

The proof of Theorem 5.5 is based on a surprising use of the maximum principle. The main step is given by the following lemma.

Lemma 5.6. *Let*

$$\eta = \omega_{xy} - \frac{\cosh \omega}{\sinh \omega} \omega_x \omega_y$$

in R *where* ω *satisfies*

$$\Delta\omega + \sinh \omega \cosh \omega = 0 \quad \text{in } R$$
$$\omega > 0 .$$

Then $$\Delta\eta = \left(1 + \frac{2 |\nabla\omega|^2}{\sinh^2 \omega} \right) \eta \quad \text{in } R .$$

Proof. The proof proceeds by careful computation. We compute

$$\Delta\eta = \Delta\omega_{xy} - \frac{\cosh \omega}{\sinh \omega} \left(\omega_x \Delta\omega_y + \omega_y \Delta\omega_x + 2\omega_{xy} \Delta\omega \right) \tag{48}$$
$$- \omega_x \omega_y \left(- \frac{1}{\sinh^2 \omega} \Delta\omega + 2 \frac{\cosh \omega}{\sinh^3 \omega} |\nabla\omega|^2 \right)$$
$$+ \frac{2}{\sinh^2 \omega} \left(\omega_x \omega_y \Delta\omega + |\nabla\omega|^2 \omega_{xy} \right) .$$

To simplify (48), we use

$$\Delta\omega_x + \left(\sinh^2\omega + \cosh^2\omega\right)\omega_x = 0$$
$$\Delta\omega_y + \left(\sinh^2\omega + \cosh^2\omega\right)\omega_y = 0$$

$$\Delta\omega_{xy} + \left(\sinh^2\omega + \cosh^2\omega\right)\omega_{xy} + 4\sinh\omega\cosh\omega\,\omega_x\omega_y = 0\ .$$

Then

$$\Delta\eta = -\left(\sinh^2\omega + \cosh^2\omega\right)\omega_{xy} - 4\sinh\omega\cosh\omega\,\omega_x\omega_y$$

$$-\frac{\cosh\omega}{\sinh\omega}\left(-2\left(\sinh^2\omega + \cosh^2\omega\right)\omega_x\omega_y - 2\sinh\omega\cosh\omega\,\omega_{xy}\right)$$

$$-\,\omega_x\omega_y\left(3\frac{\sinh\omega\cosh\omega}{\sinh^2\omega} + 2\frac{\cosh\omega}{\sinh^3\omega}|\nabla\omega|^2\right)\ .$$

Simplifying further, using $\cosh^2\omega = 1 + \sinh^2\omega$,

$$\Delta\eta = \left(1 + 2\frac{|\nabla\omega|^2}{\sinh^2\omega}\right)\omega_{xy}$$

$$+\left(-4\sinh\omega\cosh\omega + 2\frac{\cosh\omega}{\sinh\omega}\left(2\sinh^2\omega + 1\right) - 3\frac{\cosh\omega}{\sinh\omega} - 2\frac{\cosh\omega}{\sinh^3\omega}|\nabla\omega|^2\right)\omega_x\omega_y$$

$$= \left(1 + 2\frac{|\nabla\omega|^2}{\sinh^2\omega}\right)\omega_{xy} - \frac{\cosh\omega}{\sinh\omega}\left(1 + 2\frac{|\nabla\omega|^2}{\sinh^2\omega}\right)\omega_x\omega_y$$

$$= \left(1 + 2\frac{|\nabla\omega|^2}{\sinh^2\omega}\right)\eta$$

proving the lemma.

Lemma 5.6 is the key to the proof of Theorem 5.4. All that is required is

Lemma 5.7. *The function*

$$\eta = \omega_{xy} - \frac{\cosh\omega}{\sinh\omega}\omega_x\omega_y$$

of Lemma 5.6 is continuous in $\overline{R}$ *and* $\eta = 0$ *on* ∂R.

Proof. Since solutions of (47) extend to all of R^2 by odd reflection across ∂R, ω is real analytic in R^2 and we can expand ω in a power series centered at a point $P = (x_0, y_0)$ on ∂R. The proof breaks down into two cases.

Case i) P is *not* a corner point of ∂R. Then $|\nabla\omega| \neq 0$ at P by the Hopf boundary point lemma. It suffices (by symmetry) to assume that P is an interior point of the top side of R. Then $\omega_y(P) \neq 0$ and

$$\omega_x = (y - y_0)\,\omega_{xy}(P) + (y - y_0)O(r)$$

$$\sinh\omega = (y - y_0)\,\omega_y(P) + (y - y_0)O(r)$$

where $r^2 = (x - x_0)^2 + (y - y_0)^2$. Hence,

$$\lim_{\substack{(x,y)\to P \\ (x,y)\in R}} \frac{\omega_x}{\sinh\omega} = \frac{\omega_{xy}(P)}{\omega_y(P)}$$

and so

$$\lim_{(x,y)\to P} \eta(x,y) = 0 \quad .$$

Case ii) P is a corner point of ∂R. Now $|\nabla\omega(P)| = 0$ so we must make use of the subtler form of the Hopf lemma, Lemma S of [GNN]. This implies that $\omega_{xy}(P) \neq 0$. Then

$$\omega_x = (y - y_0)\,\omega_x(P) + (y - y_0)O(r)$$

$$\omega_y = (x - x_0)\,\omega_{xy}(P) + (x - x_0)O(r)$$

$$\sinh\omega = (x - x_0)(y - y_0)\omega_{xy}(P) + (x - x_0)(y - y_0)O(r)$$

and so,

$$\lim_{(x,y)\to P} \frac{\omega_x\omega_y}{\sinh\omega} = \omega_{xy}(P) \quad .$$

Therefore $\lim_{(x,y)\to P} \eta = 0$ proving the lemma.

Theorem 5.4 now follows directly from Lemmas 5.6 and 5.7 and the Hopf maximum principle. Namely, η cannot have a positive interior maximum or a negative interior minimum. Since $\eta = 0$ on ∂R it follows that $\eta \equiv 0$ in R or equivalently,

$$\sinh\omega\;\omega_{xy} - \cosh\omega\;\omega_x\omega_y = 0 \quad .$$

References

[Ab] **U. Abresh,** Constant mean curvature tori in terms of elliptic functions, preprint.

[A] **A.D. Alexandrov,** Uniqueness theorems for surfaces in the large, I. Vestnik Leningrad Univ. 11, No. 19 (1956), 5-17 (in Russian). A.M.S. transl. (Series 2) 21, 412-416.

[CR] **M. Crandall and P. Rabinowitz,** Bifurcation from simple eigenvalues, J. Funct. Anal. 8 (1971), 321-340.

[E] **L.P. Eisenhart,** A treatise on the differential geometry of curves and surfaces, Dover Reprint 1960.

[ES] **A. Eydeland and J. Spruck,** The inverse power method for semilinear elliptic equations, to appear.

[GNN] **B. Gidas, Wei-Ming Ni and L. Nirenberg,** Symmetry and related properties via the maximum principle, Comm. in Math. Physics 68 (1979), 209-243.

[GT] **D. Gilbarg and N.S. Trudinger,** Elliptic partial differential equations of second order, Springer-Verlag 1977.

[H] **H. Hopf,** Lectures on differential geometry in the large, Stanford Lecture Notes 1955; reprinted in Lecture Notes in Math. 1000, Springer-Verlag 1984.

[R] **P. Rabinowitz,** A global theorem for nonlinear eigenvalue problems and applications. Contributions to Nonlinear Functional Analysis, Academic Press (1971), 11-36.

[S] **J. Serrin,** A symmetry problem in potential theory, Arch. Rat. Mech. 43 (1971), 304-318.

[W] **H. Wente,** Counterexample to a conjecture of H. Hopf, Pacific J. 121 (1986), 193-243.

LECTURE NOTES IN MATHEMATICS

Edited by A. Dold and B. Eckmann

Some general remarks on the publication of proceedings of congresses and symposia

Lecture Notes aim to report new developments - quickly, informally and at a high level. The following describes criteria and procedures which apply to proceedings volumes. The editors of a volume are strongly advised to inform contributors about these points at an early stage.

§1. One (or more) expert participant(s) of the meeting should act as the responsible editor(s) of the proceedings. They select the papers which are suitable (cf. §§ 2, 3) for inclusion in the proceedings, and have them individually refereed (as for a journal). It should not be assumed that the published proceedings must reflect conference events faithfully and in their entirety. Contributions to the meeting which are not included in the proceedings can be listed by title. The series editors will normally not interfere with the editing of a particular proceedings volume - except in fairly obvious cases, or on technical matters, such as described in §§ 2, 3. The names of the responsible editors appear on the title page of the volume.

§2. The proceedings should be reasonably homogeneous (concerned with a limited area). For instance, the proceedings of a congress on "Analysis" or "Mathematics in Wonderland" would normally not be sufficiently homogeneous.

One or two longer survey articles on recent developments in the field are often very useful additions to such proceedings - even if they do not correspond to actual lectures at the congress. An extensive introduction on the subject of the congress would be desirable.

§3. The contributions should be of a high mathematical standard and of current interest. Research articles should present new material and not duplicate other papers already published or due to be published. They should contain sufficient information and motivation and they should present proofs, or at least outlines of such, in sufficient detail to enable an expert to complete them. Thus resumes and mere announcements of papers appearing elsewhere cannot be included, although more detailed versions of a contribution may well be published in other places later.

Surveys, if included, should cover a sufficiently broad topic, and should in general not simply review the author's own recent research. In the case of surveys, exceptionally, proofs of results may not be necessary.

"Mathematical Reviews" and "Zentralblatt für Mathematik" require that papers in proceedings volumes carry an explicit statement that they are in final form and that no similar paper has been or is being submitted elsewhere, if these papers are to be considered for a review. Normally, papers that satisfy the criteria of the Lecture Notes in Mathematics series also satisfy this

.../...

requirement, but we would strongly recommend that the contributing authors be asked to give this guarantee explicitly at the beginning or end of their paper. There will occasionally be cases where this does not apply but where, for special reasons, the paper is still acceptable for LNM.

§4. Proceedings should appear soon after the meeeting. The publisher should, therefore, receive the complete manuscript within nine months of the date of the meeting at the latest.

§5. Plans or proposals for proceedings volumes should be sent to one of the editors of the series or to Springer-Verlag Heidelberg. They should give sufficient information on the conference or symposium, and on the proposed proceedings. In particular, they should contain a list of the expected contributions with their prospective length. Abstracts or early versions (drafts) of some of the contributions are very helpful.

§6. Lecture Notes are printed by photo-offset from camera-ready typed copy provided by the editors. For this purpose Springer-Verlag provides editors with technical instructions for the preparation of manuscripts and these should be distributed to all contributing authors. Springer-Verlag can also, on request, supply stationery on which the prescribed typing area is outlined. Some homogeneity in the presentation of the contributions is desirable.

Careful preparation of manuscripts will help keep production time short and ensure a satisfactory appearance of the finished book. The actual production of a Lecture Notes volume normally takes 6 -8 weeks.

Manuscripts should be at least 100 pages long. The final version should include a table of contents and as far as applicable a subject index.

§7. Editors receive a total of 50 free copies of their volume for distribution to the contributing authors, but no royalties. (Unfortunately, no reprints of individual contributions can be supplied.) They are entitled to purchase further copies of their book for their personal use at a discount of 33.3 %, other Springer mathematics books at a discount of 20 % directly from Springer-Verlag. Contributing authors may purchase the volume in which their article appears at a discount of 33.3 %.

Commitment to publish is made by letter of intent rather than by signing a formal contract. Springer-Verlag secures the copyright for each volume.

LECTURE NOTES

ESSENTIALS FOR THE PREPARATION OF CAMERA-READY MANUSCRIPTS

The preparation of manuscripts which are to be reproduced by photo-offset requires special care. Manuscripts which are submitted in technically unsuitable form will be returned to the author for retyping. There is normally no possibility of carrying out further corrections after a manuscript is given to production. Hence it is crucial that the following instructions be adhered to closely. If in doubt, please send us 1 - 2 sample pages for examination.

Typing area. On request, Springer-Verlag will supply special paper with the typing area outlined.

The CORRECT TYPING AREA is 18 x 26 1/2 cm (7,5 x 11 inches).

Make sure the TYPING AREA IS COMPLETELY FILLED. Set the margins so that they precisely match the outline and type right from the top to the bottom line. (Note that the page-number will lie outside this area). Lines of text should not end more than three spaces inside or outside the right margin (see example on page 4).

Type on one side of the paper only.

Type. Use an electric typewriter if at all possible. CLEAN THE TYPE before use and always use a BLACK ribbon (a carbon ribbon is best).

Choose a type size large enough to stand reduction to 75%.

Word Processors. Authors using word-processing or computer-typesetting facilities should follow these instructions with obvious modifications. Please note with respect to your printout that
i) the characters should be sharp and sufficiently black;
ii) if the size of your characters is significantly larger or smaller than normal typescript characters, you should adapt the length and breadth of the text area proportionally keeping the proportions 1:0.68.
iii) it is not necessary to use Springer's special typing paper. Any white paper of reasonable quality is acceptable.
IF IN DOUBT, PLEASE SEND US 1-2 SAMPLE PAGES FOR EXAMINATION. We will be glad to give advice.

Spacing and Headings (Monographs). Use ONE-AND-A-HALF line spacing in the text. Please leave sufficient space for the title to stand out clearly and do NOT use a new page for the beginning of subdivisions of chapters. Leave THREE LINES blank above and TWO below headings of such subdivisions.

Spacing and Headings (Proceedings). Use ONE-AND-A-HALF line spacing in the text. Start each paper on a NEW PAGE and leave sufficient space for the title to stand out clearly. However, do NOT use a new page for the beginning of subdivisions of a paper. Leave THREE LINES blank above and TWO below headings of such subdivisions. Make sure headings of equal importance are in the same form.

The first page of each contribution should be prepared in the same way. Therefore, we recommend that the editor prepares a sample page and passes it on to the authors together with these ESSENTIALS. Please take

.../...

the following as an example.

MATHEMATICAL STRUCTURE IN QUANTUM FIELD THEORY

John E. Robert
Fachbereich Physik, Universität Osnabrück
Postfach 44 69, D-4500 Osnabrück

Please leave THREE LINES blank below heading and address of the author. THEN START THE ACTUAL TEXT OF YOUR CONTRIBUTION.

Footnotes. These should be avoided. If they cannot be avoided, place them at the foot of the page, separated from the text by a line 4 cm long, and type them in SINGLE LINE SPACING to finish exactly on the outline.

Symbols. Anything which cannot be typed may be entered by hand in BLACK AND ONLY BLACK ink. (A fine-tipped rapidograph is suitable for this purpose; a good black ball-point will do, but a pencil will not). Do not draw straight lines by hand without a ruler (not even in fractions).

Equations and Computer Programs. Equations and computer programs should begin four spaces inside the left margin. Should the equations be numbered, then each number should be in brackets at the right-hand edge of the typing area.

Pagination. Number pages in the upper right-hand corner in LIGHT BLUE OR GREEN PENCIL ONLY. The final page numbers will be inserted by the printer.

There should normally be NO BLANK PAGES in the manuscript (between chapters or between contributions) unless the book is divided into Part A, Part B for example, which should then begin on a right-hand page.

It is much safer to number pages AFTER the text has been typed and corrected. Page 1 (Arabic) should be THE FIRST PAGE OF THE ACTUAL TEXT. The Roman pagination (table of contents, preface, abstract, acknowledgements, brief introductions, etc.) will be done by Springer-Verlag.

Corrections. When corrections have to be made, cut the new text to fit and PASTE it over the old. White correction fluid may also be used.

Never make corrections or insertions in the text by hand.

If the typescript has to be marked for any reason, e.g. for TEMPORARY page numbers or to mark corrections for the typist, this can be done VERY FAINTLY with BLUE or GREEN PENCIL but NO OTHER COLOR: these colors do not appear after reproduction.

Table of Contents. It is advisable to type the table of contents later, copying the titles from the text and inserting page numbers.

Literature References. These should be placed at the end of each paper or chapter, or at the end of the work, as desired. Type them with single line spacing and start each reference on a new line. Please ensure that all references are COMPLETE and PRECISE.

Vol. 1173: H. Delfs, M. Knebusch, Locally Semialgebraic Spaces. XVI, 329 pages. 1985.

Vol. 1174: Categories in Continuum Physics, Buffalo 1982. Seminar. Edited by F.W. Lawvere and S.H. Schanuel. V, 126 pages. 1986.

Vol. 1175: K. Mathiak, Valuations of Skew Fields and Projective Hjelmslev Spaces. VII, 116 pages. 1986.

Vol. 1176: R.R. Bruner, J.P. May, J.E. McClure, M. Steinberger, H_∞ Ring Spectra and their Applications. VII, 388 pages. 1986.

Vol. 1177: Representation Theory I. Finite Dimensional Algebras. Proceedings, 1984. Edited by V. Dlab, P. Gabriel and G. Michler. XV, 340 pages. 1986.

Vol. 1178: Representation Theory II. Groups and Orders. Proceedings, 1984. Edited by V. Dlab, P. Gabriel and G. Michler. XV, 370 pages. 1986.

Vol. 1179: Shi J.-Y. The Kazhdan-Lusztig Cells in Certain Affine Weyl Groups. X, 307 pages. 1986.

Vol. 1180: R. Carmona, H. Kesten, J.B. Walsh, École d'Été de Probabilités de Saint-Flour XIV – 1984. Édité par P.L. Hennequin. X, 438 pages. 1986.

Vol. 1181: Buildings and the Geometry of Diagrams, Como 1984. Seminar. Edited by L. Rosati. VII, 277 pages. 1986.

Vol. 1182: S. Shelah, Around Classification Theory of Models. VII, 279 pages. 1986.

Vol. 1183: Algebra, Algebraic Topology and their Interactions. Proceedings, 1983. Edited by J.-E. Roos. XI, 396 pages. 1986.

Vol. 1184: W. Arendt, A. Grabosch, G. Greiner, U. Groh, H.P. Lotz, U. Moustakas, R. Nagel, F. Neubrander, U. Schlotterbeck, One-parameter Semigroups of Positive Operators. Edited by R. Nagel. X, 460 pages. 1986.

Vol. 1185: Group Theory, Beijing 1984. Proceedings. Edited by Tuan H.F. V, 403 pages. 1986.

Vol. 1186: Lyapunov Exponents. Proceedings, 1984. Edited by L. Arnold and V. Wihstutz. VI, 374 pages. 1986.

Vol. 1187: Y. Diers, Categories of Boolean Sheaves of Simple Algebras. VI, 168 pages. 1986.

Vol. 1188: Fonctions de Plusieurs Variables Complexes V. Séminaire, 1979–85. Edité par François Norguet. VI, 306 pages. 1986.

Vol. 1189: J. Lukeš, J. Malý, L. Zajíček, Fine Topology Methods in Real Analysis and Potential Theory. X, 472 pages. 1986.

Vol. 1190: Optimization and Related Fields. Proceedings, 1984. Edited by R. Conti, E. De Giorgi and F. Giannessi. VIII, 419 pages. 1986.

Vol. 1191: A.R. Its, V.Yu. Novokshenov, The Isomonodromic Deformation Method in the Theory of Painlevé Equations. IV, 313 pages. 1986.

Vol. 1192: Equadiff 6. Proceedings, 1985. Edited by J. Vosmansky and M. Zlámal. XXIII, 404 pages. 1986.

Vol. 1193: Geometrical and Statistical Aspects of Probability in Banach Spaces. Proceedings, 1985. Edited by X. Fernique, B. Heinkel, M.B. Marcus and P.A. Meyer. IV, 128 pages. 1986.

Vol. 1194: Complex Analysis and Algebraic Geometry. Proceedings, 1985. Edited by H. Grauert. VI, 235 pages. 1986.

Vol. 1195: J.M. Barbosa, A.G. Colares, Minimal Surfaces in $\mathbb{R}^3$. X, 124 pages. 1986.

Vol. 1196: E. Casas-Alvero, S. Xambó-Descamps, The Enumerative Theory of Conics after Halphen. IX, 130 pages. 1986.

Vol. 1197: Ring Theory. Proceedings, 1985. Edited by F.M.J. van Oystaeyen. V, 231 pages. 1986.

Vol. 1198: Séminaire d'Analyse, P. Lelong – P. Dolbeault – H. Skoda. Seminar 1983/84. X, 260 pages. 1986.

Vol. 1199: Analytic Theory of Continued Fractions II. Proceedings, 1985. Edited by W.J. Thron. VI, 299 pages. 1986.

Vol. 1200: V.D. Milman, G. Schechtman, Asymptotic Theory of Finite Dimensional Normed Spaces. With an Appendix by M. Gromov. VIII, 156 pages. 1986.

Vol. 1201: Curvature and Topology of Riemannian Manifolds. Proceedings, 1985. Edited by K. Shiohama, T. Sakai and T. Sunada. VII, 336 pages. 1986.

Vol. 1202: A. Dür, Möbius Functions, Incidence Algebras and Power Series Representations. XI, 134 pages. 1986.

Vol. 1203: Stochastic Processes and Their Applications. Proceedings, 1985. Edited by K. Itô and T. Hida. VI, 222 pages. 1986.

Vol. 1204: Séminaire de Probabilités XX, 1984/85. Proceedings. Edité par J. Azéma et M. Yor. V, 639 pages. 1986.

Vol. 1205: B.Z. Moroz, Analytic Arithmetic in Algebraic Number Fields. VII, 177 pages. 1986.

Vol. 1206: Probability and Analysis, Varenna (Como) 1985. Seminar. Edited by G. Letta and M. Pratelli. VIII, 280 pages. 1986.

Vol. 1207: P.H. Bérard, Spectral Geometry: Direct and Inverse Problems. With an Appendix by G. Besson. XIII, 272 pages. 1986.

Vol. 1208: S. Kaijser, J.W. Pelletier, Interpolation Functors and Duality. IV, 167 pages. 1986.

Vol. 1209: Differential Geometry, Peñíscola 1985. Proceedings. Edited by A.M. Naveira, A. Ferrández and F. Mascaró. VIII, 306 pages. 1986.

Vol. 1210: Probability Measures on Groups VIII. Proceedings, 1985. Edited by H. Heyer. X, 386 pages. 1986.

Vol. 1211: M.B. Sevryuk, Reversible Systems. V, 319 pages. 1986.

Vol. 1212: Stochastic Spatial Processes. Proceedings, 1984. Edited by P. Tautu. VIII, 311 pages. 1986.

Vol. 1213: L.G. Lewis, Jr., J.P. May, M. Steinberger, Equivariant Stable Homotopy Theory. IX, 538 pages. 1986.

Vol. 1214: Global Analysis – Studies and Applications II. Edited by Yu.G. Borisovich and Yu.E. Gliklikh. V, 275 pages. 1986.

Vol. 1215: Lectures in Probability and Statistics. Edited by G. del Pino and R. Rebolledo. V, 491 pages. 1986.

Vol. 1216: J. Kogan, Bifurcation of Extremals in Optimal Control. VIII, 106 pages. 1986.

Vol. 1217: Transformation Groups. Proceedings, 1985. Edited by S. Jackowski and K. Pawalowski. X, 396 pages. 1986.

Vol. 1218: Schrödinger Operators, Aarhus 1985. Seminar. Edited by E. Balslev. V, 222 pages. 1986.

Vol. 1219: R. Weissauer, Stabile Modulformen und Eisensteinreihen. III, 147 Seiten. 1986.

Vol. 1220: Séminaire d'Algèbre Paul Dubreil et Marie-Paule Malliavin. Proceedings, 1985. Edité par M.-P. Malliavin. IV, 200 pages. 1986.

Vol. 1221: Probability and Banach Spaces. Proceedings, 1985. Edited by J. Bastero and M. San Miguel. XI, 222 pages. 1986.

Vol. 1222: A. Katok, J.-M. Strelcyn, with the collaboration of F. Ledrappier and F. Przytycki, Invariant Manifolds, Entropy and Billiards; Smooth Maps with Singularities. VIII, 283 pages. 1986.

Vol. 1223: Differential Equations in Banach Spaces. Proceedings, 1985. Edited by A. Favini and E. Obrecht. VIII, 299 pages. 1986.

Vol. 1224: Nonlinear Diffusion Problems, Montecatini Terme 1985. Seminar. Edited by A. Fasano and M. Primicerio. VIII, 188 pages. 1986.

Vol. 1225: Inverse Problems, Montecatini Terme 1986. Seminar. Edited by G. Talenti. VIII, 204 pages. 1986.

Vol. 1226: A. Buium, Differential Function Fields and Moduli of Algebraic Varieties. IX, 146 pages. 1986.

Vol. 1227: H. Helson, The Spectral Theorem. VI, 104 pages. 1986.

Vol. 1228: Multigrid Methods II. Proceedings, 1985. Edited by W. Hackbusch and U. Trottenberg. VI, 336 pages. 1986.

Vol. 1229: O. Bratteli, Derivations, Dissipations and Group Actions on C*-algebras. IV, 277 pages. 1986.

Vol. 1230: Numerical Analysis. Proceedings, 1984. Edited by J.-P. Hennart. X, 234 pages. 1986.

Vol. 1231: E.-U. Gekeler, Drinfeld Modular Curves. XIV, 107 pages. 1986.

Vol. 1232: P.C. Schuur, Asymptotic Analysis of Soliton Problems. VIII, 180 pages. 1986.

Vol. 1233: Stability Problems for Stochastic Models. Proceedings, 1985. Edited by V.V. Kalashnikov, B. Penkov and V.M. Zolotarev. VI, 223 pages. 1986.

Vol. 1234: Combinatoire énumérative. Proceedings, 1985. Edité par G. Labelle et P. Leroux. XIV, 387 pages. 1986.

Vol. 1235: Séminaire de Théorie du Potentiel, Paris, No. 8. Directeurs: M. Brelot, G. Choquet et J. Deny. Rédacteurs: F. Hirsch et G. Mokobodzki. III, 209 pages. 1987.

Vol. 1236: Stochastic Partial Differential Equations and Applications. Proceedings, 1985. Edited by G. Da Prato and L. Tubaro. V, 257 pages. 1987.

Vol. 1237: Rational Approximation and its Applications in Mathematics and Physics. Proceedings, 1985. Edited by J. Gilewicz, M. Pindor and W. Siemaszko. XII, 350 pages. 1987.

Vol. 1238: M. Holz, K.-P. Podewski and K. Steffens, Injective Choice Functions. VI, 183 pages. 1987.

Vol. 1239: P. Vojta, Diophantine Approximations and Value Distribution Theory. X, 132 pages. 1987.

Vol. 1240: Number Theory, New York 1984–85. Seminar. Edited by D.V. Chudnovsky, G.V. Chudnovsky, H. Cohn and M.B. Nathanson. V, 324 pages. 1987.

Vol. 1241: L. Gårding, Singularities in Linear Wave Propagation. III, 125 pages. 1987.

Vol. 1242: Functional Analysis II, with Contributions by J. Hoffmann-Jørgensen et al. Edited by S. Kurepa, H. Kraljević and D. Butković. VII, 432 pages. 1987.

Vol. 1243: Non Commutative Harmonic Analysis and Lie Groups. Proceedings, 1985. Edited by J. Carmona, P. Delorme and M. Vergne. V, 309 pages. 1987.

Vol. 1244: W. Müller, Manifolds with Cusps of Rank One. XI, 158 pages. 1987.

Vol. 1245: S. Rallis, L-Functions and the Oscillator Representation. XVI, 239 pages. 1987.

Vol. 1246: Hodge Theory. Proceedings, 1985. Edited by E. Cattani, F. Guillén, A. Kaplan and F. Puerta. VII, 175 pages. 1987.

Vol. 1247: Séminaire de Probabilités XXI. Proceedings. Edité par J. Azéma, P.A. Meyer et M. Yor. IV, 579 pages. 1987.

Vol. 1248: Nonlinear Semigroups, Partial Differential Equations and Attractors. Proceedings, 1985. Edited by T.L. Gill and W.W. Zachary. IX, 185 pages. 1987.

Vol. 1249: I. van den Berg, Nonstandard Asymptotic Analysis. IX, 187 pages. 1987.

Vol. 1250: Stochastic Processes – Mathematics and Physics II. Proceedings 1985. Edited by S. Albeverio, Ph. Blanchard and L. Streit. VI, 359 pages. 1987.

Vol. 1251: Differential Geometric Methods in Mathematical Physics. Proceedings, 1985. Edited by P.L. García and A. Pérez-Rendón. VII, 300 pages. 1987.

Vol. 1252: T. Kaise, Représentations de Weil et GL_2 Algèbres de division et GL_n. VII, 203 pages. 1987.

Vol. 1253: J. Fischer, An Approach to the Selberg Trace Formula via the Selberg Zeta-Function. III, 184 pages. 1987.

Vol. 1254: S. Gelbart, I. Piatetski-Shapiro, S. Rallis. Explicit Constructions of Automorphic L-Functions. VI, 152 pages. 1987.

Vol. 1255: Differential Geometry and Differential Equations. Proceedings, 1985. Edited by C. Gu, M. Berger and R.L. Bryant. XII, 243 pages. 1987.

Vol. 1256: Pseudo-Differential Operators. Proceedings, 1986. Edited by H.O. Cordes, B. Gramsch and H. Widom. X, 479 pages. 1987.

Vol. 1257: X. Wang, On the C*-Algebras of Foliations in the Plane. V, 165 pages. 1987.

Vol. 1258: J. Weidmann, Spectral Theory of Ordinary Differential Operators. VI, 303 pages. 1987.

Vol. 1259: F. Cano Torres, Desingularization Strategies for Three-Dimensional Vector Fields. IX, 189 pages. 1987.

Vol. 1260: N.H. Pavel, Nonlinear Evolution Operators and Semigroups. VI, 285 pages. 1987.

Vol. 1261: H. Abels, Finite Presentability of S-Arithmetic Groups. Compact Presentability of Solvable Groups. VI, 178 pages. 1987.

Vol. 1262: E. Hlawka (Hrsg.), Zahlentheoretische Analysis II. Seminar, 1984–86. V, 158 Seiten. 1987.

Vol. 1263: V.L. Hansen (Ed.), Differential Geometry. Proceedings, 1985. XI, 288 pages. 1987.

Vol. 1264: Wu Wen-tsün, Rational Homotopy Type. VIII, 219 pages. 1987.

Vol. 1265: W. Van Assche, Asymptotics for Orthogonal Polynomials. VI, 201 pages. 1987.

Vol. 1266: F. Ghione, C. Peskine, E. Sernesi (Eds.), Space Curves. Proceedings, 1985. VI, 272 pages. 1987.

Vol. 1267: J. Lindenstrauss, V.D. Milman (Eds.), Geometrical Aspects of Functional Analysis. Seminar. VII, 212 pages. 1987.

Vol. 1268: S.G. Krantz (Ed.), Complex Analysis. Seminar, 1986. VII, 195 pages. 1987.

Vol. 1269: M. Shiota, Nash Manifolds. VI, 223 pages. 1987.

Vol. 1270: C. Carasso, P.-A. Raviart, D. Serre (Eds.), Nonlinear Hyperbolic Problems. Proceedings, 1986. XV, 341 pages. 1987.

Vol. 1271: A.M. Cohen, W.H. Hesselink, W.L.J. van der Kallen, J.R. Strooker (Eds.), Algebraic Groups Utrecht 1986. Proceedings. XII, 284 pages. 1987.

Vol. 1272: M.S. Livšic, L.L. Waksman, Commuting Nonselfadjoint Operators in Hilbert Space. III, 115 pages. 1987.

Vol. 1273: G.-M. Greuel, G. Trautmann (Eds.), Singularities, Representation of Algebras, and Vector Bundles. Proceedings, 1985. XIV, 383 pages. 1987.

Vol. 1274: N. C. Phillips, Equivariant K-Theory and Freeness of Group Actions on C*-Algebras. VIII, 371 pages. 1987.

Vol. 1275: C.A. Berenstein (Ed.), Complex Analysis I. Proceedings, 1985–86. XV, 331 pages. 1987.

Vol. 1276: C.A. Berenstein (Ed.), Complex Analysis II. Proceedings, 1985–86. IX, 320 pages. 1987.

Vol. 1277: C.A. Berenstein (Ed.), Complex Analysis III. Proceedings, 1985–86. X, 350 pages. 1987.

Vol. 1278: S.S. Koh (Ed.), Invariant Theory. Proceedings, 1985. V, 102 pages. 1987.

Vol. 1279: D. Ieşan, Saint-Venant's Problem. VIII, 162 Seiten. 1987.

Vol. 1280: E. Neher, Jordan Triple Systems by the Grid Approach. XII, 193 pages. 1987.

Vol. 1281: O.H. Kegel, F. Menegazzo, G. Zacher (Eds.), Group Theory. Proceedings, 1986. VII, 179 pages. 1987.

Vol. 1282: D.E. Handelman, Positive Polynomials, Convex Integral Polytopes, and a Random Walk Problem. XI, 136 pages. 1987.

Vol. 1283: S. Mardešić, J. Segal (Eds.), Geometric Topology and Shape Theory. Proceedings, 1986. V, 261 pages. 1987.

Vol. 1284: B.H. Matzat, Konstruktive Galoistheorie. X, 286 pages. 1987.

Vol. 1285: I.W. Knowles, Y. Saitō (Eds.), Differential Equations and Mathematical Physics. Proceedings, 1986. XVI, 499 pages. 1987.

Vol. 1286: H.R. Miller, D.C. Ravenel (Eds.), Algebraic Topology. Proceedings, 1986. VII, 341 pages. 1987.

Vol. 1287: E.B. Saff (Ed.), Approximation Theory, Tampa. Proceedings, 1985–1986. V, 228 pages. 1987.

Vol. 1288: Yu. L. Rodin, Generalized Analytic Functions on Riemann Surfaces. V, 128 pages, 1987.

Vol. 1289: Yu. I. Manin (Ed.), K-Theory, Arithmetic and Geometry. Seminar, 1984–1986. V, 399 pages. 1987.